Creating Shapes in Civil and Naval Architecture

History of Science and Medicine Library

VOLUME 11

Creating Shapes in Civil and Naval Architecture

A Cross-Disciplinary Comparison

Edited by

Horst Nowacki and Wolfgang Lefèvre

BRILL

LEIDEN • BOSTON
2009

On the cover: Hamburg Set. See also Appendix C, Fig. C.5, p. 384. (Cover image courtesy of Wichmann GmbH)

This book is printed on acid-free paper.

Library of Congress Cataloging-in-Publication Data

Creating shapes in civil and naval architecture : a cross-disciplinary comparison / edited by Horst Nowacki and Wolfgang Lefevre.
p. cm. — (History of science and medicine library, ISSN 1872-0684 ; v. 11)
Includes bibliographical references and index.
ISBN 978-90-04-17345-3 (hardback : alk. paper) 1. Naval architecture—Design and construction—History. 2. Hulls (Naval architecture)—Design and construction—History. 3. Shape theory (Topology) 4. Structural optimization. 5. Shipbuilding—History. I. Nowacki, H. (Horst) II. Lefèvre, Wolfgang, 1941– III. Title. IV. Series.
VM156.C74 2009
623.8'1—dc22

2009016176

ISSN 1872-0684
ISBN 978 90 04 17345 3

PRINTED IN THE NETHERLANDS

CONTENTS

MIDDLE AGES

MODERN ERA

APPENDICES

LIST OF FIGURES

Colour plates I–LII can be found in a separate section at the back of this volume

Figures 1–16 (belonging to Chapter One by Horst Nowacki):

Figures 1–14 (belonging to Chapter Seven by Ole Crumlin-Pedersen):

Figures 1–13 (belonging to Chapter Eight by Enrique Rabasa Díaz and José Calvo López):

Figures 1–25 (belonging to Chapter Eleven by Ulrich Alertz):

Figures 1–4 (belonging to Appendix B by Jens Høyrup):

Figures 1–8 (belonging to Appendix C by Jobst Lessenich):

PREFACE

This book is based on the essential results of a Workshop held on December 7 to 9, 2006 at the Max Planck Institute for the History of Science (MPIWG) in Berlin on the subject of "Creating Shapes in Civil and Naval Architecture: A Cross-Disciplinary Comparison". It presents the final version of 14 articles presented there and 4 relevant Appendices.

The meeting in Berlin was held to lay certain foundations in a comparative study of the knowledge history in two related disciplines, Civil Architecture and Naval Architecture. Civil Architecture deals with the conception, design and realization of terrestrial buldings, Naval Architecture treats the corresponding tasks for floating, maritime structures. The two disciplines throughout their history have been sharing certain organizational objectives, technical methods and tools. Both aspire to create complex, multifunctional objects, though for different sets of functional requirements and operational environments.

It is therefore legitimate to ask how much common ground do the two architectures share also in their knowledge histories. Which are their sources of knowledge genesis, how did their knowledge grow and migrate, how did they exchange any knowledge, did they influence each other or inherit common knowledge from other sources, was there any significant dialogue, both with respect to practical processes and methodical concepts? In short, with respect to their knowledge domains, were they more like twin sisters or more like distant cousins?

The meeting was also motivated by two earlier developments from which it resulted. A team at MPIWG together with international partners had been involved in a research study on the "Knowledge History in Architecture". This project encompassed many facets of architecture history and spanned a time frame from the neolithic age through the modern era up to 1700. Simultaneously a group of naval historians and archaeologists, Richard Barker (Derby), Eric Rieth (Paris) and Horst Nowacki (Berlin, Visiting Scholar at MPIWG) had been in communication since an earlier MPIWG Workshop in 2001 concerning the development and migration of knowledge on shipbuilding and ship design practices over major time spans since antiquity. In the course of their discussions a paper published by Eric Rieth in September 2004

("*Des mots aux pratiques techniques: gabarits et architecture navale au Moyen Age*"), whose English translation by Richard Barker is included in this book as Appendix A, stirred much new interest: It addressed the technical, functional and linguistic commonalities of craftsmen, especially stonemasons and carpenters, in the transition from late Roman antiquity to the French and English Middle Ages. The comparison included especially the terminologies which they shared, also with shipbuilders. This discussion also triggered lively interest in holding a joint meeting with a group of civil architects.

Thus it was resolved to organize a workshop for cross-disciplinary discussions and to invite a group of international experts from both architectures, well balanced with historians, archaeologists, practicing architects and engineers, mathematicians and information scientists, to convene for a three-day meeting in Berlin. A total of 29 persons actively participated in the Workshop, 18 papers were presented and discussed. A documentation of all original contributions as presented to this workshop was first published in Preprint No. 338 of the Max Planck Institute for the History of Science, Berlin, a precursor to this book.

The subject of *shape creation* was chosen as a focal point of interest for the comparisons between the two disciplinary knowledge domains because it played a decisive role in object developments in both disciplines, it permitted direct comparisons of geometric information contents and corresponding practices, and the issue of whether there was much or only limited commonality was not trivial or obvious. Clearly this concentration limited the scope of the comparisons to only a subset of the many aspects which the two architectures might share or differ in.

The presentations and discussions presented at this meeting and consequently the results reported in this book set themselves the following main objectives:

Collection and documentation of sources (literary, archaeological, by artifacts) of shape creation knowledge in the two architectures where suitable for comparisons in their knowledge histories.

Classification of practical and procedural knowledge types in both disciplines by a suitable reference framework in order to structure the comparative discussions.

Enumeration of milestone events at which historically mutual communication or transfer of pertinent knowledge may have occurred and may have been particularly inspiring, motivating and effective.

- Priorities placed on knowledge concerning the design and construction of complex shape objects.
- Historical changes in shape design styles and methodologies between antiquity and the early modern era (up to 1700).
- Comparisons of shape design methods and construction practices for complex shapes.
- Shape representation methods and tools (marking, lofting, drafting, modelmaking, measuring, controlling of quality).
- Terminologies and language clues.
- Application case studies.

The articles submitted were encouraged to be dealing with both architectures or at least with shape creation processes of generic enough type to form the basis for comparisons of different approaches to complex shape creation. A particular interest was in the conception and geometric modelling of complex shapes, but also in the practical realization of such objects.

The media of representation of shapes changed several times in the course of history, from sketches in sand, on ceramics and in rock, from carved wooden or clay models and assemblies via lofting practices on level ground and chalked floors to drawings on paper and other printable materials. Both architectures were influenced by each transition in representation medium in their conceptual approaches and building practices. The discussions were intended also to compare the effects of such media changes in parallel disciplines. Incidentally the recent transition to computer based modeling of geometric shapes has taught us how deeply the change of medium may influence the thought process that leads to shape creation. Was this also true at certain earlier milestones in history?

The agenda of the meeting did not impose any restrictions on the kind of knowledge on shape creation that was to be presented. On the contrary all kinds of knowledge were encouraged to be addressed, whether practical or theoretical, formal or informal, simple or complex, based on tools or procedures, scientific or pre- or non-scientific. Such knowledge would consist of any structured sets of facts and rules.

The meeting and the results presented here were intended as an opening move, the first of several steps toward a historical review of shape creation knowledge. Initially there did not exist a sufficient body of sources, literary or archaeological, to achieve a systematic comparison of the related disciplines. The near future goal was to initiate a documentation of any such available sources, which was begun here. The more distant goal remains to identify patterns and stages of development in the processes of shape creation in different eras, regions and applications. We hope that the material collected here will offer enough food for thought and inspiration for further stages of comparison in the historical development of geometric modeling, design and building.

We want to thank Jürgen Renn, head of Department I of the Max Planck Institute for the History of Science, whose unceasing support enabled us to organize the preceding workshop as well as the edition of this volume.

Horst Nowacki
Wolfgang Lefèvre

INTRODUCTION

CHAPTER ONE

SHAPE CREATION KNOWLEDGE IN CIVIL AND NAVAL ARCHITECTURE

Horst Nowacki

Abstract

This opening contribution will serve to define a few concepts and comparative methodologies for shape creation in the two disciplines of civil (terrestrial) and naval (maritime) architecture. It will thus help to set a platform for more specific individual case studies. Initially it will seek to find an understanding of what is "shape" in this cross-disciplinary context and how different types of shape might be classified. Then it will examine "the shape creation process" of conceptual design, construction and fabrication in terms of the knowledge required at each stage. This knowledge is intimately linked with the media of concept documentation and with the methodical tools of construction, fabrication and assembly. Thirdly, it will proceed to review important historical milestones at which by particular events or turning points the proximity in shape creation knowledge between the two disciplines became apparent and should be more closely investigated, i.e., it will identify candidates for more specific case studies. This survey will range from antiquity to the early modern age as does the whole book.

Overall this chapter will provide a first glance at the theme of the book in terms of three axes of comparison for historical shape creation developments:

- The time axis (genesis of shape creation knowledge)
- The geometric complexity and functionality axis (which shape knowledge when?)
- The axis of tools and methods of shape creation (which approaches when?)

Thus in summary this opening chapter will create a list of questions and try to systematize and classify the issues studied in this book. Its purpose will not and cannot be to provide all the finished answers.

1. Introduction

Buildings and ships are among the earliest artifacts conceived, created and shaped by humans in team work. The beginnings of major construction projects in both categories are certainly prehistorical. Watercraft e.g. were apparently used by the first human occupants of New Guinea and Australia some 40000 years ago to transport their people and goods when they crossed over the deep and wide trench then separating the South East Asian mainland from Australia/New Guinea [1]. Houses and ships date back to the earliest sessile human societies in prehistory in the Fertile Crescent, Egypt and elsewhere. Prestigious building projects, but also waterborne vehicles are known from many early societies. They often were prominent landmarks of their era and are thus a telling testimony to the level of advanced construction knowledge achieved in their societies. An historical comparison of the evolution of knowledge in these two applications of human construction skills can thus yield a representative sample of human knowledge history over a very long time span. A monocultural bias can be avoided by contrasting two distinct evolutions. This ideal goal of course can be achieved only to the extent that we have sufficient comparable historical evidence from both sides.

Civil architecture and naval architecture owe their naming to the period of classical Greek history. The designations "architect" (ἀρχιτέκτων) and "naval architect" (ἀρχιτέκτων ἐπὶ τὰς ναῦς) or their immediate precursors are both encountered in Greek texts by no later than 500 B.C. (see details below). According to the Athenian constitution [2] the leading architects and naval architects for state projects were elected by the Council of 500 (Boule). By this time these professions already had achieved a high public prestige and were carrying much responsibility. The words have remained in usage until today, although somewhat varying in scope and substance with time and place.

Civil architecture deals with the realization of building objects on firm ground (terrestrial structures). Naval architecture is concerned with the same creation process for floating objects (maritime structures). Both processes encompass the stages of conception, construction and fabrication of the structures to be created. This similitude in practical objectives has been further reinforced by several organisational commonalities:

- Building and shipbuilding projects are usually of such magnitude that they require a major workforce with many specialized skills and hence a strong division of labor. This workforce needs coordination and synchronisation.
- Ambitious projects involve major investments and financial risks. Structural and operational safety is of significant concern.
- Advance planning, material procurement and fabrication site logistics are of high priority.
- Thus strong supervisory control and centralized project responsibility are generally required throughout the project. These are generally exerted by the leading architect or naval architect.

Many of these parallels between the two architectures lie in the organisation of the entire creation process. The dissimilarities whose existence cannot be denied either stem from the different functional requirements, the distinct character of shape and intended structure, and the differences in the operational environments. Thus to a large degree each architecture has provided its own answers to its spectrum of questions.

This study will focus on the shape creation process in the two architectures. This is of course only one special aspect in a large spectrum of tasks in both architectures. As Fig. 1.1 illustrates, the intended function plays a central role in the conception and realization of buildings and ships. Shape is only one of several properties that determine the character of the object. Other important properties of the object are its functionality, its physical structure, its internal subdivision, safety against numerous risks, feasibility of practical production, and last but not least the appeal of its style. All of these properties and requirements interact with each other. Thus focussing the attention on shape creation may suppress the discussion of other strong commonalities. It is entirely possible that all other parallels alone are sufficient to justify the homonym "architecture" for both disciplines, which will not be disputed here. But whether the shape creation process, too, shows strong or weak commonalities is a key question addressed in this study. The answer is not trivial, but requires a thorough investigation. This question should be examined from the viewpoints of design methodology, constructive geometry, functional design, design criteria, esthetic criteria and others.

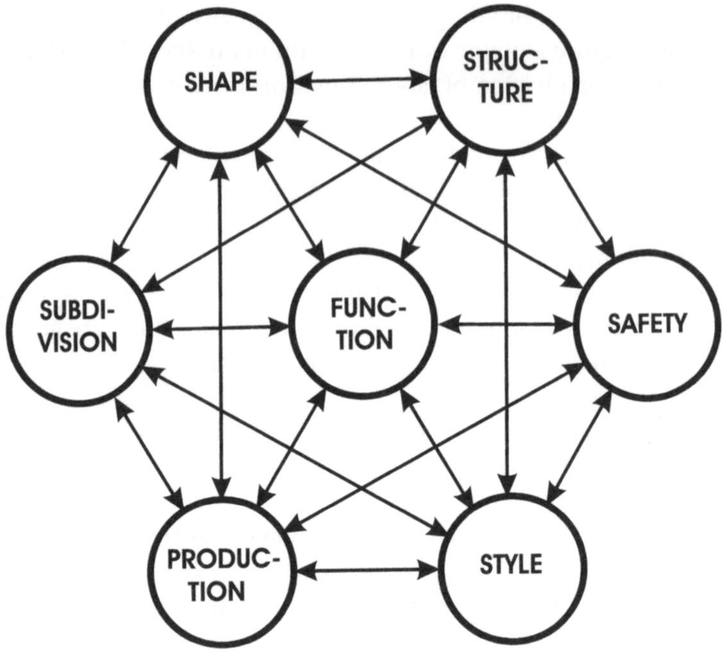

Fig. 1.1: Properties of the shape creation object.

Fig. 1.2 describes the scope of the studied issues. The knowledge on shape creation, civil and naval architecture overlaps in the central shaded area where the tasks, methods and results of the geometric shaping processes have certain central themes in common with the two application disciplines ("CANA shape"). The history of shape genesis in each individual discipline is better understood and has been intensively studied (white areas). But comparative studies between civil architecture (CA) and naval architecture (NA) are scarce, above all in the area of shape design. Thus the first task the study must address is the collection of source material for performing such comparisons.

Some of the pertinent open questions are:

- Were there any commonalities in practical and theoretical knowledge?
- If yes, did the two disciplines share such knowledge and communicate it across disciplinary boundaries?
- Did one discipline consciously benefit from or at least tacitly adopt any knowledge from the other?
- Or did they both rely on an accumulated reservoir of common, perhaps intuitive knowledge prevalent in their region and era?

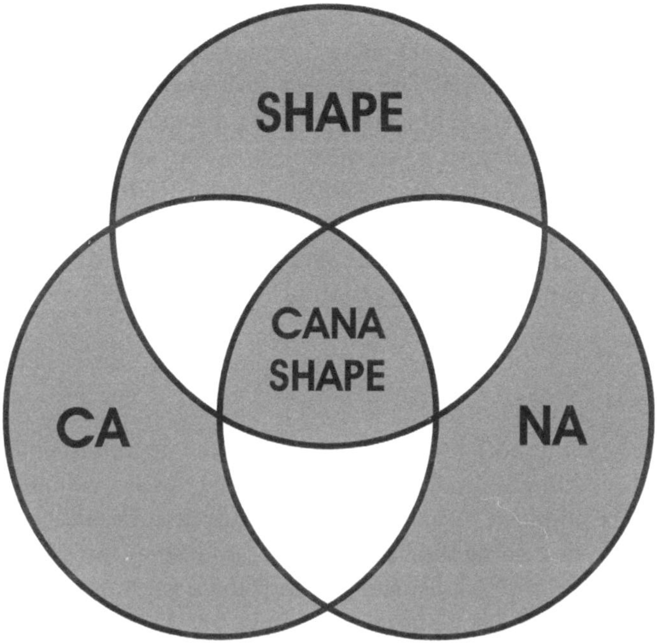

Fig. 1.2: Focus of investigation on CANA shape creation.

- If there were also dissimilarities in the approach to similar tasks at any given time and place, why did they occur and how long did they last?
- Were there any barriers to communication?

Even if we do not detect any strong and frequent direct interdependencies between the two professional bodies of knowledge, it will be of interest to compare the development in the professional expertise of the two sister disciplines when exposed to similar external influences. E.g., how were they affected by the cultural influences of the Renaissance, the rebirth of classical knowledge, the resurrection of scientific traditions from antiquity? Did such influences make a similar or distinctly different impact?

Thus the two architectures may be compared by viewing themselves through the mirror of their sister discipline in order to develop a deeper understanding of their own knowledge histories. This may also shed some light on the general mechanisms of knowledge genesis, transfer, sharing and evolution.

A valuable supplement to this introductory chapter was presented by Richard Barker during the workshop and his article was included in the original workshop report, which has appeared as an MPI Preprint [56]. His contribution gives an extensive overview of authentic and relevant historical source material on both architectures from the classical to the early modern era. The article further discusses several issues of methodology, which the two architectures historically had in common, ranging from geometric design and modeling to drafting in various media, mathematical methods, material technology, mensuration and fabrication. The interested reader should refer to the MPI Preprint or to any forthcoming internet version of Richard Barker's article.

2. Definitions

Shape creation in both architectures aims at the definition and realization of three-dimensional physical objects that possess volume properties and are bounded by exterior (and often interior) surfaces. We are interested in describing the shape of the bounding surfaces and thereby the object shapes. It will be useful to introduce a terminology that can be applied in both disciplines.

Shape

Civil architecture and naval architecture deal with the creation of physical objects, buildings and ships, whose shape belongs to their most characteristic features. Human shape perception is based on the human viewing and tactile sensing of objects, hence on human interpretation and judgment of shape impressions. Thus human notions of shape are influenced by subjective elements which are not unique so that it is not trivial to define what constitutes a shape.

However I ought to explain what I mean by *shape* in the context of this chapter and in the broader context of shape comparisons between objects in the two architectural disciplines. Here we are mainly interested in the shape of physical volume objects, such as buildings or ships, which are bounded by surfaces. In short we are mainly addressing body shapes and surface shapes, although the shape of curves and point sets which lie in the surfaces can be derived therefrom.

Essentially we need to agree on a terminology by which we want to describe the properties of a shape. Following practices in constructive geometry let us strictly distinguish between *morphological* and *geometric* attributes of a shape. *Morphological attributes* are related to those properties which characterize a shape by the general pattern of arrangement, connectivity and contiguity of shape elements. Thus they are related to the topological organisation of the elements, to the element types and to the shape character of each type. E.g. a cube has the following *morphological* properties: A cube is a singly connected volume domain, bounded by six planar faces. Each face is bounded by four edges. Faces meet at their edges which they share with their neighboring faces. The edges extend between their end vertices. Three faces share a corner at a corner vertex. A corresponding data structure can be built for the figure of a cube in which the faces, edges and vertices are properly connected with their neighbors by suitable pointers.

Note that this morphological description of a shape is kept free of any positional information. The same morphological shape description applies to a cube of any size and still holds for a parallelepiped or for a corresponding shape with unequal edge lengths or with curved faces. All of these figures share the morphology of a cube. Some terminologies call this class of shapes "objects of cube morphology".

Geometric properties of shapes by contrast are expressed by positions in a suitable reference frame. These properties are required to dimension and locate a body shape and to measure its size, volume, surface etc. In conjunction with a morphological structure of the shape they make the definition of body shape explicit and apt to be fabricated.

This discussion has illustrated that a suitable set of morphological attributes is sufficient to characterize a *class of shapes*. In order to describe an *individual shape* both morphological and geometric attributes are necessary.

In a comparative study on shape creation in two related disciplines the methodologies and technologies for creating certain classes of shapes are of more general interest than those for individual objects. The main issue thus is whether the presumed conceptual methods, concept documentation tools and fabrication technologies are suitable for creating the classes of product shapes occurring in some historical context.

Let us note on the side that our terminology for "shape", which I am adopting here for pragmatic reasons, resembles that utilized by Aristotle ([3], Physika II2, 193b) who on different grounds distinguishes between the "shape" (Greek: "μορφή", German: "Gestalt") of a physical object, thereby denoting the common properties of a class of objects, and the "form" (Greek: "εἶδος"), which refers to the specific characteristics of an individual object. The Greek language is fortunate in offering two distinct words for these distinct concepts.

Simple and complex shapes

The literature abounds of references to *simple shapes* of solids that are intuitively known to everyone: Sphere, cylinder, cone, torus, cube, pyramid, polyhedra and others. These shapes have unique morphological properties with which mankind has been familiar since very early history. Each simple shape can be declared to serve as a primitive element in a building block system to create more complex shapes by composition. Each class of these primitives is characterized by a simple and unique curvature distribution in its surface: The faces are either planar (zero curvature) or have typical Gaussian curvature distributions. Let the Gaussian curvature be denoted by

$$k_G = k_1 k_2,$$

where the principal curvatures k_1, k_2 are the maximum and minimum curvatures at a surface point. Then we have the following curvature properties (Table I, Figs. 1.3a and 1.3b):

Table I: Curvature properties of simple shapes.

Shape	k_1	k_2	k_G	Surface type
Sphere	1/R	$k_2 = k_1$	$k_1 k_2 > 0$	elliptic
Cylinder	1/R	0	0	parabolic
Cone	1/r	0	0	parabolic
Torus	$1/r_1$	$1/r_2$ or $-1/r_2$	> 0 or < 0	elliptic/hyperbolic
Cube	0	0	0	planar

Complex shapes of solids differ from simple shapes in one or both of the following ways:

- They may be composed of more than one simple shape by union or difference or both: *Composite shapes* (Constructive Solid Geometry).
- Their faces may possess curvature distribution patterns of any desired variation: *Freeform shapes*.

In the latter case a surface may consist of several domains of different signs of Gaussian curvature, e.g., elliptic domains (convex, convex principal curvatures) and hyperbolic domains (convex, concave principal curvatures), bounded and separated by parabolic lines or domains (where $k_G = 0$). In Figs. 1.3c and 1.4 the color table for visualizing the

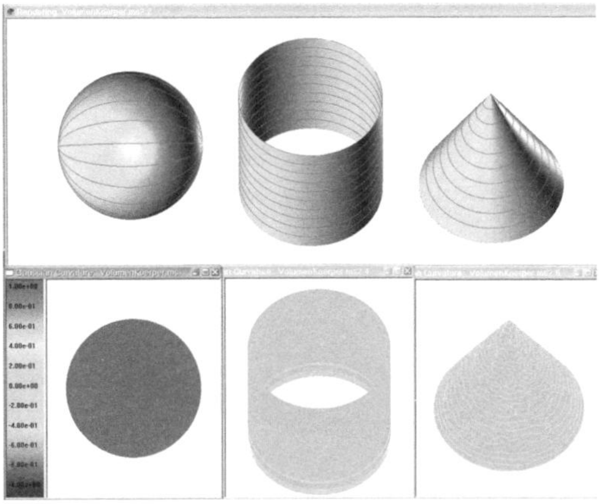

Fig. 1.3a [Col. Pl. I]: Simple Shapes: Sphere, cylinder, cone: Perspective views, top, and Gaussian curvature bottom row.

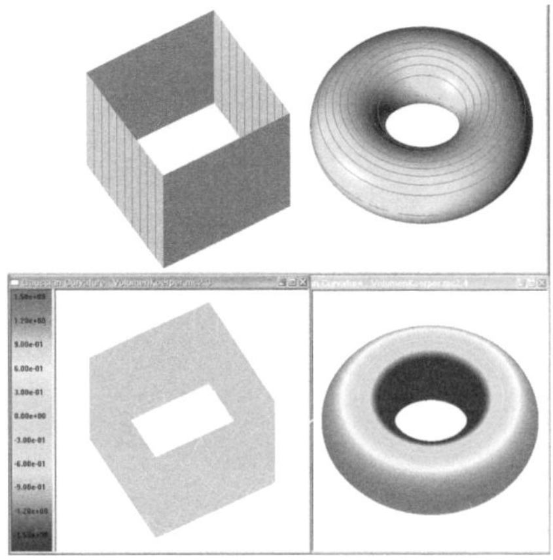

Fig. 1.3b [Col. Pl. I]: Simple Shapes: Cube, torus: Perspective views and Gaussian curvature.

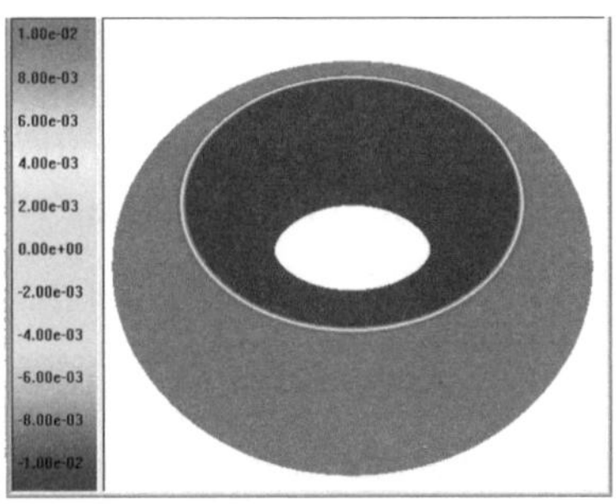

Fig. 1.3c [Col. Pl. I]: Torus, Gaussian curvature, color table reset.

Gaussian curvatures has been set with such extreme values of the color band that elliptic regions appear red, hyperbolic regions blue and a narrow parabolic separation zone is green. Thus the gross character of the shape or its classification can be described in terms of its pattern of red and blue zones.

In Fig. 1.4 the same visualization technique is applied to the complex forebody shape of a modern tanker with a protruding bulbous bow. The more complex pattern of red (elliptic) and blue (hyperbolic) zones indicates a map of red hillscapes (or concave-concave grooves) embedded in regions of a blue saddle surface.

The pattern of arrangement of elliptic and hyperbolic domains can thus be used to serve for a *classification* of complex surface shapes. Civil

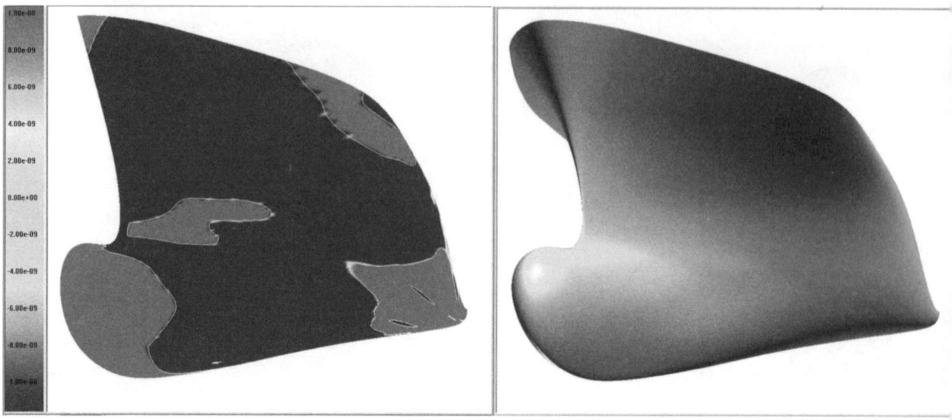

Fig. 1.4 [Col. Pl. II]: Tanker forebody with bulbous bow: Gaussian curvature map and perspective view.

and naval architecture are rich in such complex surface shapes whose conception and fabrication may pose increasing difficulties in advanced shape creation. Artistic sculptures can exhibit even more complex surface curvature patterns. Yet human imagination and perception are well trained in recognizing the intended shape character even in very complex curvature distributions.

Shape creation

The shape creation process leads from the earliest idea of a newly conceived shape to the final realization of a fabricated object shape. This process can be structured into several stages in order to identify certain evolutionary steps of the shape model (Fig. 1.5).

The process is delineated in a sequential way for the ease of discussion, but in practice may be iterative since the conceptual freedom in the earlier stages may well depend on the feasible domain in the later decisions. Yet it is useful to distinguish several levels of maturity and completeness of the shape model.

Shape conception is a mental process of imagination and judgment. It results in a *mental model* (*shape concept*) in the designer's mind. Initially this conceptual model need not be complete or even consistent and free of contradictions. Gradually it will gain maturity and accuracy. It is often influenced by experience from earlier, similar tasks, but must be tailored to comply with a new set of requirements,

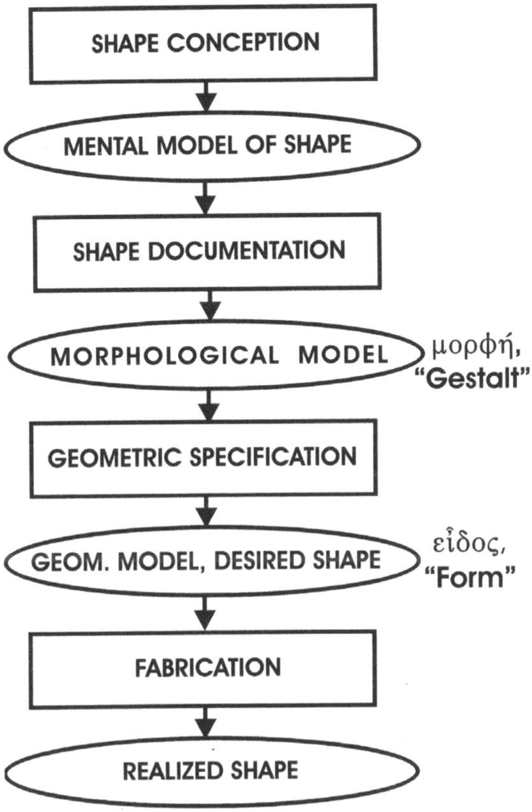

Fig. 1.5: Shape creation process.

at least in orientation. In the beginning this model is still confined to a single person's mind. It must be documented in some medium of expression to be communicated to other persons.

For this purpose the shape idea can be described textually or by a sketch in terms of the morphological character of the shape. This results in a communicable morphological shape model. This model is not yet dimensioned.

Once the geometric features of the shape are specified in terms of dimensions and positions, a *geometric model* of the desired shape is completed. The object can now be fabricated according to this information yielding a *realized shape*. The fabrication process includes the transfer of geometric information from the documentation medium of the geometric model to physical parts used in fabrication by the stages of marking, lofting, template cutting, part preparation, subassembly and assembly, outfitting and finishing, usually ending in trials and delivery.

It is evident that shape creation thus proceeds from initially very limited information on the intended shape to an elaborate, complete description and fabrication of the whole object. The methodologies of design, construction and fabrication are responsible for providing the missing information and obtaining the desired product.

In its cross-disciplinary comparisons this investigation must therefore examine the following questions:

- Which methods of design and above all constructive geometry were used by the two architectures historically at any given time and place?
- Which shapes were achievable by these methods and how do these shapes and their construction methods compare between the two architectures?

Knowledge

Knowledge is a mental model of our perception of objects and processes in the real world. It is usually derived by observation, interpretation, abstraction and validation as a result of reflection and judgment.

Knowledge must be communicable to be of practical relevance. For communication knowledge requires a representation, i.e., a description in a representation medium.

Communicable knowledge can be described as consisting of facts and rules (as in Knowledge Based Systems). Facts are represented by data, rules by "if-then"-constructs. Facts can be derived from rules and vice versa. Thus facts and rules are to a certain extent interchangeable.

We are interested in shape creation knowledge. Here the data describing the shapes serve as facts. The methods for creating the shapes are composed of rules, e.g., those of computational geometry. Such rules and procedures are our main interest in this study.

Knowledge may be of many types, all of them of historical interest:

- Intuitive, practical, theoretical knowledge
- Informal vs. formal
- Simple vs. complex

- Scientific (based on axioms, proofs, validation) vs. pre- or non-scientific
- Based on tools and practical procedures.

The etymological origins of "architect" and "architecture"

The role of an architect, involved in some major construction project, whether for buildings on land or ships afloat, is probably very ancient. This type of function became mandatory as soon as construction projects reached such magnitude that the construction process required a strong division of labor, both in sequential and in parallel subtasks. This is always the case when a major workforce must be coordinated toward an ambitious goal. The function of the architect then encompasses the conceptual design, the supervision of the construction and the contractual responsibility for the results of the project. This form of organisation and coordination has most likely occurred in many construction projects in early history, e.g., in the Near East and in Egypt.

However the designation "architect" is of Greek origin. Affluent urban societies in Greek antiquity developed the resources and desires for ambitious public buildings and also large ships. Beautiful, awe inspiring temples and powerful naval fleets are among the earliest testimonies. Leading architects or naval architects were appointed by state authorities to carry the overall responsibility for such ambitious tasks. What does the root of the word "architect" tell us about the substance of this task and the scope of the architect's responsibilities?

The designation τέκτων (tekton) is found in very early sources and has the meaning [4]:

ὁ τέκτων = a) craftsman (carpenter, shipbuilder,...) in hard materials (wood, stone, metal,...)
b) originator, producer, master, artist

This word is related to similar, sometimes earlier words in other indo-european languages (e.g. Sanskrit), also to Latin texere = to shape, to build. The existence of such early roots and relationships illustrates the importance of building tasks and the challenge of working hard materials, also in earlier pre-Greek societies. In ancient Greek τίκτω is a verb of similar root, meaning "to create".

An early usage of τέκτων in connection with ships and naval architecture occurs in the Odyssey IX, 126/127 [5] where Homer describes the lack of shipbuilding skills on the island of the Cyclopes:

> ...οὐδ' ἄνδρες νηῶν ἐνὶ τέκτονες οἵ κε κάμοιεν
> νῆας εὐσσέλμους, αἵ κεν τελέοιεν ἕκαστα...
>
> ...and there are no shipbuilders who can finish
> well covered ships which can reach any destination....

This quotation "ἄνδρες...νηῶν τέκτονες" as shipbuilders stems from the 7th c. B.C., when the Odyssey was recorded. It may not much predate the derived form of ἀρχιτέκτων (architekton) whose dictionary definition [4] evolved into:

> ὁ ἀρχιτέκτων = Architect, chief builder, initiator, originator, conceptual designer, construction supervisor, responsible contractor.

In Athens, according to the Athenian constitution (section XLVI.1, lines 5/6), as narrated by Aristotle [2] or one of his students, the custom was to appoint architects and naval architects by vote of the Council (Boule):

> ...χειροτονεῖ δ' ἀρχιτέκτονας ὁ δῆμος ἐπὶ τὰς ναῦς.
>
> ...but the People's Council votes by hand on the naval architects (ἀρχιτέκτονας ἐπὶ τὰς ναῦς).

To my knowledge this is the earliest literary reference to the designation "naval architect" in Greek. The architects for public building projects in Athens were appointed by a similar process.

Thus from the time of the coining of the Greek words for architect and naval architect the two professions were regarded as something closely related by societal status, by public responsibility and professional function. The two types of architects' roles in their professional environments may have been very similar although their required knowledge and experience may have differed.

Sources of evidence

The historical evidence for a comparison of the two architectures stems principally from three kinds of sources:

- *Literary*: Any contemporary documents with textual or graphical information.
- *Based on artifacts*: Specifically on sculptures, models, paintings and other images.
- *Archaeological*: Stemming from excavations of historical prototypes, terrestrial ruins or underwater shipwrecks, conserved wholly or in part.

Eric Rieth in [6], e.g., has given a thorough overview of some of the earliest archaeological finds and literary sources on historical ships, especially from mediaeval sources in Mediterranean countries. Other overviews related to ship hull shape developments are presented in Rieth [7], [8], Pomey [9], Basch [10], [11], Crumlin-Pedersen [12], Steffy [13], Barker [14], Hoving [15], [16], Hocker et al. [17] and by Barker [56].

It appears from this body of information that suitable literary source material for comparisons between civil and naval architecture is very scarce in written sources up until the high Middle Ages (c. 1300), and where earlier material does exist it is often only indirectly related to shape creation, open to arguable interpretation. Artifacts, in particular artistic sculptures and monuments, do date back to antiquity in both architectures, but are often not very accurate and revealing for the shape creation process. Thus any archaeological evidence available on historical buildings and ships or shipwrecks is probably the most direct and accurate information we can get on the earlier periods of shape genesis in history. Historians must make the best of the rather meager material on the origins of shape creation knowledge in civil and naval architecture. Several chapters in this book will address this matter in more detail.

Confining myself to a few linguistic pieces of evidence I will only illustrate the situation by three example themes, mainly picked up from Eric Rieth [6], pertinent to word histories for concepts used in both architectures:

1) *From modulus to mold*:
As the American Heritage Dictionary [18] suggests, one can trace the etymology of the word "mold" (American spelling), which today inter alia means "to shape in or by a mold", via the following stations:

> *Modulus* (Latin) → *modle* (old French) → *moles* (mediaeval French, Gimpel [19]) → *molde* (Anglo-Norman and middle-English, Gimpel: 1445) → *mould* (modern British English) → *mold* (American spelling).

As Gimple says and du Colombier [20] attests in more detail, the word "*moles*" in mediaeval cathedral buildings denoted a wooden (sometimes metal), planar, thin, but stiff enough template used by the stonemasons for the marking of the level faces of stone blocks. By means of these instruments they were lifting off curve data from drawings made in chalk or gypsum on the floor of "la chambre aux traits", a special room in cathedral building lots where the shape of certain contours for stone parts was developed and inscribed in the floor drawing. (The naval architect may be tempted to believe that this technique was an ancestor of later mold lofting practices for ships).

As Rieth [6] explains, also citing du Colombier [20], such "*moles*" are first attested to have been used in 1174 for the marking of stones in the quarries of Caen destined for the cathedral of Canterbury. In shipbuilding for comparison Rieth [6] quotes an unpublished building document from 1318–1320 for the construction of galleys at Narbonne in which in Latin the use of "*moduli*", presumably for the marking of ship parts, is mentioned. These events in connection with the word history are very convincing evidence for the use of "*moulds*" as planar marking templates in both architectures. But this only confirms their analogous geometric purpose. Their functional role in shape creation, however, here for marking stones for volume cutting, there for marking wooden planar boards, may have been quite different. It remains an open question whether *moulds* in this sense were not used as marking templates by carpenters or stonemasons much earlier than attested so far.

2) *From sextus to sesto*:
Another fascinating story surrounds the words "*sextus*" (Latin) and "*sesto*" (Venetian, Italian). According to Rieth [6], as mentioned in a mediaeval shipbuilding contract specification from around 1275, the word *sextus* is used to denote the "*master template*" of the ship from which the shape of any derived frames or transverse sections at the desired longitudinal stations can be deduced by certain rules on the basis of the "*Mediterranean method*" (Rieth [30]) or the equivalent Venetian method of "scorrer del sesto" [33]. This approach to geometry generation will be explained in more detail later. This technique of ship surface construction is unique for naval architecture and is not known in civil architecture. It is a fundamental method of "*skeleton first*" shipbuilding. The example illustrates again how earlier Latin words for concepts of craftsmanship were adopted and converted to modern languages.

3) *From navis to nave*:
The Latin word "*navis*" for ship has assumed a further special meaning in the Middle Ages in the architecture of a church or cathedral as the high central part of the building extending from the narthex to the transept crossing. This meaning, which is found in Latin texts of the 12th c., is encountered in several modern European languages soon after that: *Nef* (French), *navata* (Italian), *nave* (English, Spanish),

Kirchenschiff (German), can also be dated as of the high Middle Ages. Since the Latin origin of this usage is evident, it is of interest in our context to know how and when the word "*navis*" acquired this meaning in Latin and whether any resemblance between ship and church caused this transition.

The conjecture that perhaps the naming stems from the observation that the church roof looks like a ship turned upside down appears too far-fetched. Actually neither the geometry nor the structure of the roof in its carpentry have any close similarity to ships.

Nor does the phonetic resemblance between Greek η ναῦς (ship) and ὁ ναὸς (temple cella or core) offer any convincing explanation.

The most plausible reason for the origin of the likeness between church and ship seems to be at the symbolic level rather than in appearance or construction. The German Encyclopedia of the Ancient World [21] says in this spirit:

> "As a Christian symbol *the ship* is found in the catacombs (as already earlier in antiquity) as a reference to the course of life (Goal: A lighthouse or a monograph of Christ). The mainmast was regarded as the symbol of the cross; therefore frequently found with accompanying text '*in pace*' (also on glyptographic gems and rings). In biblical scenes (Noah, Jonah) the motif of life saving predominates. Under the influence of secular symbols (Roman coins before and into the 4th c. A.D. carried the symbol of the ship with Isis or Victoria at the helm) the ship developed into the symbol for the church with Christ at the helm." This quotation establishes a very early reference to the ship as a symbol for a safe haven and for the church. It alludes to a continuity of meaning from pre-Christian to Christian times. In fact, Isis was the goddess of the oceans and of lucky fortunes, hence an ideal helmsman. Thus Christ was ideally suited to take over those roles for the church. This explains how and why the ship had an important religious meaning to Christians which was related to the purpose of a church as a building.

The symbolic closeness between ship and church is underscored also in "*Schiffe der Völker*" [22] where the origin of the meaning of "*navis*" in architecture is investigated and after discarding other conjectures The Fathers of the Church in early Christianity and in late antiquity are quoted seven times depicting the church as a safe vessel running into a secure haven through the dangerous ocean of time.

In conclusion these plausible explanations do testify to a symbolic proximity between the concepts of ship and church, but do not present any new arguments for structural or geometric interdependencies between civil and naval architecture.

3. Stages of development in ship hull shape creation

In this section a compact overview will be given on the development of ship hull forms during some major historical eras. The purpose will be to analyze the characteristic shape features, the underlying conception and fabrication processes and the knowledge that forms the background for the practised decisions on hull shape. This brief survey cannot intend to be historically complete or to cover the whole range of design aspects in any of the ship types cited as examples. Rather it will limit itself to information relevant to shape creation knowledge.

1) *Early History (Prehistory, Near East and Egypt, Asia, Europe)*

Types: Raft, skin boat, bark boat, logboat, outrigger, first river and ocean cargo ships.

Shape: Tendency to elongated (high L/B), often slender shapes, but wide enough to ensure stability. Not seldom blunt bows and sterns. The shape in this era expresses human experience and intuitive understanding of shape performance.

Knowledge: Basic practical knowledge on buoyancy, stability, maneuverability, steering and propulsion of boats or small ships. Propulsion by sail, paddle, oar. Limits in size and speed.

Fabrication: Early primitive methods. Promising line: "keel boats". Keel log with planks on the sides, connected by mortise and tenon or sewn together, caulking.

"Shell first" principle, "passive" transverse floors and beams. Carvel planking with mortise and tenon predominant in the Mediterranean, clinkered construction in Northern Europe.

Tools (for shape definition): Cord, compass, ruler. Perhaps first beginnings with templates?

Evolutionary trends: Increasing ship size and speed. Ocean voyages and oceanborne trade become feasible.

Pictures: Figs. 1.6a–1.7d.

References: Lavery [23], Dudszus et al. [24], Greenhill et al. [25], Greenhill [26], Crumlin-Pedersen et al. [27].

2) *Classical Greece and Rome (ca. 500 B.C.–A.D. 450)*

Types: Oared vessels: Fast, slender warships: Galleys, triremes etc. Sailing ships: Large, bulky cargo ships.

Shape: Elongated, slender forms with rambows for warships vs. short, beamy, stable cargo ship forms: "Long ships" vs. "round ships".
Knowledge: First theoretical knowledge on buoyancy, stability (Archimedes). Propulsion of major ships by sail and oar.
Fabrication: Mediterranean: Generally "shell first" construction, carvel planking. Large series building programs of warship fleets in Greece and Rome. Hence copying of ships feasible.
Tools (for shape definition): Cord, compass, ruler. Probably templates?
Evolutionary trends: Increasing ship size, multi-deck galleys, heavy cargo transport (obelisk), grain ships.
Pictures: Figs. 1.8a–1.8b.
References: Lavery [23], Dudszus et al. [24], Greenhill et al. [25], Basch [11], Kemp [28].

3) *Middle Ages (ca. A.D. 450–1450)*

Types: Oared galleys, sailing merchantmen, Viking ships, early cogs (ca. A.D. 1225).
Shape: Slender, fast vs. full, beamy, slow. Transition to "skeleton first" construction in combination with carvel planking (beginning as of ca. 650 A.D.).
Knowledge: Basic practical knowledge in naval architecture. First use of stern centerplane rudder (ca. 1250 in a cog).
Fabrication: In the Mediterranean beginning of "skeleton first" construction with carvel planking. Templates and later "master templates" in use. Northern Europe: Still clinker construction (until the late Middle Ages).
Tools (for shape definition): Cord, compass, ruler. Templates and master molds (sesti, gabarits).
Evolutionary trends: Transition from "shell first" to "skeleton first" in Mediterranean countries.
Pictures: Figs. 1.9a–1.9b.
References: Steffy [29], Rieth [30], Crumlin-Pedersen [31], Unger [32].

4) *Renaissance (ca. A.D. 1450–1600)*

Types: Still galleys (at least till Lepanto, 1571), different types of galleons (Venetian, Genoese, Spanish, Portuguese). Spanish and Portuguese vessels: Não, caravel, ca. 1450).

Fig. 1.6a [Col. Pl. III]: Sailing log raft (Taiwan), [23], © National Maritime Museum, Greenwich, London.

Fig. 1.6b [Col. Pl. III]: 1.6b: Basket boat (Iraq, Vietnam), [23], © Corbis GmbH, Düsseldorf.

Fig. 1.6c [Col. Pl. III]: Bark canoe (North America), [23], © National Maritime Museum, Greenwich, London.

Fig. 1.6d [Col. Pl. III]: Indian fishing dugout (North America), [23], from "Admiranda Narratio…", engraved in 1590 by Theodore de Bry (1528–1598) after John White, © Service Historique de la Marine, Vincennes, France/The Bridgeman Art Library.

Fig. 1.7a [Col. Pl. IV]: Nile sailing boat (ca. 1900 B.C.), [23], © Dorling Kindersley, London.

Fig. 1.7b [Col. Pl. IV]: Indonesian outrigger, [23], © Dorling Kindersley, London.

Fig. 1.7c [Col. Pl. IV]: Hjortspring boat (ca. 300 B.C.), [23], © Nationalmuseet, Denmark.

Fig. 1.7d [Col. Pl. IV]: Carvel planking construction, [23], © Dorling Kindersley, London.

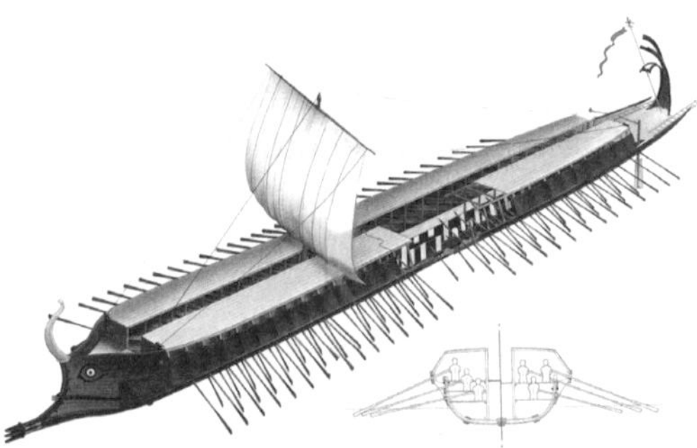

Fig. 1.8a [Col. Pl. V]: Greek trireme (ca. 480 B.C.), [28], © Little, Brown Book Group, London.

Fig. 1.8b [Col. Pl. V]: Roman merchantman (ca. A.D. 200), [23], © National Maritime Museum, Greenwich, London.

Fig. 1.9a [Col. Pl. V]: Roskilde warship, [23], © Vikingskibsmuseet, Roskilde, Denmark.

Fig. 1.9b: A cog, as reconstructed by Th. Macklin, [24], © 1995 Weltbild Verlag, Augsburg.

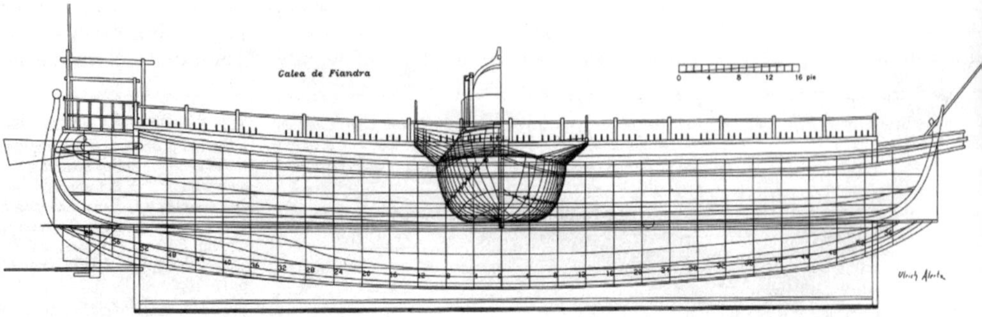

Fig. 1.10a: Venetian galley Galea de Fiandra (ca. 1410), as reconstructed by Ulrich Alertz, [53], © Ulrich Alertz.

Shape: Mediterranean method: Gabarit and sesto: Master mold and longitudinal interpolants. Round, piecewise circular arc sections.
Knowledge: Advance planning of hull shapes based on Mediterranean method. Fabrication of long series of large ships, skeleton first principle.
Fabrication: Marking of ship parts (frames) from master mold in the mold loft. Skeleton first assembly. Northern method still clinkered with the beginnings of carvel planking.
Tools (for shape definition): Mold loft, full scale lofting on the floor, marking by templates and master molds.
Evolutionary trends: Increasing ship sizes, shipboard gunnery since ca. 1500, trend toward fully rigged ships.
Pictures: Figs. 1.10a–1.10b.
References: Drachio [33], Rieth [30], Lane [34], Gardiner et al. [53].

5) *Early modern era (ca. 1600–1700)*

Types: Sailing ships of the line and frigates with heavy gunnery armament. Growth of the navies in Spain, England, France, Holland, Scandinavia. Larger cargo ships: East Indiamen.
Shape: Short and beamy with several decks and high freeboard and slender hull form in warships. Full merchantmen.
Knowledge: Rapid spreading of carvel, skeleton first construction method for larger vessels to Northwestern and Northern Europe. Solid practical shipbuilding knowledge, but largely still pre-scientific, non-analytic.

ig. 1.10b [Col. Pl. V]: A caravel, [28], © Little, Brown Book Group, London.

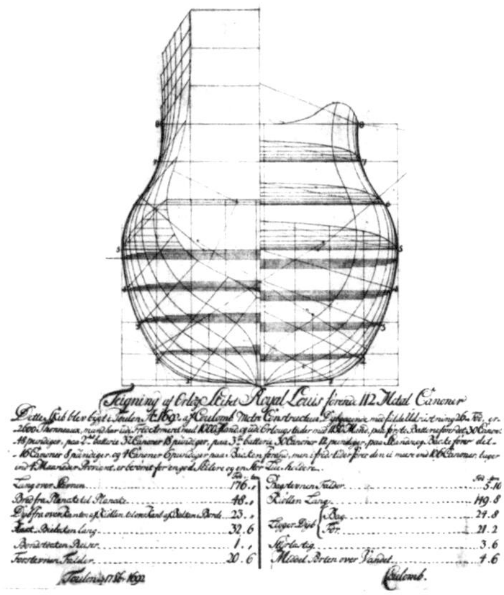

Fig. 1.11: "Royal Louis", F. Coulomb, 1692, [36], courtesy of the Musée national de la Marine, Paris.

Fabrication: Skeleton first construction methodology being accepted in Britain and Northern Europe, too.

Tools (for shape definition): Gradual introduction of scale, orthographic drawings (lines plans), and of lines fairing instruments: Bows and splines. Continuing use of mold lofting.

Evolutionary trends: Heavily armed, full rigged ships of the line and frigates. Short, beamy hull forms.

Picture: Fig. 1.11.

References: Barker [35], Hoving [15], Boudriot [36].

4. Milestones of Civil and Naval Architecture

In this section we will collect a short list of major events and in particular publications or citations that have played an important role in developing or documenting a similar consciousness in the two architectures. Some of these sources were really of incisive influence in developing an awareness of the identity of each architecture or in fact also of the interrelationships between both. Others are less significant as events, but they do show the awareness of traditional similarities. This listing does not claim to be complete, but may demonstrate a certain continuity in these relationships, though with fluctuations in their intensity. Table II presents a synopsis of some of the most remarkable milestone events and sources.

Vitruvius (2nd half of first c. B.C.), the respected Roman writer and military engineer, was the first pre-Christian author who wrote on the theory of architecture in his famous work *The Ten Books on Architecture* [37]. He created a systematic description and classification of building types and styles in accordance with his programmatic motto: "Architectura est scientia". His proposed goals of good architecture were "firmitas, utilitas, venustas", i.e., "strength, usefulness, beauty". Criteria for the esthetic assessment of architectural beauty were formulated, based on the harmony of proportions. He analyzed shape and function of architectural elements, e.g., in column orders. His influence in antiquity and the early Middle Ages was limited, but his scientific methodology in architectural theory was revived by the humanists and was fully recognized and admired during the Renaissance. His approach to the classification of shape elements and building types had lasting effects in architecture theory and may indirectly have influenced similar trends in naval architecture.

Alberti (1414–1472), Italian humanist and prolific writer, resurrected and further extended the Vitruvian theory of architecture. His influential book *"De re aedificatoria"* [38] adopted certain fundamental elements of Vitruvian style, but also added fresh, contemporaneous themes and criteria. His ranking and classification of building functions and styles took into account social-ethical aspects of the living community as well as formal-esthetical criteria, again derived from rules and proportions. He characterized the architect as a person of high responsibility and broad required practical and theoretical knowledge. The steps and instruments of architectural design were outlined. His work provided lasting guidelines for several centuries. *Alberti* is also cited as having written a book about ships ("NAVIS"), which however unfortunately is lost.

The affluent urban Renaissance societies in Italy and other countries led to many ambitious building projects and an increasing public prestige of renowned architects, also favored by their increasingly scientific public image. The profession took responsibility for a greater diversity of tasks, which was accompanied by more varied specialization. This

Table II: Historical Milestone Sources and Quotations.

Author	Source/Quotation	Date	Country	Reference
Vitruvius	De architectura libri decem	Ca. 25 B.C.	Roman Empire	[37]
Alberti	De re aedificatoria libri decem	1452, printed 1485	Italy	[38]
Francesco di Giorgio	Trattato d'architettura civile e militare. See Jacob Burckhardt [39]	1480	Italy	
Oliveira	Ars nautica	1570	Portugal	[40]
Digges	Stratiotikos, quoting incomplete "Architectura nautica"	1579	England	[41]
Lavanha	Livro Primeiro da Architectura Naval	1614–1615	Portugal	[42]
Furttenbach	Architectura Navalis	1629	Germany	[43]
Dudley	Arcano del mare, quoting "L'architettura nautica"	1646	England/Italy	[44]
Wilhelm	Architectura Civilis	1649	Germany	[45]
Deane	Deane's Doctrine of Naval Architecture	1670	England	[46]
Dassié	L'Architecture Navale	1695	France	[47]
Decker	Architectura Civilis	1711–1716	Germany	[48]
Sutherland	Quoting "...the Art of Marine Architecture"	1711	England	[49]
Chapman	Architectura navalis mercatoria	1768	Sweden	[50]

caused a conscious division of architecture into several branches, based on types of tasks. Thus apparently already in early Renaissance new terminologies are encountered which distinguish between "*civil*" and "*military*" architecture (*di Giorgio*, 1480 [39]). A century later the terms "*ars nautica*" (*Oliveira*, 1570 [40]) and "*Architectura nautica*" (*Digges*, 1579 [40]) are found.

Lavanha (ca. 1550?–1624) and *Furttenbach* (1591–1667) are the first literary sources who are using the designation "Architectura Naval" (Portuguese) or "Architectura Navalis" (Latin) in the titles of their treatises. Neither claims to have invented these names, so they might have been used orally for a while before the treatises were printed.

Lavanha, a leading scientist, mathematician and engineer on the Iberian peninsula in his era, a Portuguese scholar and writer engaged in Spanish court services for a long time, is indeed a key witness for the theme of our study. He was well educated and familiar with *Vitruvius*, *Alberti* and the architectural literature of his day. But he was equally well informed of the shipbuilding practices and geometric foundations, also in technical detail, at the turn from the 16th to the 17th c. when the Iberian countries held an important position in this industry. When he wrote his treatise "First Book of Naval Architecture" (abt. 1614–1615, [42]), which remained unfinished, it was one of his intentions to underscore the direct correspondence between civil and naval architecture, by emphasizing the analogies in objectives, design methods and instruments. He quoted Alberti in describing the tasks and knowledge background required from an architect; he postulated the same from a naval architect, in particular in arithmetic, geometry and drafting skills. He related *Alberti*'s six parts of architecture (ordinance, disposition, correspondence, adornment, decorum, distribution) to equivalent functions in naval architecture. He postulated the use of drawings with three orthogonal views and a perspective as well the demonstration by a model. And most importantly for our investigation theme, he adhered to the tripartition of architecture into *military*, *civil* and *naval* architecture. The origin of these words thus seems to lie before his treatise, but *Lavanha*, for all we know, definitely first used *Naval Architecture* in the title of a book.

Furttenbach, citizen of Ulm, merchant, architect, engineer and writer, is another prominent figure in the literary history of the architectures. As a young man he traveled extensively in Italy (Milan, Genova) and spent there more than a decade between 1607 and about 1620. He met many influential people, also architects and engineers, and developed a professional interest in architecture. He must have visited several shipyards, too, in particular in Venice, because his later writings on this topic reflect strong Italian influence. In the years 1628, 1629, 1630, back in Ulm, in quick succession he published three voluminous books under the titles of: "Architectura civilis", "Architectura navalis" and "Architectura martialis". The classical tripartition is reflected here again although he later wrote three more books and invented the titles "Architectura universalis", "…recreationis" and "…privata". There is no indication that *Furttenbach* knew *Lavanha*'s treatise, more likely he picked up his terminology for the first three architectures while in Italy. In his substantial books on these three architectures he crisply explains the unique traits of each branch, but also their common professional roots. Like *Lavanha* he sees the affinities in similar tasks and subtasks, similar organisation of the design process, similar criteria and conceptual instruments of design. It becomes very evident that the two or three architectures felt a consciousness of being closely related intellectually and in motivation.

Most of the further references in Table II are related to books or treatises that appeared in the century that followed (Dudley [44], Wilhelm [45], Deane [46], Dassié [47], Decker [48], Sutherland [49]). They are mentioned here mainly to demonstrate that the reference to Civil or Naval Architecture in many languages had become rather common and not infrequent, although of course the contents of each field developed further as technologies and knowledge changed. Whether the substance of the professions changed so as to strengthen or weaken the close ties that had existed during the Renaissance is a subject that deserves further investigation.

Chapman (1721–1808), the famous Swedish naval constructor, stands at the end of the list in Table II for two main reasons: First, by his work "Architectura navalis mercatoria" [50], a collection of several hundred lines plans with orthographic views and beautiful perspective drawings of the merchant ships of his era, compiled essentially from his own drawings, he stands at the culmination point of a development of ship description that began with the promises and postulates of Alberti and Lavanha. Second, Chapman made use of the still reasonably novel techniques of ship lines fairing by French curves and probably spline curves, drawn with elastic battens. Thereby he made himself independent of the rather rigid system of proportionate subdivision along diagonals and was able to construct free-form surfaces of any desired shape. This method was augmented by his invention of the Sectional Area Curve, a plot of underwater cross sectional areas in vertical planes against ship length, whose integration yields the ship's displacement volume and volume centroid longitudinally. This planning and control device in lines creation enabled him to hit the desired ship size and trim condition right the first time, a breakthrough made possible by the use of calculus and numerical integration. Thus *Chapman* also stands at the beginning of the modern era of hull shape creation. This is documented in his "*Treatise on Shipbuilding*" [51].

Fig. 1.12a [Col. Pl. VI]: Frederik Henrik af Chapman (1721–1808), [54], © Swedish National Museum of Fine Arts, portrait by L. Pasch the younger.

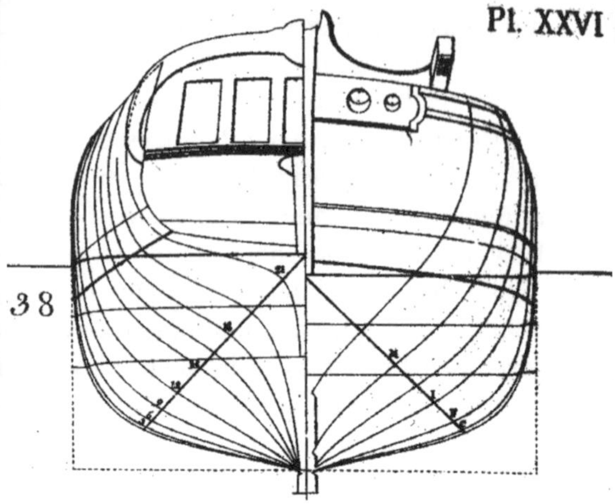

Fig. 1.12b: Body plan from "Architectura Navalis Mercatoria" [50], Chapman, 1768.

Evolution of Hull Shape Complexity

Hull shape creation certainly has multiple roots in diverse parts of the world, it cannot have originated only once. In the Mediterranean countries and in Northwestern and Northern Europe the ship and boat types with pronounced keel or strong bottom structures became the dominant shiptype in the long run, both in their "shell first" and later "skeleton first" versions. The combination of "skeleton first" molding with carvel planking (first in the Mediterranean) as of the Middle Ages constitutes the most promising line of development in many countries and certainly enabled the designers to develop methods and instruments for conceiving and fabricating repeatable shapes of varying geometries.

Generally the attainable shape variety depends essentially on the physical tools and conceptual methods for realizing any proposed shape. Thus in this section I will limit myself to a few observations on certain mainstream developments, starting with the Mediterranean method for skeleton-first, carvel planking design by means of the master mold (sesto, gabarit). The British technology of whole-moulding will not be addressed here.

I will distinguish the following stages of development in methodologies and hull shape complexity:

- The Mediterranean method
- Proportionate subdivision along diagonals
- Free subdivision along diagonals
- Free-form fairing

The discussion is intended to highlight how the geometric properties of the hull form achieved depend on the shape creation tools and methods.

1) *Mediterranean method*

There is archaeological and documentary evidence (Rieth [30], [6], Drachio [33]) that this method of hull shape design and part marking was used in several Mediterranean shipbuilding countries, especially in France, Italy (Venice) and on the Iberian peninsula, during the late Middle Ages (beginning no later than ca. 1300) and during the Renaissance in the context of skeleton-first, carvel planking ship construction.

In this method a single master mold (maître gabarit, sesto), i.e., a planar template made of wood, is used to define the frame shapes of the whole skeleton throughout the ship length except at the extreme ship ends, which are constructed separately by more pragmatic rules. The master mold contains not only the frame shape as it holds at midship section, but also carries several markings so that the shape of all other transverse sections can be derived by a geometrical process of translation, rotation and clipping, taking into account the longitudinally tapered shape of the hull. This process is also supported by two other planar marked templates, the rise of floor template (tablette, rising staff) and the narrowing of breadth template (trébuchement, breadth staff). Thus the entire hull surface shape for a great percentage of ship length is constructed from a single master curve, valid amidships, so that the other transverse sections in large measure inherit the shape features of the midship section. The shape of the midship section can be chosen arbitrarily by the designer, in practice it often consists of segments of circular arcs and straight lines, connected with tangency continuity. In this case the piecewise circular arc character of the midship section is retained over most of the ship body length.

The procedure of frame shape definition follows the sequence shown in Figs. 1.13a–1.13i (Rieth [30]) in an example based on La Madeleine [30]:

- The midship section underwater is laid out (Fig. 1.13a). It is here composed of three segments, the flat bottom (straight line), the lower circular arc with a smaller radius from the bottom to the bilge diagonal line, and the upper circular arc with a larger radius from the diagonal to the waterline.
- A pencil of straight lines is drawn (Fig. 1.13b) whose spacing varies as a parabolic function (0, 1, 4, 9, 16, 25, 36) in order to create a proportionate subdivision of certain characteristic measures of the hull shape: The "floor width" (top, "longueur du plat"), the "rise of floor" (middle, "acculment") and "narrowing of breadth" (bottom, "trébuchement") are inserted into the pencil as horizontal lines according to their designated lengths. These lines are marked for each Station (from 0 to 6) and the marks are transferred on marking staffs (floor width staff, rising staff = tablette, narrowing breadth staff = trébuchement).
- Fig. 1.13c shows how the "floor width" and "rise of floor" markings are transferred to the midship section view. A new vertical reference line, half a keel width from the centerplane, serves to define the origin for the frame construction. The "floor width" marks (red) are laid out horizontally from there, the "rise of floor" marks (blue) vertically. An enlarged view of these markings is inserted nearby (green arrow shift).
- Fig. 1.13d shows the corresponding markings and their enlargement (magenta) for the "trébuchement" operation.
- Now the example of Station 5, the next to last station forward, will illustrate the construction of station shapes. In *step* 1 (Fig. 1.13e) the midship section curve is lifted up according to the "rise of floor" mark and at the same time shifted sideways to the left to clip the floor according to the "floor width" mark.
- The next transformation (*step 2*) consists of three operations, shown in Figs. 1.13f–h. The *first* operation is another parallel upward translation of the station (Fig. 1.13f) according to the mark on the breadth staff (Fig. 1.13d).
- The *second* operation is a rotation of the station in its new position about its left-hand end point on the vertical line at the half-keel width. The station is rotated clockwise until its lowest point becomes a point

of tangency to the horizontal line of step 1 (Fig. 1.13e). Then the part of the rotated station to the left of the point of tangency is clipped off and replaced by a horizontal straight line. In the *third* operation this straight line segment is trimmed down to the applicable "floor width" (acc. To Fig. 1.13c) so that the resulting width of the station is reduced. The final section shape at Station 5 is shown in Fig. 1.13h (black).

- Fig. 1.13i shows the sections at all stations from 0 to 6, constructed accordingly.
- In Fig. 1.13j a perspective view of the surface interpolated through all stations is given in a wireframe representation.
- Fig. 1.13k presents a color rendered visualisation of this surface.
- Finally in Fig. 1.13l a color rendering of the normal curvature distribution, taken in transverse planes, is given.

The resulting surface, as is shown, inherits the shape character of the midship section, though modified by the tapering operations. The curvature pattern consists of nearly constant band widths of the given arcs. All sections remain tangency continuous. All sections are convex, unless a concave circular arc were given. The designer controls the shape only by the choice of midship section shape and of the longitudinal interpolant (here parabolic).

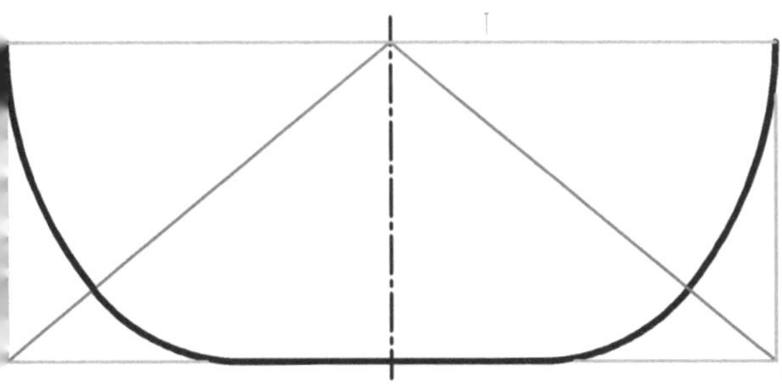

Fig. 1.13a [Col. Pl. VII]: Midship section underwater (example by La Madeleine, 1712, from Rieth [30]), courtesy of the Musée national de la Marine, Paris.

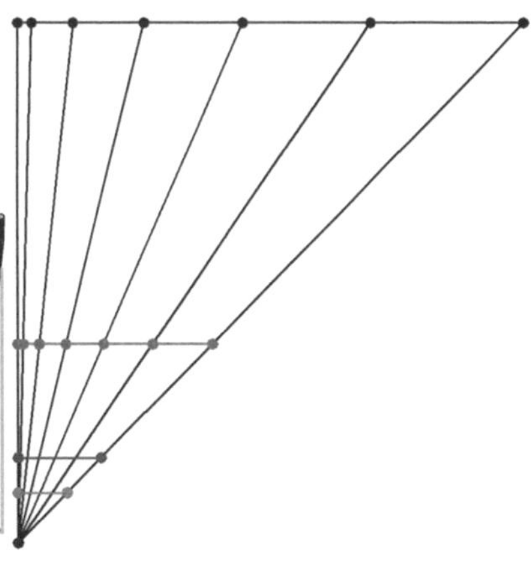

Fig. 1.13b [Col. Pl. VII]: Creation of proportionate subdivision for "floor width", "rise of floor"and "trébuchement".

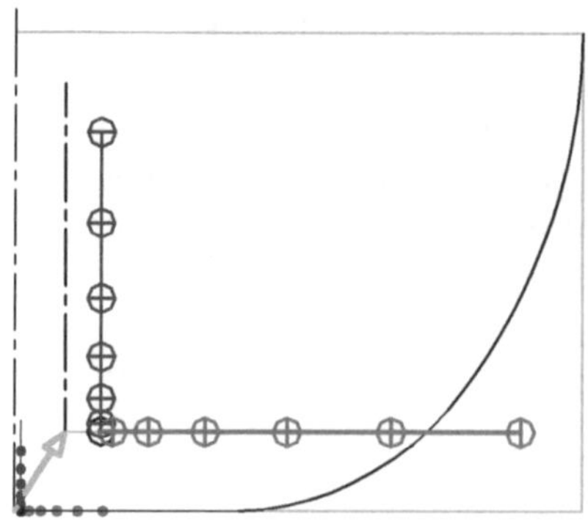

Fig. 1.13c [Col. Pl. VII]: Marking of "floor width" (red, "longueur du plat") and "rise of floor" (blue, "acculment").

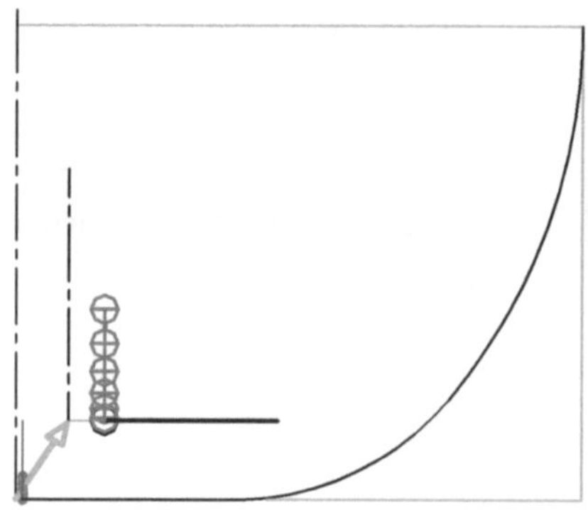

Fig. 1.13d [Col. Pl. VII]: Marking (magenta) for translation/rotation/clipping (trébuchement).

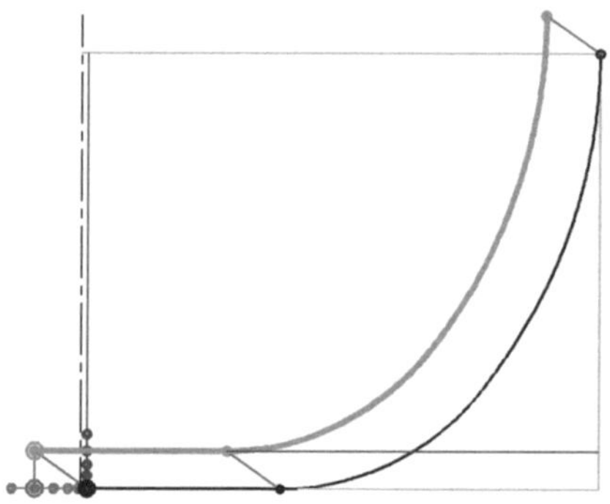

Fig. 1.13e [Col. Pl. VII]: Creation of Station 5 fwd., translation, upward shift and floor width reduction (step 1).

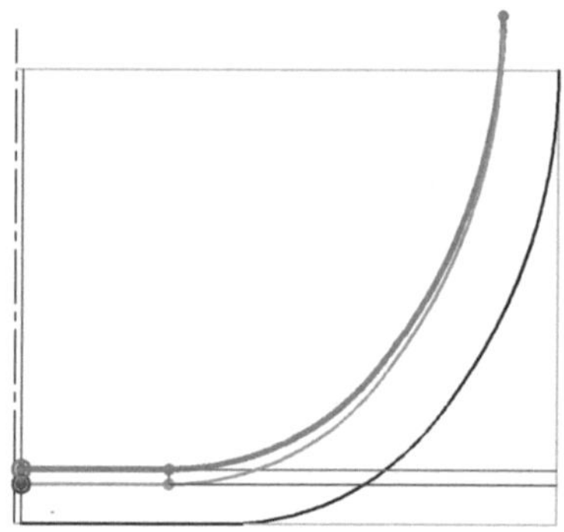

Fig. 1.13f [Col. Pl. VII]: Station 5 fwd., upward shift (step 2, trébuchement, operation 1).

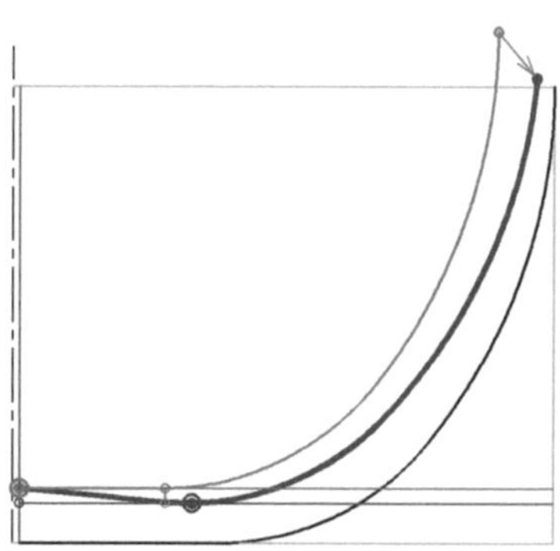

Fig. 1.13g [Col. Pl. VIII]: Station 5 fwd., rotation about magenta end point of blue curve until point of tangency on horizontal blue line is reached (step 2, trébuchement, operation 2).

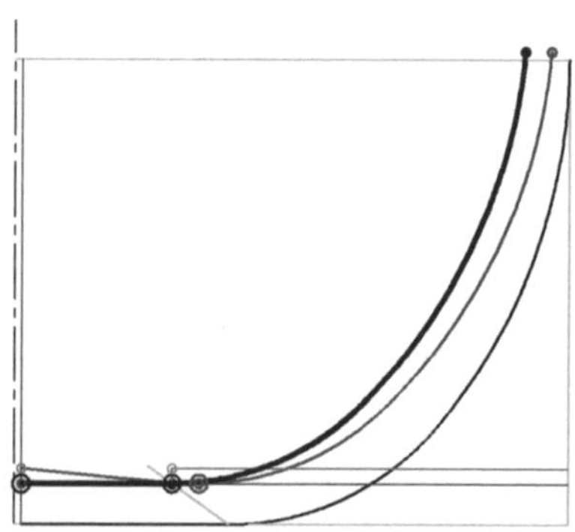

Fig. 1.13h [Col. Pl. VIII]: Clipping of left segment of blue curve, replacement by black line segment, clipped to floor width. Final curve: Black. Step 2, trébuchement, operation 3.

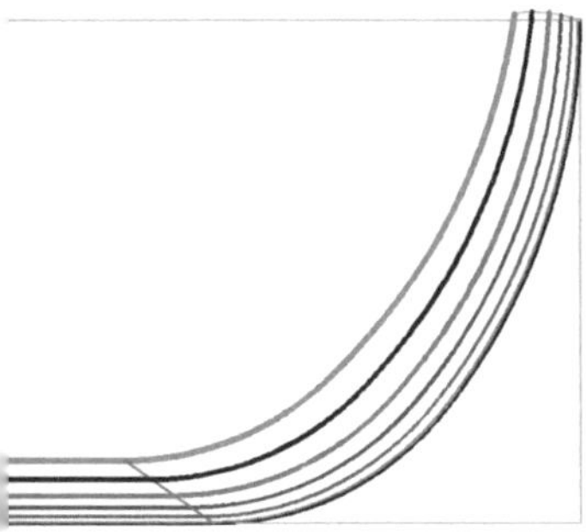

ig. 1.13i [Col. Pl. VIII]: Sections for all Stations 0 to 6.

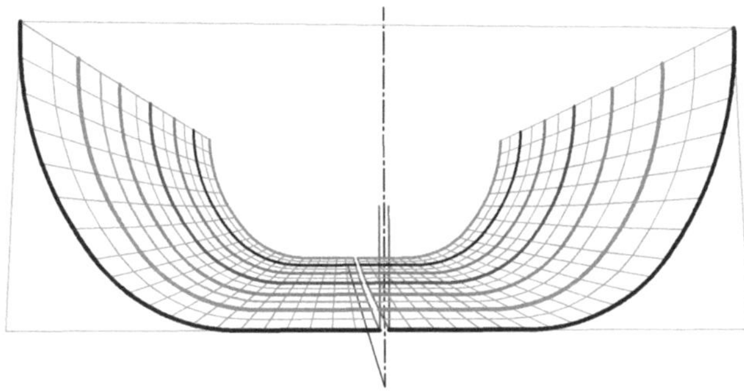

Fig. 1.13j [Col. Pl. VIII]: Wireframe perspective view of interpolated surface.

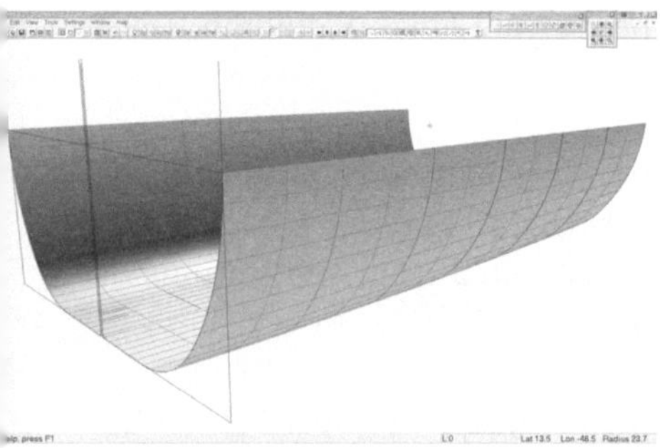

ig. 1.13k [Col. Pl. VIII]: Color rendering of surface.

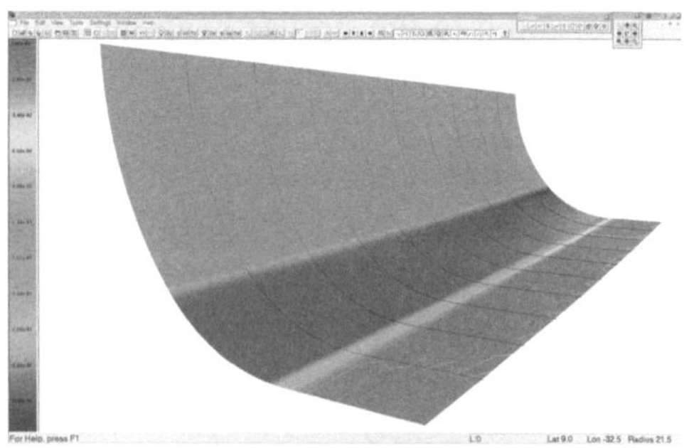

Fig. 1.13l [Col. Pl. VIII]: Rendering of normal curvatures, taken in transverse planes.

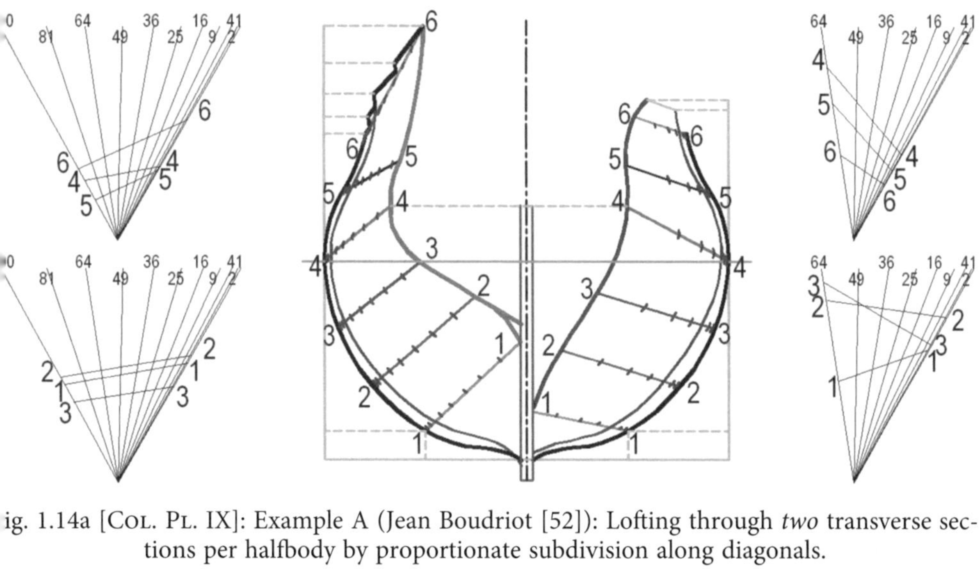

ig. 1.14a [Col. Pl. IX]: Example A (Jean Boudriot [52]): Lofting through *two* transverse sections per halfbody by proportionate subdivision along diagonals.

2) *Proportionate subdivision along diagonals*

This method, dating from the 17th c., is a further development of the Mediterranean method, dropping some of its restrictions. Here the designer can freely choose the shape of *two* transverse sections per half-body, say, for the midship section and a section near the end. Then intermediate transverse section shapes are constructed by interpolation, using a longitudinal interpolation operator that is formed by the same proportionate subdivision applied to each of the diagonal sections (Fig. 1.14b). Thus a uniform proportionate "stretching" occurs in the hull surface along all diagonals. The steps are illustrated in Figs. 1.14a–1.14i, referring to an Example Ship A from Jean Boudriot [52], and to the similar Example Ship B, the "Royal Louis" (Coulomb, 1692, Fig. 1.11):

- Example A: Choice of two end sections per half-body, proportionate subdivision, uniform for all diagonals (Fig. 1.14a)
- Lofted surface, forward and aft, reconstructed from end stations by parabolic interpolation (Fig. 1.14b)
- Results, Example A (Figs. 1.14c–1.14e)
- Similarly for "Royal Louis", Example B (Figs. 1.11 and 1.14f–1.14i)

The results illustrate that this method has gained additional freedoms by allowing two independent choices of end sections. The Gaussian curvature map shows a slight narrowing of the convex domain band width toward the ends (Fig. 1.14i). This is controled by the designer.

3) *Free subdivision along diagonals*

If further the restriction of *uniform* proportionate subdivision along *all* diagonals is dropped, then the designer can make a free choice of the interpolation operator for each diagonal. Of course he will still favor a smooth, convex, monotonically varying function for his interpolants. Yet he can design sharper tapering or more complex curvature variations into his surface. Figs. 1.15a–1.15c show a corresponding set of results for the example ship "Dauphin Royal" (Blaise Ollivier, 1735). The bandwidth of elliptic (convex-convex) Gaussian curvature narrows quite sharply towards the ends.

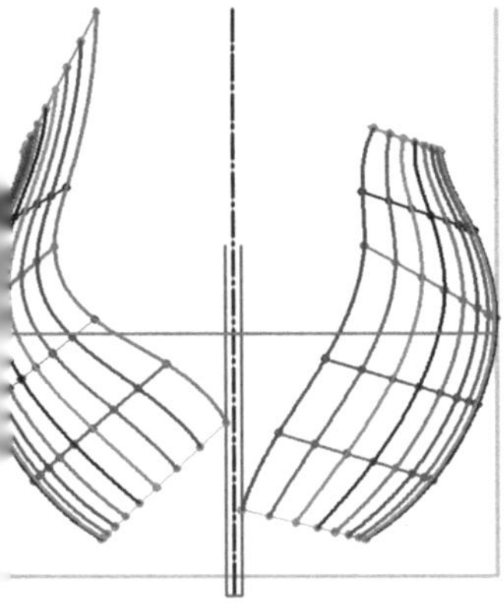

g. 1.14b [Col. Pl. IX]: Example lofted surface, forward and aft.

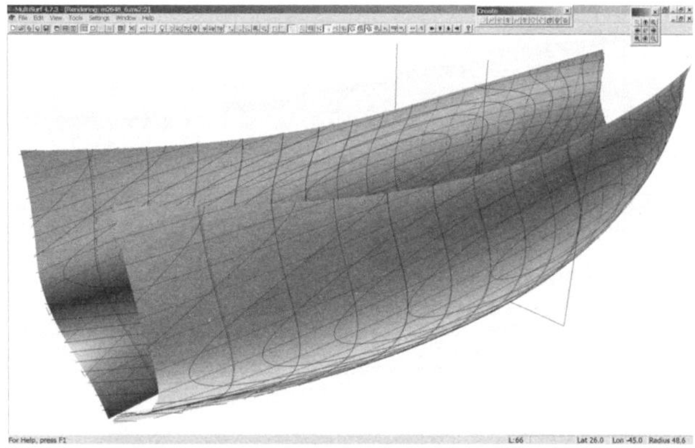

Fig. 1.14c [Col. Pl. IX]: Example A, rendered surface, perspective view.

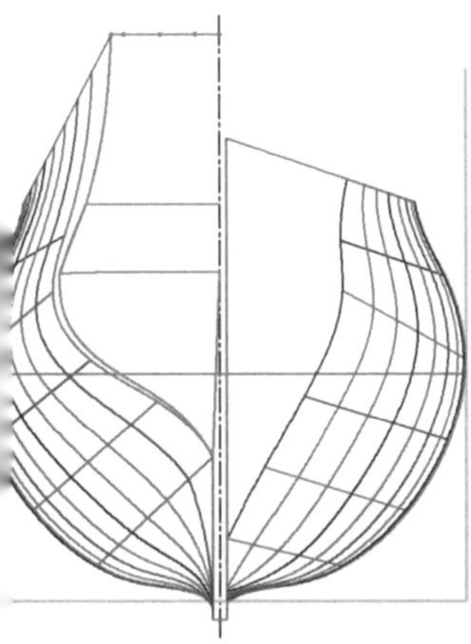

g. 1.14d [Col. Pl. IX]: Complete body plan, reconstructed.

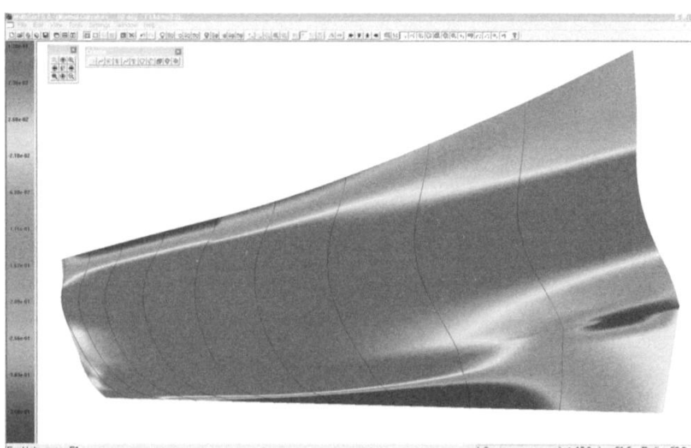

Fig. 1.14e [Col. Pl. IX]: Normal curvature distribution, taken in transverse planes, afterbody.

4) *Free-form fairing*

With the introduction of ship lines plans and lines fairing methods by elastic splines and French curves hull shape creation was freed of many previous restrictions. In essence it became feasible to create any desired shape. Thereby the designer received less guidance and more responsibility for creating fair hull shapes. Yet the ensuing problems were mastered by experienced designers and skillful loftsmen in the mold loft.

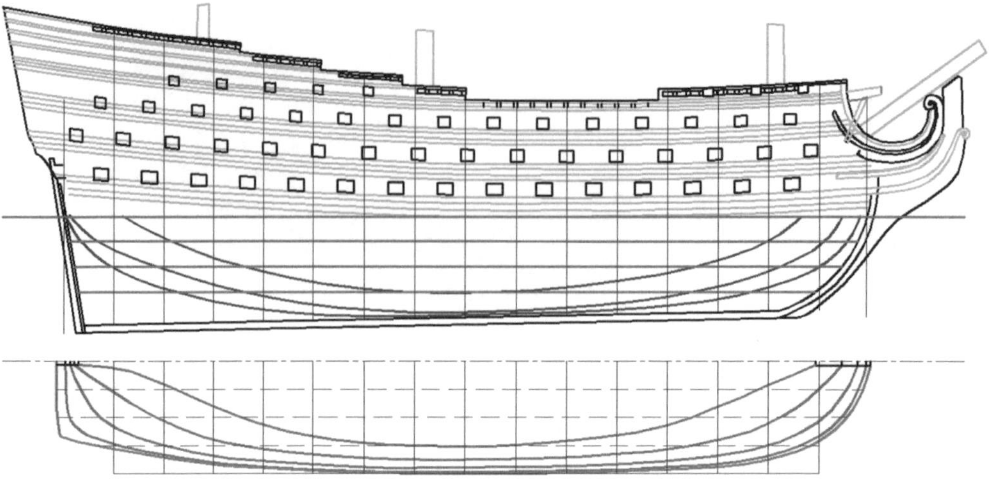

Fig. 1.14f [Col. Pl. X]: Example B, "Royal Louis" (1692), reconstructed by J. Sonntag. Source: Musée national de la Marine, Paris.

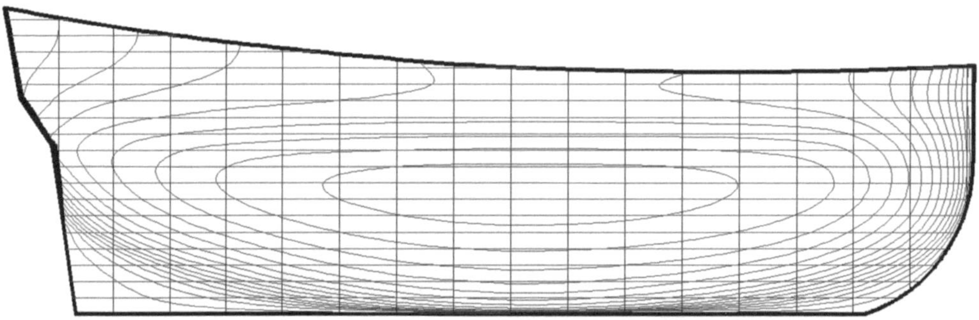

Fig. 1.14g [Col. Pl. X]: "Royal Louis", reconstructed elevation, wireframe view, with longitudinal planar intersections (buttock lines).

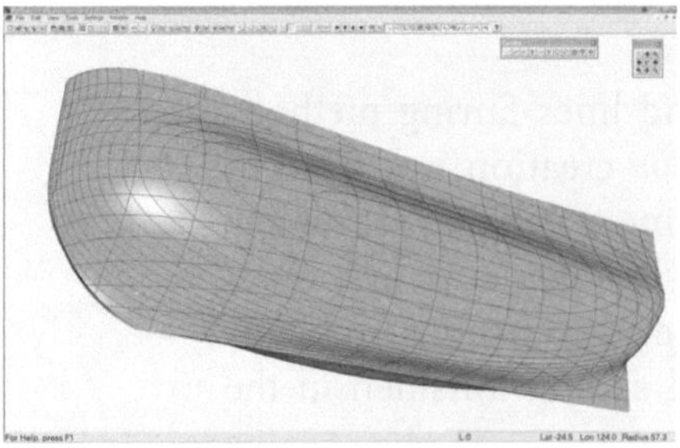

Fig. 1.14h [Col. Pl. X]: "Royal Louis", color rendering.

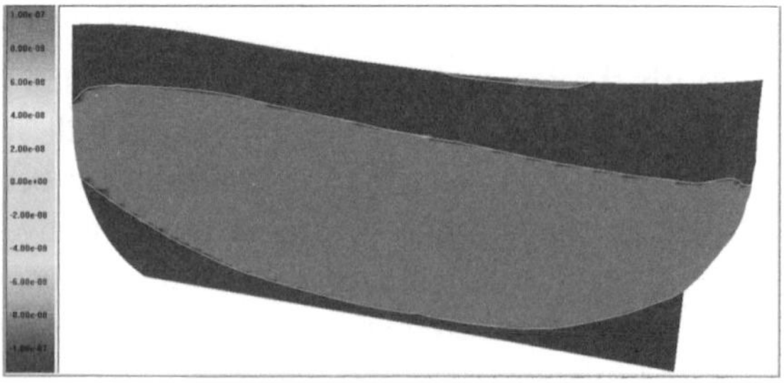

Fig. 1.14i [Col. Pl. X]: "Royal Louis", Gaussian curvature distribution.

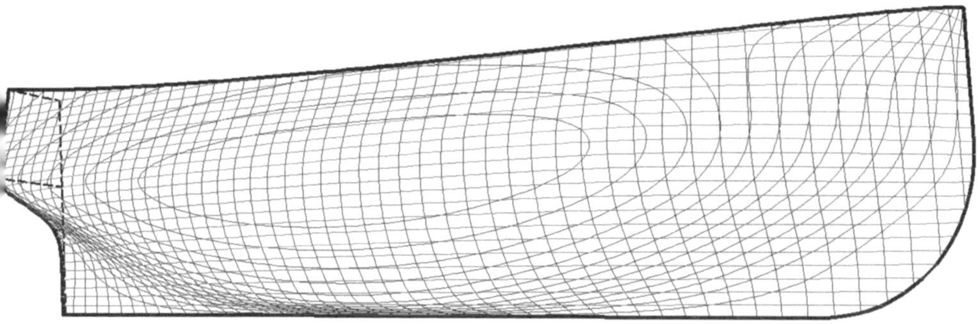

Fig. 1.15a [Col. Pl. XI]: "Dauphin Royal", Blaise Ollivier, 1735, wireframe perspective view.

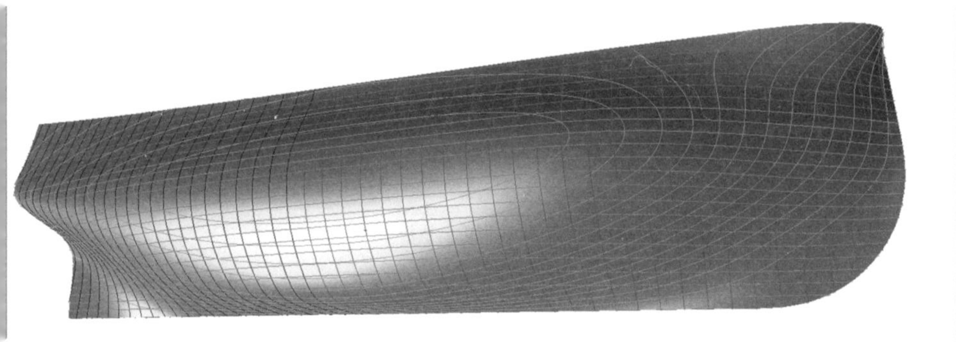

Fig. 1.15b [Col. Pl. XI]: "Dauphin Royal", color rendered view.

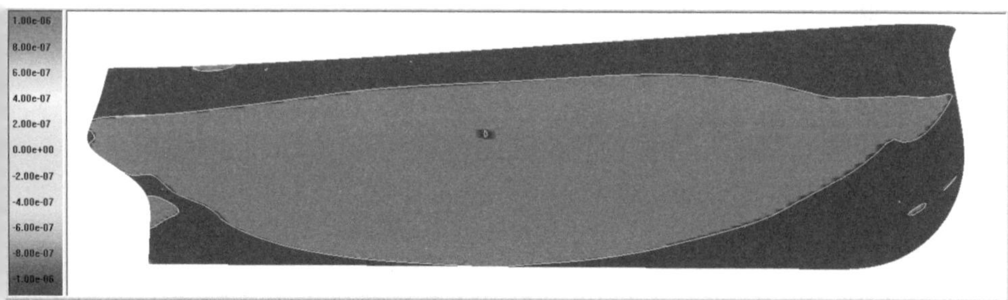

Fig. 1.15c [Col. Pl. XI]: "Dauphin Royal", Gaussian curvature rendering.

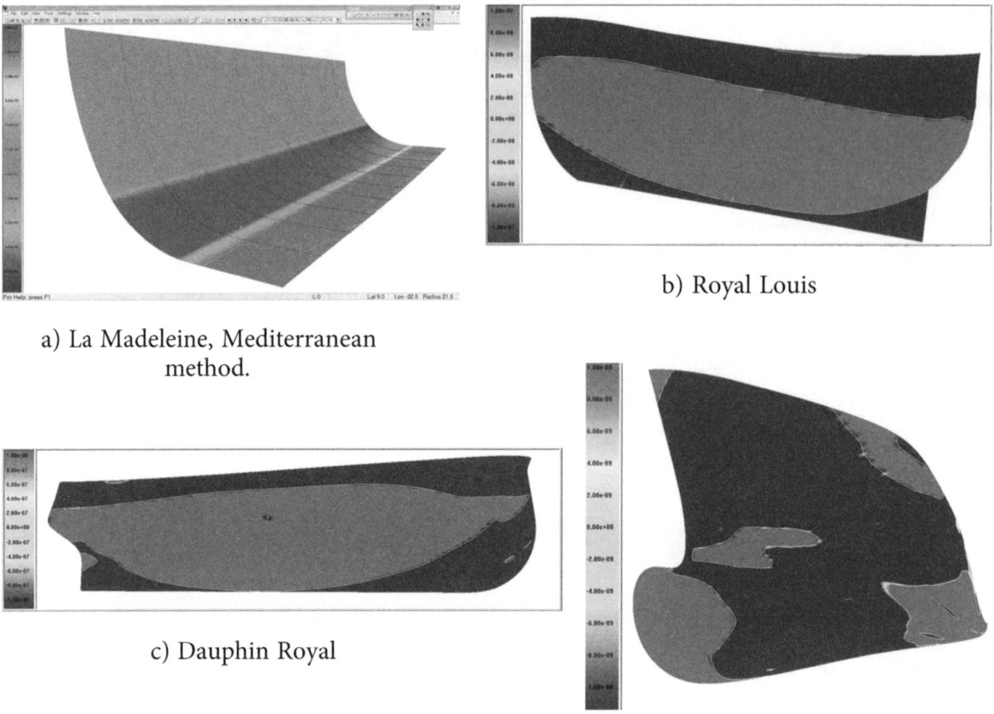

a) La Madeleine, Mediterranean method.

b) Royal Louis

c) Dauphin Royal

d) Modern Tanker

Fig. 1.16 [Col. Pl. XII]: Summary comparison of Gaussian curvature maps for various design methods.

Naval architecture fairing tools and methods for free-form design in the following centuries became a much imitated model in other industries, e.g., in the geometric design of railway cars, automobile body shapes and aircraft fuselages. Free-form design by corresponding computer-based methods is still prevalent today. For now, as an illustration, just one of Chapman's hull form creations is shown in Fig. 12b.

Fig. 1.16 compares the four principal methods of hull form creation discussed earlier and demonstrates the increasing degree of curvature control exerted from one method to the next one: a) Mediterranean method, nearly constant curvature value band width, b) Equal proportionate subdivision of diagonals, mildly narrowing curvature bands, c) Variable proportionate subdivision, strongly narrowing bandwidth, d) Free-form design, arbitrary curvature patterns. These developments in methods and tools took several centuries.

5. On Fairness

It seems to be a human trait since the dawn of history that the shape of a well-designed ship is regarded as a thing of beauty. This can be illustrated by many quotations and artifacts throughout the course of history. But how is the beauty of a ship judged and how is it achieved?

Naval architects have developed the concept of "fairness" for the desired harmony in a hull form. Fairness is a bit elusive to define and subject to personal tastes. Technically the absence of flaws and geometric degeneracies, the presence of a smooth, gradual development of ship lines and surfaces is sometimes considered as the minimum necessary requirement for a fair hull shape. The overall proportions of a hull shape certainly play a role, too. Mathematically one can play with the minimization of "noise" or "acceleration" in a surface. One can define mathematical fairness measures and thus quantify "fairness". This will not lead to a single unique answer, but generally to quite acceptable solutions. This methodology has been the subject of intensive recent studies, which are synoptically reported by Nowacki et al. in [55].

In practice, however, fortunately the methods and tools of hull shape design in many historical periods have generally tended to favor "fair shapes" as a result. The use of planking, clinkered or especially carvel, almost naturally produces a "fair surface" The lofting principles of the Mediterranean method and of the proportionate subdivision techniques by virtue of their built-in smooth interpolants tend to provide fair shapes. Likewise free-form fairing by means of splines and French curves in the hands of an experienced designer will naturally facilitate the achievement of a fair ship shape. Hydrodynamic functional advantages may provide further incentives.

Architects have the same strong desire to create things of beauty. They judge beauty by different criteria and with different objectives. But in the end this innate desire for "beauty" or "fairness" is another strong motivation that the two architectures are sharing.

6. Conclusions

Civil Architecture and Naval Architecture have shared a long and interrelated history. It is arguable whether the two disciplines have always mutually and substantially benefitted from this basic affinity. But it is not disputable that Naval Architecture at least has had advantages by considering itself as a branch of architecture and sharing with it certain traditions and the general consciousness of responsibility for the quality of shape creation.

Acknowledgments

I am pleased to express my gratitude for the help received from several persons and institutions. First, I acknowledge with thanks the support received from Richard Barker and Eric Rieth by a continuing communication during the last several years on matters related to the theme of this book, especially on subjects of naval architecture history. This was both motivating and inspiring. Both colleagues in the recent years also helped in preparing the workshop held at the Max Planck Institute on this subject in 2006 by valuable advice and orientation. Further, through Eric Rieth the Musée national de la Marine, Paris also helped and deserves credit for providing geometric data on several examples of historical ships. The active support contributed by Jürgen Sonntag, VWS Berlin, in the computer-based reconstruction and geometric modeling of the majority of historical hull forms and their visualisation, also for the simple shapes, is also gratefully acknowledged. The hull form curve and surface reconstructions shown in Section 6, Figs. 1.13–1.15, were performed by means of the AEROHYDRO MultiSurf software system. Last, but not least I am indebted for and gratefully recognize the encouragement and help I have received by the Max Planck Institute for the History of Science, Berlin where I have been a Visiting Scholar for the last eight years. I have much enjoyed the free scientific atmosphere and open dialogue with many colleagues there. I am also grateful for several useful editorial comments by my friends Harry Benford, Walter Debler and Jobst Lessenich.

References

1. Diamond, Jared: "Guns, Germs, and Steel", W. W. Norton Co., New York-London, 1997.
2. Aristotle: "The Athenian Constitution", ca. 325 B.C., translation by H. Rackham, Loeb Classical Library, Aristotle vol. XX, Section XLVI, pp. 128–129, Harvard University Press, Cambridge, MA, reprinted 1992.
3. Aristotle: "Physika", Book II2 193b, Loeb Classical Library, Harvard University Press, Cambridge, MA, reprinted 1995.

4. Menge, Hermann: "Langenscheidt's Large Dictionary Greek-German", Langenscheidt Publ., 22nd ed., Berlin-Munich, 1973.
5. Homer: "Odyssey", IX 126/127, translated by A. T. Murray, Loeb Classical Library, Harvard University Press, Cambridge, MA, reprinted 1995.
6. Rieth, Eric: "Des mots aux pratiques techniques: gabarits et architecture navale au Moyen Age", Chronique d'Histoire Maritime, no. 56, pp. 13–34, 2004.
7. Rieth, Eric (ed.): "Concevoir et construire les navires: De la trière au picoteux", Technologie/Idéologie/Pratiques, Editions Erès, Ramonville Saint-Agne, 1998.
8. Rieth, Eric: "Une tradition médiévale et méditerranéenne de conception des navires construits à franc-bords 'membrure première'", article in [7].
9. Pomey, Patrice: "Conception et réalisation des navires dans l'Antiquité méditerranéenne", article in [7].
10. Basch, Lucien: "Ancient wrecks and the archaeology of ships", The International Journal of Nautical Archaeology and Underwater Exploration, vol. 1, pp. 1–58, 1972.
11. Basch, Lucien: "Le musée imaginaire de la marine antique", Hellenic Institute for the Preservation of the Nautical Tradition, Athens, 1987.
12. Crumlin-Pedersen, Ole: "Nordic Clinker Construction", Chapter 4 in [17], 2004.
13. Steffy, J. Richard: "Ancient Scantlings: the projection and control of Mediterranean hull shapes", in H. Tzalas (ed.), TROPIS III, 3rd International Symposium on Ship Construction in Antiquity, pp. 417–428, Athens, 1995.
14. Barker, Richard: "English Shipbuilding in the Sixteenth Century: Evidence for the Processes of Conception and Construction", article in [7].
15. Hoving, Ab: "Nikolaes Witsens Scheeps-Bouw-Konst Open Gestelt" ("Nikolaes Witsen's Shipbuilding Art openly Presented"), Uitgeverij Van Wijnen, Franeke, 1994.
16. Hoving, Ab, and Lemmers, A. A.: "In Tekening Gebracht: De achtiende eeuwse scheepsbouwers en hun ontwerpsmethoden", De Bataafse Leeuw, Amsterdam, 2001.
17. Hocker, Fred, and Ward, Cheryl (eds.): "The Philosophy of Shipbuilding", Texas A&M Univ. Press, College Station, 2004.
18. Morris, William (ed.): "The American Heritage Dictionary", American Heritage Publ. Co., New York, 1971.
19. Gimpel, Jean: "Les bâtisseurs des cathédrales", Editions du Seuil, 1980.
20. Colombier, Pierre du: "Les chantiers des cathédrales", A.&J. Picard, Paris, 1953, reprinted 1973.
21. Bartels, Klaus, and Huber, Ludwig (eds.): "Lexikon der Alten Welt" ("Encyclopedia of the Ancient World"), Patmos Publ., Albatros Publ., 1990.
22. Lächler, Paul, and Wirz, Hans (eds.): "Die Schiffe der Völker" ("The Ships of the Peoples"), Walter-Verlag, Olten-Freiburg, 1962.
23. Lavery, Brian: "SHIP: 5000 Years of Maritime Adventure", Dorling Kindersley Ltd., National Maritime Museum, London, 2004.
24. Dudszus, Alfred; Henriot, Ernest: Köpcke, Alfred; Krumrey, Friedrich: "Das grosse Buch der Schiffstypen" ("The Great Book of Ship Types"), Weltbild-Verlag, Augsburg, 1995.
25. Greenhill, Basil, and Morrison, John: "The Archaeology of Boats and Ships: An Introduction", Conway Maritime Press, London, 1976 and 1995.
26. Greenhill, Basil: "The Evolution of the Wooden Ship", Facts on File, New York-Oxford, 1988.
27. Crumlin-Pedersen, Ole: "Hjortspring: A Pre-Roman, Iron-Age Warship in Context", Viking Ship Museum, Roskilde, 2003.
28. Kemp, Peter: "The History of Ships", Macdonald & Co. Publ. Ltd., Black Cat, London, 1978 and 1988.

29. Steffy, J. Richard: "The Mediterranean shell to skeleton transition. A Northwest European parallel", in Reinder Reinders and Kees Paul (eds.), Fifth International Symp. on Boat and Ship Archaeology, Amsterdam, 1984, Oxbow Monograph 12, Oxford, 1991.
30. Rieth, Eric: "Le Maître-Gabarit, la Tablette et le Trébuchet: Essai sur la conception non-graphique des carènes du Moyen Âge au XXe siècle", Comité des Travaux Historiques et Scientifiques, Paris, 1996.
31. Crumlin-Pedersen, Ole: "Viking-Age Ships and Shipbuilding in Hedeby/Haithabu and Schleswig", Ships and Boats of the North, vol. 2, Wikinger Museum Haithabu/ Viking Ship Museum, Schleswig and Roskilde, 1997.
32. Unger, Richard (ed.): "Cogs, Caravels and Galleons", Conway Maritime Press, Conway's History of the Ship, London, 1994 and 2000.
33. Drachio, Baldessara: "Visione di Drachio", Venetian manuscript, ca. 1593.
34. Lane, Frederic Chapin: "Venetian Ships and Shipbuilders of the Renaissance", Greenwood Press Publ., Westport, CT, 1934 and 1975.
35. Barker, Richard: "Design in the Dockyards, About 1600", in 4th International Symposium for Boat and Ship Archaeology, Porto, 1985, O. L. Figueira (ed.), BAR Int'l Series 438, Oxford, 1988.
36. Lemineur, Jean Claude: "Les Vaisseaux du Roi Soleil", Collection Archéologie Française, Editions OMEGA, Nice, 1996.
37. Vitruvius Pollio: "De architectura libri decem", ca. 25 B.C., translated into English by Frank Granger, Loeb's Classical Library, Harvard University Press, Cambridge, MA, reprinted 1995.
38. Alberti, Leon Battista: "De re aedificatoria libri decem", 1452, printed 1485, translated into German and edited by M. Theuer: "Zehn Bücher über die Baukunst", Wien and Leipzig 1912, reprinted Darmstadt, 1991.
39. Burckhardt, Jacob: "Geschichte der Renaissance in Italien" ("History of the Renaissance in Italy"), 7th ed., Esslingen, Paul Neff Verlag (Max Schreiber), 1924.
40. Oliveira, Fernando: "Ars Nautica", manuscript, 1579, in Barata, João da Gama Pimentel, Memórias do Centro de Estudos de Marinha, Lisbon, 1972.
41. Digges, Leonard, completed by his son Thomas: "Stratiotikos, an Arithmeticall Militare Treatise, compendiously teaching the Science of Numbers, as well in Fractions as Integers,...", London, 1579. Reprinted by Theatrum Orbis Terrarum Ltd., Amsterdam, and Da Capo Press, New York, 1968.
42. Lavanha, João Baptista: "Livro Primeiro da Architectura Naval", ca. 1614–1615, reprinted by the Academia de Marinha, Lisbon, with English translation by Richard Barker, 1996.
43. Furttenbach, Joseph: "Architectura Navalis, das ist: Von dem Schiff-Gebäw, auff dem Meer und Seeküsten zu gebrauchen..." ("...i.e., on the ship construction, to be used on the ocean and sea coasts..."), Ulm, 1629. Reprinted by Georg Olms Verlag, Hildesheim-New York, 1971/1975.
44. Dudley, Roberto: "Dell'Arcano del Mare", Book IV: Shipbuilding, Florence, 1646.
45. Wilhelm, Johann: "Architectura Civilis, oder Beschreibung und Verreissung vieler vornehmer Dachwerck..." ("...or Description und Documentation of many prominent roof constructions..."), Nürnberg, 1649.
46. Deane, Sir Anthony: "Deane's Doctrine of Naval Architecture", 1670, newly edited by Brian Lavery, Naval Institute Press, Annapolis/ Conway Maritime Press, London, 1980/81.
47. Dassié, Charles: "L'Architecture navale, contenant la manière de construire les Navires, Galères & Chaloupes,..." ("...containing the manner of constructing the Ships, Galleys & Longboats..."), Jean de la Caille, Paris, 1677, Laurent d'Houry, Paris, 1695, facsimile reprint by Ancre, Nice, 1994.

48. Decker, Paulus: "Architectura Civilis" ("Princely Architecture or: Civil Architecture"), two parts, Augsburg, 1711–1716.
49. Sutherland, William: "The Ship-Builder's Assistant: or some Essays Towards Compleating the Art of Marine Architecture", A. Bell and A. Smith, London, 1711.
50. Chapman, Frederik Henrik af: "Architectura Navalis Mercatoria", Stockholm, 1768. Reprinted by Prager Publ., New York, 1968.
51. Chapman, Frederik Henrik af: "Traktat om Skeppsbyggeriet", 1775, French translation by Vial du Clairbois, Paris, 1781, German translation from the French by G. Timmermann, Hamburg, 1972.
52. Boudriot, Jean: "Formation et Savoir des Constructeurs Royaux", article in [7].
53. Gardiner, Robert; Morrison, John (eds.): "The Age of the Galley: Mediterranean Oared Vessels since pre-Classical Times", Conway's History of the Ship, Conway Maritime Press, London, 1995, reprinted 2000.
54. Harris, D. G.: "F. H. Chapman: The First Naval Architect", Conway Maritime Press, London, 1989.
55. Nowacki, Horst, Geir Westgaard and Justus Heimann: "Creation of Fair Surfaces Based on Higher Order Fairness Measures with Interpolation Constraints", in H. Nowacki, P. D. Kaklis (eds.): "Creating Fair and Shape Preserving Curves and Surfaces", B. G. Teubner, Stuttgart, Leipzig, 1998.
56. Barker, Richard: "Two architectures—a view of sources and issues", in Horst Nowacki and Wolfgang Lefèvre (eds.): "Creating Shapes in Civil and Naval Architecture—A Cross-Disciplinary Comparison", Preprint No. 338, pp. 41–133, Max Planck Institute for the History of Science, Berlin, 2007.

ANTIQUITY

CHAPTER TWO

ON THE USE OF DESIGN IN ANCIENT MEDITERRANEAN SHIP CONSTRUCTION

Patrice Pomey

The question of the use of design in ancient Mediterranean ship construction has been discussed for a long time. In the 1960s, J. Rougé in his famous thesis "Recherches sur l'organisation du commerce maritime en Méditerranée sous l'Empire romain" [1] raised the question about a text of Plautus (*Miles Gloriosus*, 915–919) in which there is a metaphor concerning a naval architect and ship construction. In his interpretation, J. Rougé suggested that the function of the naval architect (*architekton / architectus navalis*) was to calculate and to draw the plans of the ship, whereas the function of the shipwright (*naupegos / faber navalis*) was to build the ship under the supervision of the naval architect. Therefore, according to this hypothesis, J. Rougé assumed the existence of designs and hull plans in ancient Mediterranean shipbuilding, but without any further specifics about it. However, the question was often raised again afterwards, particularly by L. Basch [2], F. Salviat [3] and myself [4, 5] with conflicting opinions. While, at the time, I was following partly the point of view of J. Rougé, considering the design documentation more as some sort of construction diagrams than actual ship lines plans, L. Basch was opposed to the idea of plans altogether, but in favor of the existence of sketches. F. Salviat, emphasizing the importance of the architect, was agreeing with the idea of plans.

Today, the continuous advance of knowledge—due, on one hand, to the development of nautical archaeology and the analysis of new archaeological data, and, on the other hand, to the research about ship conception—allows a new approach to this problem.

First, we know that ancient Mediterranean shipbuilding was based on a "longitudinal shell conception", in which the hull shapes were determined by a longitudinal concept given by the planking and in which the mechanical structure is provided by the shell planking [6, 7, 8, 9, 10]. And this was the practice until the end of the ancient world, whatever the process of building with or without active frames (or templates) may have been [11]. In this approach everyone agrees that there was no use of traditional hull lines plans with vertical cross sections, vertical and horizontal longitudinal sections (body plan, buttock plan, waterline plan). Indeed, such plans were not in use before the end of the XVIIth century [12]. Moreover the longitudinal conception of the ancient ship is in conflict with such type of plans based on a "transversal skeleton conception" of the ship. For the same reason, the use of "construction diagrams", like those used in medieval times, or the use of rules based on the "gabarit", is quite unlikely.

Secondly, we have to take into consideration two different types or two different levels of ship construction as pointed out by L. Basch [13]. The first one is the private construction on which common shipbuilding relies. Such usual constructions were done on private shipyards by *naupegoi* or *fabri navales* whose knowledge was empirical and based on practical experience. The second type is State construction, under the rule of the governing power, which was concerned mainly with ships for the war fleets or with unconventional ships. This State construction, performed in State shipyards, disposed of considerable resources and was probably under the supervision of *architecti navales.*[1] Their knowledge belonged to science and allowed the use of drawings and calculations.[2] It is very likely that the tools and processes of ship conception were different between these two types of construction.

It is in this light that we have to consider the question of the use of design in ancient Mediterranean ship construction.

The first document that I want to present is also the oldest one and is, maybe, the most surprising. It is a letter, on a clay tablet, sent by the Pharaoh Ramses II to the Hittite King Hattusili III, probably after the Hittite peace, around 1259 B.C. The tablet (ref. KUB III 82) written in Accadian, the diplomatic language of the time, belonged to the archives of Bogahazköy, the capital city of the Hittite kingdom, and was published by Elmar Edel in 1994 [17]. The document was quoted for

[1] The idea of two levels of ship construction was proposed, for the first time, by J. Rougé [1] who distinguishes the construction of little boats by shipwrights and the construction of regular, large size ships by architects.

[2] About the status and the practice of the architect in the ancient societies and about the use of drawings in ancient architecture, see the Symposium "Architecture et Société" held in 1980 at the Ecole française de Rome [14] and particularly the paper given by P. Gros [15], and the Symposium of Strasbourg 1984 "Le dessin d'architecture dans les sociétés antiques" [16].

79. KUB III 82
(H 4)

Vs.
[0′] [.. ù ša LUGAL KUR Ḫa-at-ti]
1′ [iq-bu-ú a-na ŠEŠ-šu um-ma-a .... ME]Š-⌈šu⌉-nu
2′ [..]-šu-nu
3′ [...LUGAL KUR Ḫa-at-ti kán^{an}-na] iq-ta-ba-aš-šu
4′ [um-ma-a a-na ŠEŠ-ia-ma aš-šum ^{GIŠ}MÁ an-ni-ti a-nu-ma al-ta-ap-ra-]ak-ku ṭe$_4$-e-ma la-qè-ša
5′ [a-na ka-a-ša ù al-tap-ra DUMU.MEŠ.KIN-ri-ia a-na LUGAL KUR A-mur-ri li-]il-qú-ni-iš-ši
6′ [ù iq-ta-bu-ni-iš-šu um-ma-a li-qè-ši a-na LUGAL KUR Ha-at-ti ká]n^{an}-na iq-ta-bu-ni-iš-šu
7′ [......... a-nu-ma a-mur a-na-ku ul-te-bi-la-ak-ku ^{GI}]^{Š}MÁ 1-en ù ša-ni-ta
8′ [^{GIŠ}MÁ a-šap-pa-ra-ak-ku i-na ša-ni-ti MU-ti ù ^{LÚ}NAGAR.MEŠ-k]a GABA.RI li-ṣi-ru-šu-n[u]
9′ [a-na me-ḫe-er ^{GIŠ}MÁ.MEŠ an-na-ti ša ú-še-bi-la-ak-ku ....]ù li-ṣi-ru-šu-nu
10′ [........ ù li-pu-šu ^{GIŠ}MÁ.MEŠ an-na-ti ù ŠEŠ-ia ú-ša-ak-k]a-la kạ-am-ma-ti ku-up-ra
11′ [li-ik-pu-ru-ši-na-ti i-na ki-di ù i-na ŠÀ-bi .... a-na l]a-a i-re-bi
12′ [A.MEŠ i-na ^{GIŠ}MÁ.MEŠ an-na-ti a-na la-a na-da-ni a-na ṭe-be-e-ši-na i-na Š]À-bi A.AB.BA
13′ [...... ú-ṣur-ta ša ^{GIŠ}MÁ an-ni-ti ša LUGAL ú-še-bi-la-ak-ku i-na ṭup-p]í il-ta-kán-ši
14′ [...ú-ṣ]ur-ta
15′ [...i-/li-]pu-šu
16′ [.. -]aš-ši
17′ [.. e]n$_6$-ša-ti
18′ [...m]a-ad
19′ [...x x x [...

Rs. abgebrochen

Fig. 2.1: Ramses' letter: Accadian text.

the first time in a nautical context by D. Meeks in 1997 [18] and then studied by myself from a ship conception point of view [7, 19].

Unfortunately, the text is gravely incomplete and has been restored by Edel (Figs. 2.1, 2.2, 2.3). According to this restoration Ramses wrote a letter to inform Hattusili about the sending of two ships (one first and a second one later) by the intermediary of the king of the Amurru kingdom in order that the Hittite king's shipwrights draw a copy of the ship. For that purpose they had to make a drawing to build a replica. He recommended also to make frames (?) and to use asphalt to caulk the ship. The ship drawing had to be made on a tablet. But this restoration is quite hypothetical and raises a lot of questions about the circumstances and the conditions to carry out the operation. However, the remaining original part of the text which is not damaged is nevertheless clear enough to understand that it is the issue to send a boat and then another and to make a drawing of them. There is also the question of

79. Ramses an Ḫattušili
(H 4)

Vs.

[0′] [.. Und was der König des Landes Ḫatti]

1′ [zu seinem Bruder sagte, wie folgt: „ ....] ihre [...... [Plu]][r.?]

2′ [..] ihr/sie[a)]

3′ [..." so] sagte [der König des Landes Ḫatti] zu ihm.

4′ [So (sprich) zu meinem Bruder: Was dieses Schiff angeht, so habe ich] dir [nunmehr] den Bescheid [mitgeteilt], es [dir] zu bringen

5′ [und ich sandte meine Boten zum König des Landes Amurru, damit] sie es bringen,

6′ [und sie sagten zu ihm, wie folgt: „Bringe es zum König des Landes Hatti!" – s]o sagten sie zu ihm.

7′ [.................................Siehe, ich habe dir nunmehr] ein Schiff [gesandt] und ein zweites

8′ [Schiff werde ich dir im nächsten Jahr senden. Dei]ne [Zimmerleute], die sollen sich eine Kopie zeichnen

9′ [entsprechend diesen Schiffen, die ich dir bringen lasse ....] und sie sollen sich eine Zeichnung machen

10′ [... und sie sollen diese Schiffe nachbauen, und mein Bruder soll] die Spanten (?) [kunstvoll anferti]gen [lassen]. Mit Erdpech

11′ [sollen sie die Schiffe verpichen von außen und von innen ...., damit k]ein [Wasser] eindringt

12′ [in diese Schiffe (und) um nicht zuzulassen, daß sie untergehen inm]itten des Meeres!

13′ [... Den Bauplan für dieses Schiff, das dir der König bringen ließ, – auf einer Taf]el hat er ihn aufgeschrieben.

14′ [.. B]au(plan)

15′ [..]sie machten/mögen machen/das sie machten

16′ [..]es/ihm (= das/dem Schiff?)

17′ [... sch]wache(?) [....]

18′ [..s]ehr [...

19′ [..] ... [...

Rs. abgebrochen

orum/eos oder *earum/eas*

Fig. 2.2: Ramses' letter: Edel's version in German.

an unknown technical word (Edel proposes the word "Spanten", i.e., "frames") and to use asphalt. As for the "tablet" on which the drawing has to be made, it is a restoration by Edel, but based on a remaining character. For that reason the restoration is likely.

According to the comment by Edel [17, II, p. 283] the substantive *usurtu*, used in the letter, is well attested in the meaning of "drawing", especially when speaking about the drawing of a building. In that same meaning, Edel, followed by Meeks, quoted another Accadian text, but a more recent one and dated from the time of Nabonide, king of Babylon (c. 555–539 B.C.). The text was published by Lambert and Millard [20] and concerned the Babylonian story of the flood (Fig. 2.4). In the text, Atra-Hasis (Noah) who had never built a ship asked Ea (God) to draw the design of a boat on the ground in order to build it. The same word *usurtu* is used here with the same meaning of "drawing".

In both texts we are confronted with the same process: the use of a design drawing in order to build a ship. The situation in the second text is more theoretical and demonstrative, but corresponds to a common

[0'] [................................et ce que le roi du pays Hatti]

1' [a dit à son frère, comme suit: "................] leur [......]

2' [..] lui / elle

3' [..................................." ainsi] lui dit [le roi du pays Hatti].

4' [Ainsi (parle) à mon frère : en ce qui concerne ce navire, je] t'[ai communiqué]
[à présent] ma décision, de [te] l'apporter

5' [et j'ai envoyé mes messagers au roi du pays Amurru, afin] qu'ils l'apportent,

6' [et ils lui ont dit, ceci : "apporte-le au roi du pays Hatti !" – c']est ce qu'ils lui ont dit.

7' [....................... vois, je t'ai à présent] [envoyé] un navire et un deuxième

8' [navire sera envoyé l'an prochain. T]es [charpentiers] doivent se dessiner une copie

9' [correspondant à ces navires, ceux que je t'ai fait porter....] et ils doivent se faire
un dessin

10' [... et ils doivent construire une réplique de ces navires, et mon frère doit]
[faire avec art] les couples (?). Avec l'asphalte

11' [ils doivent poisser les navires, à l'extérieur et à l'intérieur....,] afin que l'[eau]
n'entre pas

12' [dans ces navires (et) pour éviter qu'ils ne sombrent et ne coulent au mi]lieu
de la mer !

13' [...le plan de construction de ce navire, celui que le roi a fait apporter, – sur
une tab]lette il l'a fait tracer.

14' [...] (plan) de construction

15' [...] ils ont fait / ils souhaitent faire faire /
ce qu'ils ont fait

16' [..] il / lui (= le / les navires ?)

....

Fig. 2.3: Ramses' letter: French translation of Edel's version.

13 ....

14 [*ina qaq*]-*qa-ri e-ṣir ú*-[*ṣur-tu*]

15 [*ú-ṣur*]-*tu lu-mur-ma* $^{GI\check{S}}$*MÁ* [*lu-pu-uš*]

13 Never yet have I constructed a ship

14 draw the design on the ground

15 that I may see [the design] and [built] the boat.

13 Je n'ai encore jamais construit de navire ;

14 trace le dessin sur le sol

15 que je puisse voir [le dessin] et [construire] le bateau.

Fig. 2.4: Babylonian text (Noah): Accadian, English and French translation.

practice still in use today.[3] But the first text corresponds to a real and concrete event in the context of a diplomatic mission. Apparently, it is the oldest evidence about a "technology transfer" and about the use of a design drawing in shipbuilding. Whatever the type of drawing, its use in ancient ship construction is undisputedly well attested, at least, since the middle of the thirteenth century B.C. Such a piece of evidence, unexpected until now, is very important for the history of ship construction, the history of design mentality and the history of technology. That is the reason why we have to consider the question of what sort of design documentation may have been used.

First, we have to remember that we have stated in the beginning of this paper that, in the conception of ancient ships, there were no construction plans.

The second remark is that the drawing concerns the copy of an existing ship and not the creation of a new one, as we are familiar with doing it today. The copying of a previous ship was a common practice in ancient shipbuilding, well attested by numerous examples. E.g., the famous *Syracusia*, according to Athenaeus (V, 206d–209b), was built based on the model of an "*eikosoros*", a ship of rank "twenty" [7, 21].

The third remark is that the design probably drawn on a tablet (or on the ground) must have been very simple and basic and not too elaborate. In short, it must have been some kind of sketch. But, to be useful, this sketch must have included some information like—by hypothesis—basic lines, principal dimensions and proportions. In brief, a quoted sketch with the elements which are considered as characteristics of the ship and which are to be controlled when building the boat or the replica. Something like the sketch given by Marzari to give the main dimensions of a Venetian ship from the 15th c. [22, p. 189] (Fig. 2.5).

Now the questions to be addressed are if such a sketch as quoted was possible at the time of Ramses II from a conceptual point of view, and whether the shipwrights of the time had the tools to control a ship construction.

Fortunately, we know several examples of such drawings or sketch-plans used by the Egyptians. Most of them belong to the New Kingdom but some date back to the Ancient Kingdom. They are drawn with ink

[3] L. Basch [2, 36] quotes a shipwright from Kalymnos, working without any plan, but drawing a sketch of a sailing boat in the sand (cf. "The Greek Merchant Marine (1453–1850)", National Bank of Greece, Athens, 1972, fig. 27).

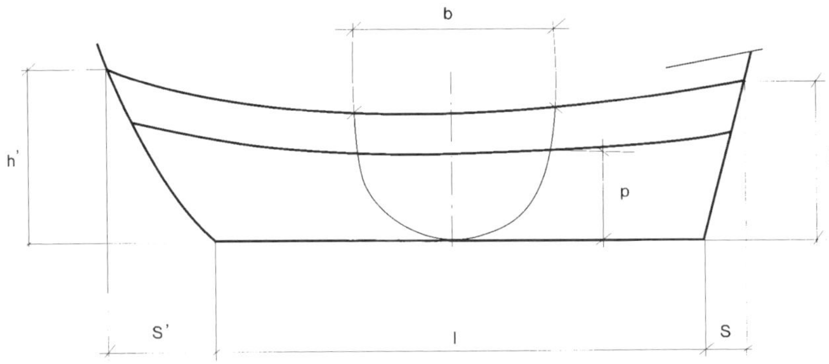

Fig. 2.5: Sketch of a Venetian ship.

on limestone chips (ostraka) or on papyrus. [23, 24]. Among them we can quote for example:

- An ostrakon (British Museum) from Deir el-Bahari (end of the XVIIIth dynasty or beginning of the XIXth) with the sketch-plan of a tomb with the main dimensions [25] (Fig. 2.6).
- A papyrus (Turin museum), plan of the tomb of Ramses IV (Valley of the Kings at Thebes, New Kingdom) with dimensions in cubits, palms and digits, and comments (in hieratic writing) giving the names of the various parts and also indications for the decoration [23, p. 49] (Fig. 2.7).
- An ostrakon (Cairo museum) found in Saqqâra and dated back to the IIIrd dynasty at the time of Djoser (beginning of the third millennium) This very interesting sketch is considered as an architect's diagram defining a curve by co-ordinates [26] (Fig. 2.8).

According to specialists, the Egyptians were able to draw ground plan views, side and end elevation and to use grid systems. But these sketches or diagrams are not architect's plans, but rather preliminary drafts or sketches for the guidance of the workmen or even for demonstrating objects [23, 24].

So, if we take in account these sketch plans in use in Ancient Egypt, we may suppose that the ship designs mentioned in the Ramses letter to Hattusili may have been of the same type, not construction plans, but sketches with the main characteristics of the ship, or some important details, for the guidance of the workmen or for the visualization of the project. Such sketch plans were surely not sufficient to build the ship,

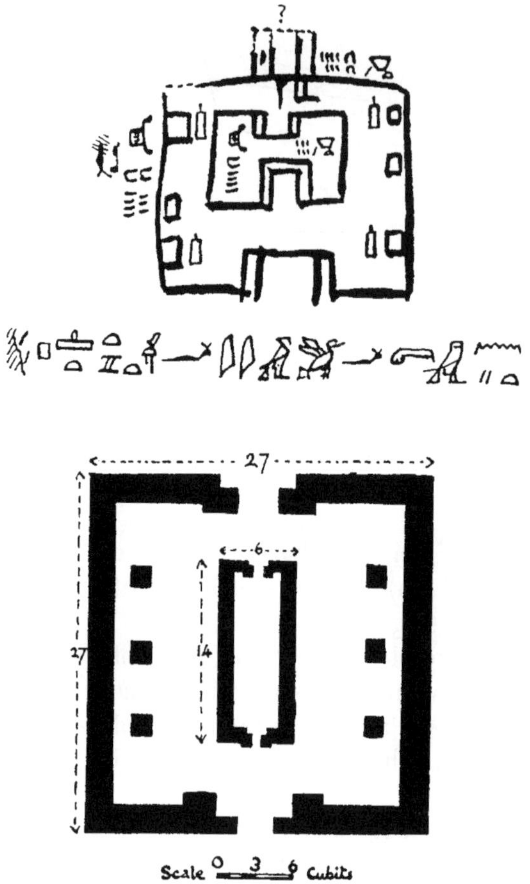

Fig. 2.6: Ostrakon from Deir el Bahari.

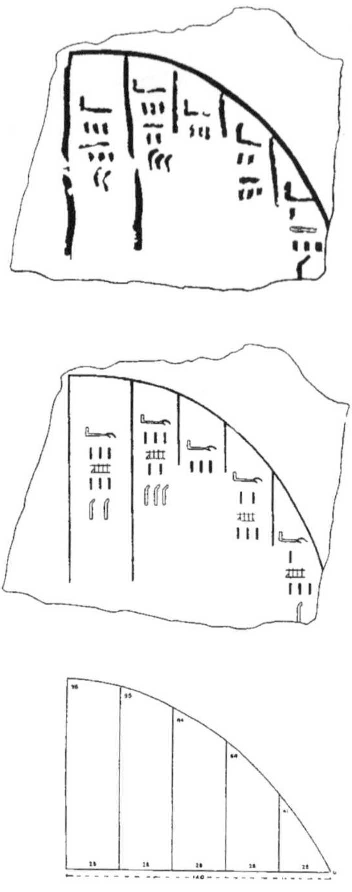

Fig. 2.8: Ostrakon from Sakkara.

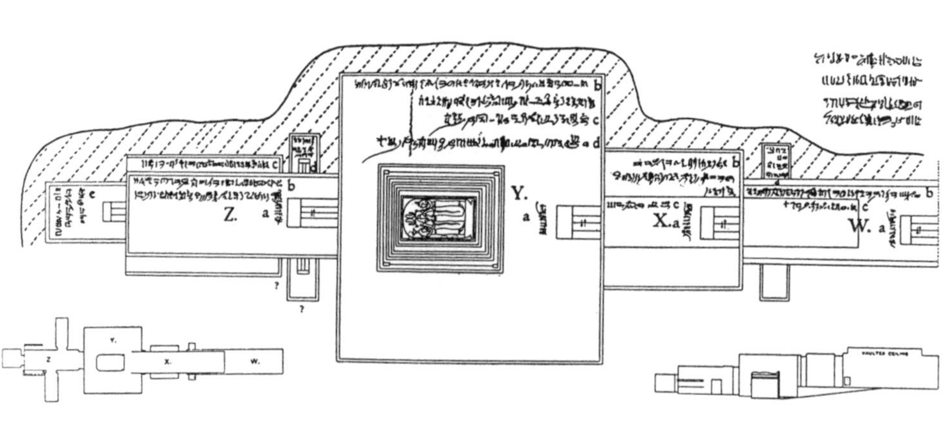

Fig. 2.7: Papyrus sketch plan with the tomb of Ramses IV.

but they offer enough information to describe the characteristics of the ship and to introduce the project in order to understand it, judge it or discuss it . If we accept the diplomatic context of the letter from Ramses to Hattusili, even if the circumstances are not clear, then it is obvious that we are considering the question of the use of ship designs in the context of a state construction. Such sketch drawings were certainly not common and were probably reserved to some specialists, in state shipyards, to define the architectural project or, in the case of our letter, to describe and visualize the characteristics of the original ship to be copied.[4]

About the control tools which are necessary to build a replica, we know that Egyptian builders, as early as the Ancient Kingdom, used measuring instruments, checking systems (marks, pegs, string lines…) and were able to measure slopes and curves [24]. These systems were applicable to shipbuilding, and several tombs from Saqqâra show the use of such instruments and checking processes applied to ships. For example:

- On the Tomb of Ti (Vth dynasty): the master shipwright is holding in his hand a measuring staff or a sort of ruler and a plumbline [27] (Fig. 2.9, top).
- Tomb of Mererouka (VIth dynasty): workmen are holding a long staff along the longitudinal axis of the boat while the master shipwright is measuring the depth with a plumbline [28] (Fig. 2.9, bottom).

Fortunately we have also archaeological evidence for this hull control. On the Cheops boat (c. 2600 B.C.) there are numerous marks (over 1100) for the assembly of the different pieces, but also, according to Lipke [29], special marks on the axial sheer line of the boat probably for the checking of the hull during its construction.[5]

So, as we can see, even if the context and the circumstances of the operation of shipbuilding contained in the letter from Ramses to Hattussili are not clear, we have enough elements to think that the drawing of a ship in order to make a copy for building a replica made sense

[4] I have emphasized the function of drawings in ancient naval architecture, for the materialization of the project in my paper about "Conception et réalisation des navires dans l'Antiquité" [7, pp. 57–58].

[5] All these hull checking methods may have been very similar to those used in Scandinavian shipbuilding, as in the construction of the *Skuldelev 3* replica [30, 31].

Fig. 2.9: Boatbuilding from the tomb of Ti and Mereruka.

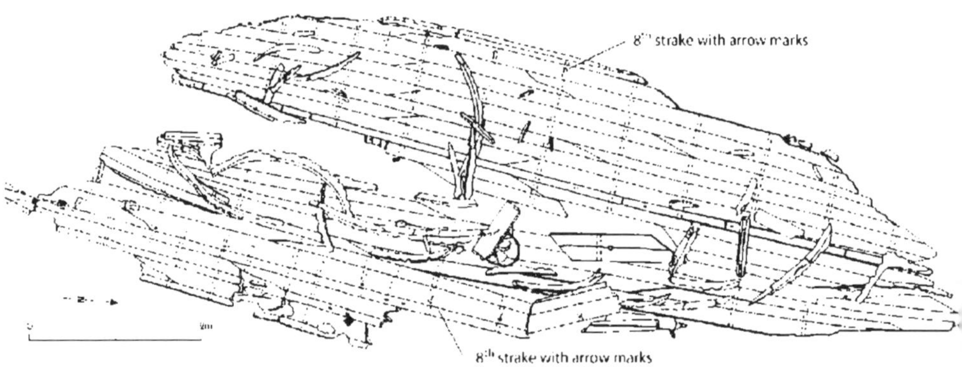

Fig. 2.10: Plan of Jules-Verne 7 wreck.

and was quite possible on the basis of the knowledge of the time in matters of drawings and checking tools. Of course, the drawing was more likely a sketch and surely not a construction plan. But whatever the type of drawing, this first attestation is so important for the history of shipbuilding that it merits to be strongly underscored.

In the Greco-Roman period, the use of drawings was probably common in civil architecture, especially for official buildings. But if the examples are very few,[6] the preserved drawings are sufficient to prove the practice [16]. So, there is no reason to think that it was not the same in ship construction. Moreover, processes of control with the help of rulers (canon), string lines, plumblines and compass used in ship construction are well testified by literary and epigraphical evidence [2, 3].

From an archaeological point of view, it has been possible to underscore such a checking process on the Greek archaic boat (end of VIth c. B.C.) *Jules-Verne 7* in Marseilles. Indeed, the asymmetry of the seven first strakes of the planking is fairly corrected at the level of the eighth strake which is larger than the others and which brings construction marks arrow-shaped (Fig. 2.10). Obviously, the eighth strake plays a role of adjustment in the construction process—that is the reason why I have called it "adjustment strake"—and a hull check was done at this level [7, pp. 63–64].[7]

Now the question is whether, beyond the existence of basic sketches, there were any more elaborate ship designs. Or, the question is: what was the role played by the naval architect (*architekton / architectus navalis*) in comparison with the shipwright (*naupegos / faber navalis*)? Besides the social distinction [15], the main technical difference seems to have been the use, by the architect, of design and calculation and, according to Vitruvius (*De architectura*) their combination with the application of numerical ratios to the design (*ratiocinatio*) [34]. The architect must know mathematics and geometry. So, we are before a "construction savante" and the main difference between the naval architect and the shipwright must have been at the level of the definition of the architectural program ("projet architectural"). In that way, we may presume the use in naval architecture of quoted designs more elaborated with geometric and regulating drawings as in civil architecture.

[6] The few architecture drawings known are preserved, engraved on stone, on Lebanese buildings from the Roman period, like the half pediment drawn on the cella wall of the temple of Bziza, near Tripoli, or the one engraved on the trilithon of the great temple of Baalbeck. For a list of these drawings with references, cf. Will [32].

[7] About the Greek archaic boat *Jules-Verne 7*, cf. [33].

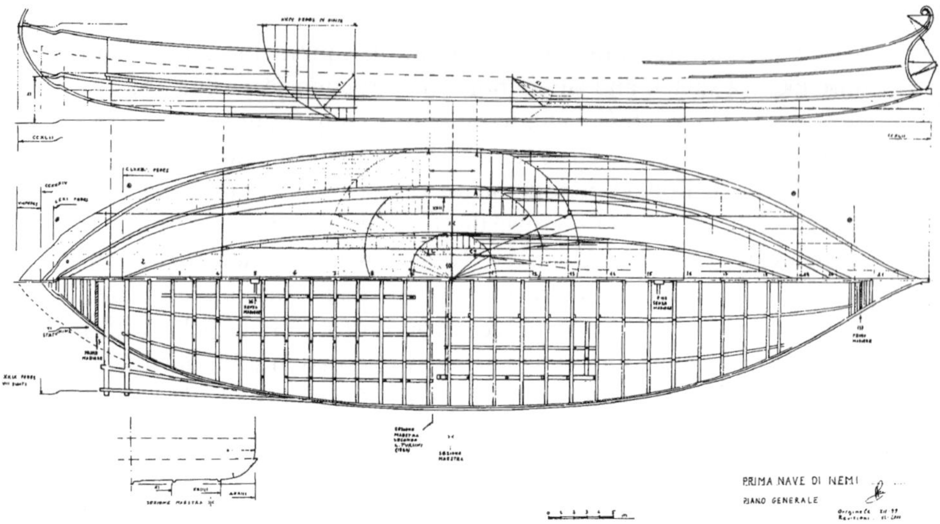

Fig. 2.11: Nemi ships with regulation drawings.

This hypothesis is supported by M. Bonino [35] for the famous Nemi ships that he considers as belonging to the Hellenistic tradition. Indeed, in the reconstruction plans of these ships, Bonino proposes, as a hypothesis, to combine the general shapes and the main curves with geometric drawings in the sense of Vitruvius (Fig. 2.11).

But if we accept this hypothesis, I consider that we have to be careful and to state that such drawings are not at all plans for ship construction, but just elaborated drawings for the presentation of the architectural program and to bring to the fore the science and the ability of the architect. These drawings correspond to the conceptual process and not to the construction phase. This role as element of presentation and discussion is, in my opinion, the main role of such drawings (or sketches at a lower level) in ship construction.[8]

A last remark concerning the construction of the famous ship *Syracusia* by Hiero II, king of Syracuse (IIIrd c. B.C.). This ship, according to Athenaeus (V, 206d–209b), was an exceptional grain carrier, the biggest ever built. Nothing was too much to make it, and the construction was done under the responsibility of the architect Archias from

[8] About the conception of the Ancient ship and the role of drawings in the process of conception see my study [7, pp. 54–59].

Corinth. He was helped by Archimedes considered as a mathematician (geometer), in charge to supervise the construction, and as an engineer (mechanic) in charge of the launching of the ship. If we consider the huge dimensions of the ship, the launching must have been very difficult and delicate. In view of the importance of this task, the nature of the problem and the intervention of Archimedes, we may think that there was a link between the launching and his theory "On Floating Bodies". So we may presume, by hypothesis, that the role of Archimedes was, precisely, to apply his theory to ensure the launching of the ship [21]. But this hypothesis implies that he was able to calculate the center of gravity and the center of buoyancy of the ship. That seems possible in principle from a theoretical point of view according to Nowacki who thinks that Archimedes may have been able to determine volumes and centroids by numerical approximation [36]. But, in fact, for that he needed to know the hull form and to have plans or sufficient drawings. But the *Syracusia* was built on the model of an "*eikosoros*" and we may suppose that it was possible for a geometer like Archimedes to make a design copy of the original hull of the "*eikosoros*", and then to make the necessary calculations, even if in an approximate, but sufficient way. Of course, it would be a question of static stability and not of dynamic stability. If we agree to this hypothesis, then we are obliged to admit the existence of some possible hull drawings in antiquity. But, I repeat this is merely an attractive hypothesis.

In conclusion, according to the testimony of the letter from Ramses to Hattusili, we may assume that some sort of designs, considered very likely like sketches, may be quoted, were in use in ancient Mediterranean ship construction since the XIIIth c. BC, at least in state shipyards. Such sketches were probably used like preliminary drafts or for the guidance of the workmen or even as demonstration drawings. For the Greco-Roman period we may think that more elaborate designs with geometric drawings were in use, especially by the architect. But this use seems to have been limited to "constructions savantes" and for more theoretical than practical purposes. But whatever the designs in use, and whatever the exceptional ship achievements made in ancient times, there were in no way construction plans.

References

1. Rougé, Jean: "Recherches sur l'organisation du commerce maritime en Méditerranée sous l'Empire romain", SEVPEN, Paris, 1966, (esp. pp. 188–192).
2. Basch, Lucien: "Eléments d'architecture navale dans les lettres grecques", L'Antiquité Classique, XLVII, 1978, pp. 5–36.
3. Salviat, François: "Sources littéraires et construction navale antique", Archaeonautica, 2, 1978, pp. 253–264.
4. Pomey, Patrice: "Plaute et Ovide architectes navals!", Mélanges de l'Ecole française de Rome, Antiquité, 85, 1973, pp. 483–515.
5. Pomey, Patrice: "La coque", in A. Tchernia, P. Pomey, A. Hesnard, "L'épave romaine de la Madrague de Giens (Var)", XXXIVe supplément à Gallia, éditions du CNRS, Paris, 1978, pp. 75–99.
6. Pomey, Patrice: "Principes et méthodes de construction en architecture navale antique", in "Navires et commerces de la Méditerranée antique, Hommage à Jean Rougé", Cahiers d'Histoire, XXXIII, 3–4, 1988, pp. 397–412.
7. Pomey, Patrice: "Conception et réalisation des navires dans l'Antiquité méditerranéenne", in E. Rieth (dir.), "Concevoir et construire les navires. De la trière au picoteux", Technologie, Idéologies, Pratiques, Revue d'anthropologie des connaissances, XIII-1, 1998, pp. 49–72.
8. Pomey, Patrice: "Principles and methods of construction in ancient naval architecture, in F. M. Hocker, C. A. Ward, "The Philosophy of Shipbuilding. Conceptual approaches to the study of wooden ships", Texas A & M University Press, College Station, 2004, pp. 25–36.
9. Steffy, J. Richard: "Wooden Ship Building and the Interpretation of Shipwrecks", Texas A & M University Press, College Station, 1994,
10. Steffy, J. Richard: "Ancient scantling: the projection and control of Mediterranean hull shapes", in H. Tzalas (ed.), "Tropis III, 3rd International Symposium on Ship Construction in Antiquity", Athens 1989 Proceedings, Athens, 1995, pp. 417–428.
11. Pomey, Patrice; Rieth, Eric: "L'archéologie navale", éditions Errance, Paris, 2005.
12. Boudriot, Jean: "Formation et savoir des constructeurs royaux", in E. Rieth (dir.), "Concevoir et construire les navires. De la trière au picoteux", Technologie, Idéologies, Pratique, Revue d'anthropologie des connaissances, XIII-1, 1998, pp. 127–149.
13. Basch, Lucien: "Construction privée et construction d'État dans l'Antiquité", in E. Rieth (dir.), "Concevoir et construire les navires. De la trière au picoteux", Technologie, Idéologies, Pratique, Revue d'anthropologie des connaissances, XIII-1, 1998, pp. 21–47.
14. "Architecture et société de l'archaïsme grec à la fin de la République romaine", Actes du Colloque international organisé par le Centre national de la recherche scientifique et l'École française de Rome, (Rome, 2–4 décembre 1980), CNRS-École française de Rome, Paris, Rome, 1983.
15. Gros, Pierre: "Statut social et rôle culturel des architectes (période hellénistique et augustéenne)", in [14], pp. 425–452.
16. "Le dessin d'architecture dans les sociétes antiques", Actes du Colloque de Strasbourg, 26–28 janvier 1984, Université de Strasbourg, Strasbourg, 1985.
17. Edel, Elmar: "Die ägyptisch-hethitische Korrespondenz aus Boghazkoï in babylonischer und hethitischer Sprache", Band I, II, Umschriften und Übersetzungen, (Abhandlungen der Rheinisch-Westfälischen Akademie der Wissenschaften,77/1, 77/2), Opladen 1994.
18. Meeks, Dimitri: "Navigation maritime et navires égyptiens : les éléments d'une controverse" in D. Garcia, D. Meeks (éd.), "Techniques et économie antiques et

médiévales. Le temps de l'innovation". Colloque d'Aix-en-Provence (mai 1996), Édition Errance, Paris, 1997, pp. 175–194.

19. Pomey, Patrice: "Le rôle du dessin dans la conception des navires antiques. À propos de deux textes akkadiens", in B. Mathieu, D. Meeks, M. Wissa (éd.), "L'apport de l'Egypte à l'histoire des techniques. Méthodes, chronologie et comparaisons", Institut français d'archéologie orientale, Le Caire, 2006, pp. 239–251.
20. Lambert, Wilfred G.; Millard, A. R.: "Atra-Hasis. The Babylonian story of the flood", Oxford, 1969.
21. Pomey, Patrice; Tchernia André: "Les inventions entre l'anonymat et l'exploit: le pressoir à vis et la Syracusia", in E. Lo Cascio, "Innovazione tecnica e progresso economico nel mondo romano", Atti degli Incontri capresi di storia dell'economia antica (Capri 13–16 aprile 2003), Edipuglia, Bari 2006, pp. 81–99.
22. Marzari, Mario: "Evolution of shipbuilding techniques and methodologies in Adriatic and Tyrrhenian traditional shipyards", in E. Rieth (dir.), "Concevoir et construire les navires. De la trière au picoteux", Technologie, Idéologies, Pratique, Revue d'anthropologie des connaissances, XIII-1, 1998, pp. 181–215.
23. Clarke, S.; Engelbach, R.: "Ancient Egyptian Construction and Architecture", New York, 1990, (esp. chap. V), (first ed. as "Ancient Egyptian Masonry", Oxford, 1930).
24. Arnold, Dieter: "Building in Egypte. Pharaonic stone masonry. New-York, Oxford, 1991 (esp; chap. 1).
25. Glanville, S. R. K.: "Working Plan for a Shrine", The Journal of Egyptian Archaeology, 16, 1930, pp. 237–239.
26. Gunn, B.: "Inscriptions from the Step Pyramide Site. II-An Architect's Diagram of the Third Dynasty", Annales du Service des Antiquités de l'Égypte, XXVI, IFAO, Le Caire 1926, pp. 197–202.
27. Wild, H.: "Le tombeau de Ti. Fascicule II. La chapelle", Mémoires de l'IFAO du Caire, tome LXV, Le Caire, 1953.
28. Duell, P., "The mastaba of Mereruka". Part II. The University of Chicago oriental Institute Publications, vol. XXXIX, Chicago, 1938.
29. Lipke, Paul, "The Royal Ship of Cheops", BAR, Int. Series 225, Oxford, 1984.
30. Andersen, Eri; Crumlin-Pedersen, Ole; Vadstrup, Soren; Viner, Max, "Roar Ege. Skuldelev 3 skibet som arkæologisk eksperiment", Vikingeskibhallen, Roskilde 1997.
31. Crumlin-Pedersen, Ole, "The 'Roar'-Project", in O. Crumlin-Pedersen, M. Vinner (ed.), "Sailing into the Past", The Viking Ship Museum, Roskilde, 1986, pp. 94–103.
32. Will, Ernest, "La maquette de l'adyton du temple de Niha (Baqa)", in [16, pp. 277–281].
33. Pomey, Patrice, "Les épaves grecques du VIe siècle av. J.-C. de la place Jules-Verne à Marseille", in P. Pomey, E. Rieth (dir.), "Construction navale maritime et fluviale. Approches archéologique, historique et ethnologique, Actes du 7e Colloque International d'Archéologie Navale—7th International Symposium on Boat and Ship Archaeology, île Tatihou, 1994", (Archaeonautica 14, 1998), CNRS Editions, Paris, 1999, pp. 147–154.
34. Frézouls, Edmond, "Vitruve et le dessin d'architecture", in [16, pp. 213–229].
35. Bonino, Marco, "Un sogno ellenistico: le navi di Nemi", Felici editore, Pise 2003.
36. Nowacki, Horst, "Archimedes and Ship Stability", Max-Planck Institute for the History of Science, Preprint 198, Berlin, 2002.

CHAPTER THREE

WERE THE HITTITES ABLE TO BUILD A REPLICA OF AN EGYPTIAN SHIP ACCORDING TO THEIR OWN DRAWINGS?

Lucien Basch

In 1994 Elmar Edel published and commented on a clay tablet inscribed in Akkadian, found somewhere in Hattusas, the capital of the Hittite kingdom (Hatti), addressed to the king of the Hittites, probably Hattusilis III, dealing with an Egyptian ship the pharaoh had sent to him (tablet KUB III 82) [1]. In 2006 Patrice Pomey [2] published a further comment where he concluded that the tablet demonstrated, at least with a high degree of probability, that its contents were sufficient to enable the subjects of the Hittite king to build a replica of this ship, though not claiming that the tablet was equivalent to a modern ship lines plan as used from the 17th c. on.

The tablet which originally contained at least 20 lines was broken at one end and what remained was severely damaged: Only one third, at the most, on the right-hand side could be read without restoration. Yet it seems that the spirit of the general meaning was that the pharaoh asked the Hittite king to have a drawing made of the ship he had sent and to build a replica of her.

Almost the whole rest of the text is the result of Edel's restoration. Restoration in epigraphy deserves much respect and Edel is without any doubt a distinguished specialist. Yet, as noted by another specialist, L. Robert, "the difficulty of restoration increases in proportion to the uniqueness of the document" [3], and this tablet is unique. Moreover a familiarity with the subject of the document is helpful. Unfortunately Edel's knowledge of Egyptian naval architecture seems to have been only limited. E.g. he writes in his comment that the "frames" (or "floor timbers") found in the Cheops ship determined the shape of the hull,[1] which is quite impossible in the system of hull construction of this

[1] [1] vol. 2, p. 284. Actually no side frames or floor timbers were found in the Cheops ship, but 16 frames in the bottom of the hull. Their role was to reinforce the bottom, and their shape was determined by the shell planking.

ship. Still in this comment Edel writes. “In dem umfassenden Buch von Ch. Boreux, Etudes de Nautique Egyptienne, (569 Seiten) steht nichts über das Abdichten von ägyptischen Schiffen” (“In the comprehensive book by Ch. Boreux... (569 pages) nothing is stated on the caulking of Egyptian ships”), whereas Boreux wrote that the seams of papyrus boats were likely caulked by “une résine ou bien une mantière bitumineuse, ou bien encore un composé de l'un et de l'autre” ([4], pp. 184, 185).

The seven first lines of the tablet are almost entirely destroyed. This does not prevent Edel from restoring them completely by a small story: The ship was first sent to the king of Amurru (a kingdom in Syria which formed the Northern boundary of Egypt), who was demanded to take her to the Southern coast of Hatti without any further specification. The fate of the Egyptian commander and his crew remained unknown,[2] This transport is completely incomprehensible, if not absurd, since no risk of interception had to be feared from Amurru, a vassal kingdom of Hatti. More importantly, the fact of sending a ship to Hatti in order to be copied is difficult to understand. Scholars agree that Hatti was never a maritime power: “The Hittites depended on Ugarit and the untrustworthy Lukka for a naval force” ([5], p. 140). Ch. Schaeffer even insists: “One knows that the Hittites, during their history, never possessed the least maritime experience, an experience which demands a long apprenticeship.”[3] If this apprenticeship nevertheless began during the reign of Hattusilis III, although there were no disputes between Hatti and Ugarit, then to begin with the building of a replica of an unfamiliar ship was a tough task. The importance of this fact has to be stressed. Once the ship had arrived somewhere in Hatti, the Hittite shipwrights (and not the Egyptians) had to make a drawing of the ship and build a replica according their own drawing. How any shipwrights might be able to achieve this demanding double task in a country without the

[2] This “small story” was probably suggested to Edel by another tablet (KUB III 51), which relates the transport from Egypt to Amurru of a medication needed by the Hittite sovereign. Ramses II suggests that the medication be sent from Amurru to Hatti by a horseman (.pethallu), a plausible solution for the delivery of a medication at the greatest possible speed. See [1] vol. 2, p. 283.

[3] [6], p. 748. The last king of Hatti was the weak Suppiluliumas II, who ascended the throne in 1190 and disappeared after a short reign. Nothing is known about his end, not even the date. Paradoxically he was the only king of Hatti who boasted that he defeated in naval battles, not just once, but three times, ships from Cyprus (Alashiya), (tablet Kbo XII 38). Schaeffer presumes, probably correctly, that these victories were won with Ugaritic ships ([6], p. 573).

least experience in shipbuilding remains a mystery. Ramses II in any case does not seem to have had a great respect for these shipbuilders since he emphatically underscores the necessity of caulking the ship. Would able shipwrights have had any doubt about this necessity?

Curiously, after the very last phase of construction, the caulking, had been addressed, a new and important element appears in line 14, where only one (reconstructed) word is left: "(u.s)urta", which normally means "drawing", but is translated by Edel as "plan of construction". (In his comment on p. 284 Edel even adds "mit Massangaben", i.e., with dimensioning information, words omitted in the translation). However in the Babylonian text quoted by Pomey ([2], p. 241), Lambert and Willard translate the same word, written and restored as "(u.sur).tu" by "design". This "design" had to be traced on the ground and for that reason cannot have been a "plan of construction", but rather a simplified sketch. If we follow Edel and his "plan of construction",[4] what might be the meaning of this somewhat obscure "design"? Pomey offers the following hypotheses:

- "a kind of sketch of the essential lines of the ship": Is "a kind of sketch" not conflicting with "essential lines"? All the more since a "sketch" (in French "esquisse") is a rather vague concept, what might be "a kind of sketch"?
- "with complementary indispensable additions" such as the overall length, the maximum beam and the depth of hold: Since the "plan of construction", always according to Edel, had been delivered aboard an Egyptian ship, all this information might easily have been obtained by measuring the ship herself.
- "the ratio of proportions between different parts of the ship: Same comment".
- "how to realize control measurements on the ship to be built".[5]

Another problem arises here: How to determine the curves of different sections of the hull shape. Pomey notes that the Egyptians since the 3rd dynasty knew how to construct a curve by a mathematical operation

[4] It must be noted that for Edel the tracing of the "plan of construction" on a clay tablet is only a hypothesis. His "Bauplan" is "*anscheinend* (apparently, underscored by the author—L.B.) on a clay table…" ([1] vol. 2, p. 283).

[5] Pomey is sure that the plan was traced on a tablet, [2], p. 243.

([2], p. 244; [7]). An open question is whether such methods were known to the Hittites. More importantly, such a method might be helpful in drafting the curve of the main section, but might not contain a practical rule for finding all the other sections between amidships and the extreme ends of the ship.

Patrice Pomey concludes that the tablet KUB III 82 is fundamental as a testimony for the use of drawings in naval architecture. To be sure, the Egyptians of the New Kingdom were certainly able to draw accurately a plan view of a ship ([8] and [9]), see Fig. 3.1, which might have been more useful for the Hittites than "a kind of sketch", but what we are asked to believe or at least to presume is that the Hittites, a people of brave soldiers and citadel builders as well as clever diplomats, but without the least maritime experience, were able to build a ship, in fact a foreign design ship, based only on

a) their own drawings (1, 8 and 9), necessarily drafted in two dimensions and not derived from orthogonal projection,
b) any Egyptian "kind of sketch", whatever it might have been.

In conclusion, while Patrice Pomey takes Elmar Edel's restored text at face value, I believe that this interesting document should be re-examined by more than one expert and within the limits of its historical context, before according it its place in the history of naval architecture.

References

1. Edel, Elmar: "Die ägyptisch-hethitische Korrespondenz aus Boghazköi in babylonischer und hethitischer Sprache" ("The Egyptian-Hittite correspondence in Babylonian and Hittite language"), I (Text), II (Kommentar), Opladen, 1994. The Accadian text and its German translation by Edel are reproduced in Patrice Pomey's paper.
2. Pomey, Patrice: "Le rôle du dessin dans la conception des navires antiques. A propos de deux textes akkadiens" ("The role of design in the conception of antique vessels. Based on two Akkadian texts"), in Mathieu, B., Meeks, D., Wissa (eds.), "L'apport de l'Egypte à l'histoire des techniques. Méthodes, chronologie et comparaisons". Institut français d'archéologie orientale, Cairo, Bibliothèque d'études, 142, 1996, pp. 239–249.
3. Robert, L.: "La restitution", in "Encyclopédie de la Pléiade. L'histoire et ses méthodes", Paris, 1961, p. 486.
4. Boreux, Ch.: "Études de nautiques égyptienne. L'art de la navigation en Égypte jusqu'à la fin de l'Ancien Empire, Cairo, 1925.
5. Sandars, N. K.: "The Sea Peoples", London, 1985.
6. Schaeffer, Cl. F. A., "Ugaritica", V.

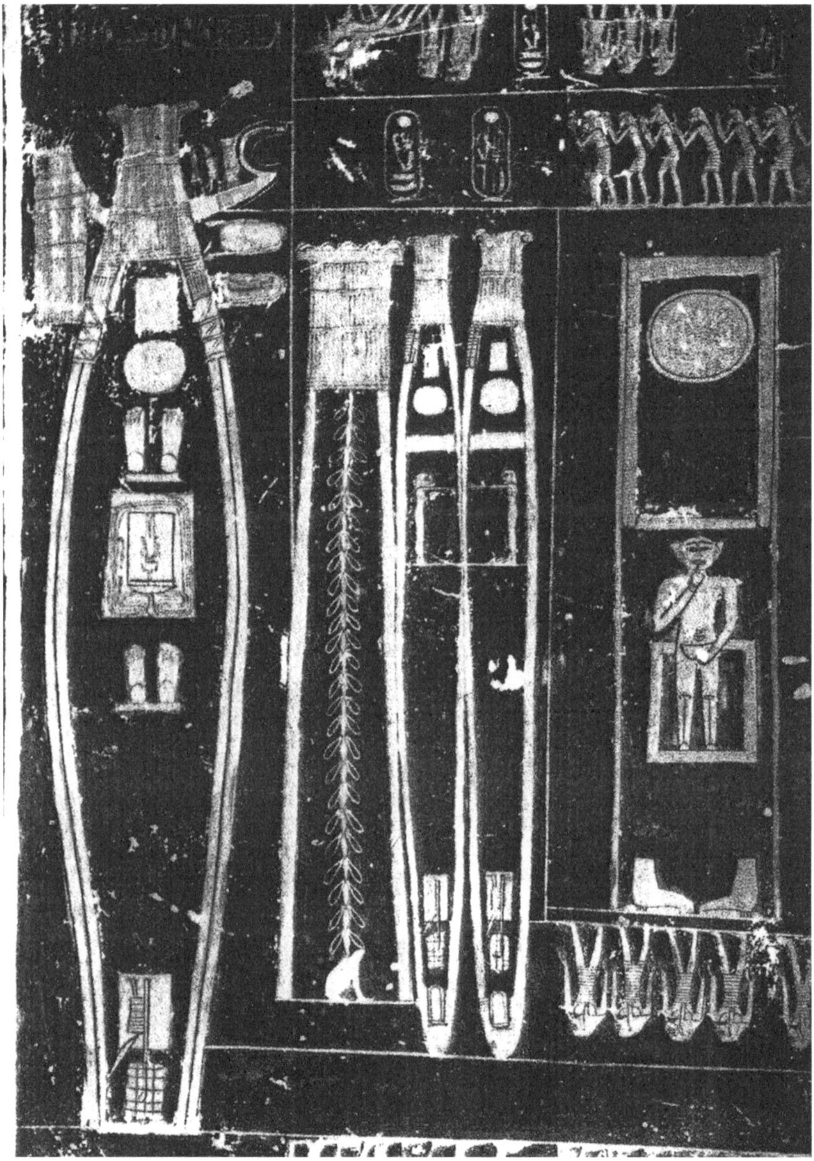

Fig. 3.1: Painting on the ceiling of the tomb of Ramses VI (ca. 1146–1143 B.C.). From PIANKOFF and MAYSTRE [14, Plate VI].

7. Gunn, B.: An Architect's Diagram of the Third Dynasty", Annales du Service des Antiquités de l'Égypte, 26, 1926, pp. 197–202.
8. Piankoff, A., and Maystre, Ch.: "Deux plafonds dans les tombes royales", Bulletin de l'Institut français d'archéologie orientale, 38, 1939, pp. 65–70.
9. Grapow, H.: "Zu dem Deckenbild im Grabe Ramses VI mit Aufrisszeichnungen der Sonnenschiffe", Zeitschrift für Ägyptische Sprache, 81, 1956, pp. 24–28.

REPLY TO LUCIEN BASCH

Patrice Pomey

The discussion of my paper by Lucien Basch is organised around two points:

The first one concerns the general context of the operation of shipbuilding based on the restoration by E. Edel. I agree with L. Basch in considering the restoration by Edel as hypothetical, it raises many questions about the circumstances and the conditions for carrying out the operation proposed by the pharaoh. What is the role of the King of Amurru? Where are the shipyards in which the ships must be copied? Who are the shipwrights? What is the type of the ship? And so on... But L. Basch criticizes Edel's restoration because he considers the transport of the ship to the Southern coast of Hatti by the King of Amurru as incomprehensible. Moreover, he regards the fact of sending a ship to Hatti in order to be copied as difficult to understand because the Hittite shipwrights have no experience in shipbuilding. Maybe. But this is a supposition of Lucien Basch about the hypothetical restoration of the text. In fact, we do not know where the shipyards are and who the shipwrights are. Nothing in the restored text says that the shipyards are in Hatti and that the shipwrights are Hittites. Edel says only that the ship has to be sent to the King of Hatti, which is not a precise location, and he speaks of the shipwrights of the King of Hatti, which does not mean that they are Hittites. The last argument used by Lucien Basch concerns the recommendation to caulk the ship which would be a proof of the inexperience of the shipwrights. But this is evidently a misinterpretation. What is important in the recommendation is not to caulk but to use asphalt to do it with.

The second point concerns the nature of the design used in the operation. Lucien Basch does not contest this point and W. Lambert writes "It can be regarded as certain that this letter concerns building sea-going boats. All the rest of the massive restorations is hypothetical and could turn out to be wrong". That is the reason why only the remaining part of the letter, whatever the uncertainty about the circumstances, has to be considered and, particularly, the question of the use of a design sketch. I consider Edel's translation "plan of construction" as impossible according to the shipbuilding technology of the time and I propose "a kind of sketch", but an annotated sketch with possible information like basic lines, dimensions and proportions. Lucien Basch considers the proposition as a "rather vague concept", but he fails to look at the illustrations that I have given and which bring us a precise idea of what kind of sketch the Egyptian were able to make. He considers also the information about dimensions and proportions as useless because they can be measured directly on the ship. But precisely, it is the use of a design sketch in shipbuilding which is new and important in this tablet. The Egyptian sketch-plans, that I have given in the example, contain the same information and sometimes more, and were used for the guidance of the workmen or even as demonstration objects. Both cases are possible, but as we do not know exactly the precise context of the operation it is difficult to decide what is the real purpose of the drawing. The use of a design sketch, as a demonstration object to document the project and to show it to the Hittite King cannot be rejected. The last objection by L. Basch concerns the difficulty of finding the sections between amidships and the extremities of the ship. But they do not need this information according to the longitudinal shell conception of the hull at the time. It was more important to know the curve of the stem and stern posts, perhaps the curve of the keel, or the curve of the bottom amidships. All that is quite possible according to the examples of Egyptian sketchplans. In conclusion, I do not take Edel's restored text at face value. I consider only the remaining part which concerns, indisputably, the use of a design sketch to build a sea-going boat. And whatever the design sketch used and its purpose, I regard this Accadian tablet as fundamental data in the history of naval architecture.

CHAPTER FOUR

MARKINGS AND PEGS: CLUES TO GEOMETRICAL PROCEDURES OF ROMAN NAVAL ARCHITECTURE?

Ronald Bockius

As examination of hundreds of ancient and prehistoric ship finds proves, shipbuilding was remarkably developed in antiquity, in particular in the Mediterranean. One of the earliest plank-built vessels known, a barge buried near Chufru's pyramid in the middle of the 3rd millennium B.C., is more than 40 m long, a size which is also encountered in seagoing freighters of the Roman period, among them some of double-planked shell. From the Hellenistic period on, war ships and special-purpose vessels, as e.g. obelisk carriers or floating palaces (respectively temples) left the dockyards, their size approaching the maximum achievable length of wooden ship construction. All of these ships apparently were built shell-first, the shipwrights being provided with traditional tools only, but having special experience and certainly the ability to improvise. It goes without saying that at least large ships whose construction was ordered by ruling bodies, by the state or rich companies, had to be designed by naval architects. As far as ancient galleys are concerned, the complex structures of oar systems might have been a particular challenge. But it is mainly the question how ships were planned in detail that remains unanswered. It can be assumed that not only boats but also large vessels of Mediterranean shipbuilding tradition were designed by eye during the construction process, the hulls being built shell-first, with framing integrated at a second stage; to arrive at the intended main dimensions. The size and shape of axial elements as keel, posts and single ribs, though not pre-erected, had to be defined by boat builders, whether based on simple sketches and the usage of ratios or otherwise. Alternatively, concepts of mixed construction are under discussion, to be understood as progressive application of individual frame elements to supplement the planking which in the geometrical sense still played a predominant role (cf. e.g. Pomey [1]). This is also reflected in Romano-Celtic ship finds (cf. Nayling and McGrail [2])—no wonder because their lack of edge-joint planking set a special accuracy standard for the construction methods. Nevertheless, it is hard to believe that such ships, which fell into the sphere of public commissions or which resulted from serial production built by large teams of craftsmen, were realized without knowledge of any principles for pre-determining shape and structure.

However, as it is clear that procedures of large-scaled ship construction not necessarily needed lines plans nor even simple drawings, it must be asked whether traces of naval architecture can be expected to be found at all on ancient ships. As long as sources and techniques of planning watercraft in antiquity remain a puzzle, we can only search for any substantial evidence potentially caused by the execution of what may be called "the organisation of the shipbuilding process" in order to discuss their meaning. Attempts to document ship finds reflecting indications of sophisticated methods of shipbuilding were made by the author some years ago (cf. Bockius, pp. 52–53 in [3]), when it appeared that a generally valid interpretation of certain practices could not be given. This still remains true, but as the analysis of the late-Roman shipwrecks from Mainz, Germany (Bockius [4]), implies, certain construction methods had been known in the 4th c. A.D. at the latest (1st) to produce and duplicate the hull geometry even for slender vessels, (2nd) to plan their internal structural elements, and (3rd) to manufacture watercraft in series:

It is necessary to remark in advance that the ship archaeological material to be treated here seems totally different from most other ancient ship finds we know. Notwithstanding the fact that the late Roman Mainz wrecks (Figs. 4.1 and 4.6) are sharing construction features with other ancient vessels, especially those from the Northern Roman provinces, their types remain unique. As substantial mute actors of amphibious warfare on inland waters (cf. Bockius [4], pp. 208–215 with further references), they do not represent ancient navigation in general. Accordingly, what can be reconstructed from individual cases of ship architecture not necessarily represents standards typical of ancient shipyards in general. On the other hand, the practical procedures applied reflect a remarkable level of technology. It seems apt to create complex hull shapes, though not in the sense of free design.

One of the most striking features of the late Roman Mainz wrecks are series of up to 3 mm wide and deep incisions cut into the keel planks by saws. To be found underneath almost all frames of vessels no. 1, 2, 3 and 5 (Fig. 4.2), these systems of grooves have been interpreted as

Fig. 4.1 [Col. Pl. XIII]: Mainz, wreck no. 5, in situ.—Bow view of the boat according to dendrochronological analysis built around A.D. 390 (Photo by O. Höckmann, Mainz).

Fig. 4.2: Mainz, wreck no. 3.—Ship according to dendrochronological analysis built after mid 3rd century A.D.—Section of the plank keel showing transversal and diagonal grooves incised underneath frames (Photo by the author).

layout markings which determined the positions where ribs had to be inserted into the pre-erected shell (cf. e.g. Höckmann, [5], pp. 18–25). This interpretation can be supported by such findings where iron nails connecting floor timbers to the keel had been driven exactly into the markings—doubtless after drilling bore holes.

The only parallels of such layout markings are known from the late Roman Guernsey wreck, the remnant of a cargo ship sunk off St. Peter Port (cf. Rule and Monaghan [6]). In this case, the occurrence of transverse grooves is limited to the keel planks as with the Mainz boats. A couple of ancient Mediterranean ship finds, e.g., Marseille-Place de Jules-Verne, Nemi, St. Gervais C, Yassi Ada B and others, also revealed incisions. But, in contrast to the Guernsey and Mainz features, markings incised into planking ran along the edge of frames or set out the position of mortise-and-tenon joints (cf. Bockius [4], pp. 195–196).

However, in the Mainz case, the sets of incisions are representing approximately uniform distances, which differ from wreck to wreck: to quote just the average, 32.5 cm on wreck no. 1 (Fig. 4.3); 46.8 cm on Mainz wreck no. 2 (Fig. 4.4). The average frame distance of the seagoing Guernsey ship was cited to be some 56 cm. These metrical distances have been suggested to be equivalents of ancient foot and cubit units (Rule and Monaghan [6], pp. 29–30). Apparently the same thing is true for the average spacings of Mainz, wrecks 1 (Fig. 4.3) and 3 (Fig. 4.5). The distances approximate the *pes Drusianus*, a unit of an ancient measuring system, well-known from Roman Gaul, but also from the Germanic provinces, in one of whom the Mainz district was situated. However, simple comparisons of ancient shipbuilding practise and modern measurements taken from archaeological remains are often misleading because of several imponderables: ancient craftsmen may have failed, and so might modern scholars when taking measurements from distorted or broken objects. And there is another problem lying in the calibration of Roman measuring instruments—tolerance ranges were remarkably large (cf. Bockius [7], pp. 524–526, fig. 4.4).

Considerations of ancient metrology would not be much help, but to discuss the problem of the constructional purpose of such markings might bring us forward. It goes without saying that such a procedure had to be based on a concept with an intellectual model in the background. So, during the process of shipbuilding, two steps were made; first, planning the scheme of the layout markings, second, the practical performance of measuring and cutting markings into the keel planks. Why so much efforts? Did late Roman boat builders actually need a

Position	to Position	Sectional Distance
A3	→ A4	32.4/32.9
A4	→ A5	32.1/32.6
A5	→ A6	32.4/32.7
A6	→ A7	32.2/32.5
A7	→ A8	33.4/33.8
A8	→ A9	32.1
A9	→ A10	ca. 32.5
A10	→ A11	ca. 32.5
A11	→ A12	32.6/32.8
A12	→ A13	32.0/32.4
A13	→ A14	32.2/32.7
A14	→ A15	31.8/32.4
A15	→ A16	32.4/32.8
A16	→ A17	32.7
		average 32.5 cm

= 1 *pes Drusianus** / **32.6 cm** (cal. 29.0 cm)
= 1 1/12 *pedes Capitolini* / **32.5 cm** (cal. 30.0 cm)
* 1 *p.D.*= 1 1/8 *pedes Capitolini*

Fig. 4.3: Mainz, wreck no. 1, plank keel: interrelated distances of setting-out marks found underneath floor-timbers and conjectural equivalents of ancient length units.—Measurements in cm taken by the author.

theoretical concept where to place ribs? As will be shown, accurately predetermined frame positions were needed; indeed, not only for the frames themselves, but also for defining the longitudinal arrangement of components of oar propulsion. Therefore, to set out the positions of frames was not the only aim of the marking systems.

Wrecks no. 1 and 5 are sharing construction elements of an oar system which was verified to be identical in detail. Both of the boats were

Position	to Position	Sectional Distance
frame A17	→ side frames A16	47.3
side frames A16	→ side frames A15	46.7
side frames A15	→ side frames A14	47.5
side frame A14	→ side frames A13	46.0
side frame A13	→ side frames A12	46.0
side frame A12	→ side frames A11	47.5

average 46.8 cm

Fig. 4.4: Mainz, wreck no. 2, plank keel: interrelated distances of setting-out marks found underneath ribs and pairs of side frames.—Measurements in cm taken by the author.

Position	to Position	Sectional Distance
M	→ A1	34.0
A1	→ A2	33.5
A2	→ A3	34.5
A3	→ A6	97.5
A6	→ A8	67.0
A8	→ A9	33.7
A9	→ A10	34.5
A10	→ A11	41.0
A11	→ A12	33.95
A12	→ A13	34.0
A13	→ A14	33.5
A14	→ A15	33.7
A15	→ A16	33.65
A16	→ A17	31.0
A17	→ A18	34.0
A18	→ A19	33.4
A19	→ A20	34.5
A20	→ A21	35.0
A21	→ A22	33.6
A22	→ A23	34.0

average 33.9 cm

= 1 *pes Drusianus** / **33.8 cm** (cal. 30.0 cm)
* 1 *p.D.* = 1 1/8 *pedes Capitolini*

Fig. 4.5: Mainz, wreck no. 3, plank keel: interrelated distances of setting-out marks found underneath and ahead frames, and conjectural equivalents of ancient length units.—Measurements in cm taken by the author.

simultaneously in use during the final decade of the 4th c. A.D., being the latest among the Mainz vessels. They evidently belong to the same type: with wreck no. 1, forward of a cross-beam, to which originally side rudders had been lashed, the section of a gunwale is preserved with a sequence of still three tholes (Fig. 4.6). The socket holes to insert tholepins are separated from each other by distances of 95 to 96 cm. Roughly the same measure is revealed from a series of recesses, up to 20 cm long and 3 to 5 cm deep, cut into the upper edge of a ceiling plank, which had been nailed to futtocks. The upper edges of such slots lay almost 20 cm below the sheer-line, i.e., 25 cm below the level where the oars rested on a specific type of a gunwale with half-rounded segments. As parts of an inner structure, remains of vertical timbers are preserved, aligned along the garboard strake on port; the longitudinal position of each of them transversely corresponds to the recesses found in the ceiling plank which can be associated with a thwart clamp. The latter originally carried benches which near the centre-line of the boat were supported by stanchions (Fig. 4.7).

As indicated by wreck no. 5, such stanchions had been arranged pairwise, one each per oarsman's room on either side of the keel, along the garboard strakes. Scattered over the propulsion sections of such boats, the stanchions had been strengthened by several planks longitudinally nailed to the vertical elements, forming fence-like structures (Fig. 4.8). Evidenced by the fragmentary finding of wreck no. 1, the vertical level of each of the boards corresponds to stringers nailed to the extremities of floor timbers and futtocks or side frames. As with the longitudinal reinforcements of the stanchions, they are provided with recesses, in each case placed in the same transverse plane. According to wreck no. 5, deck beams had been inserted into the lower slots which carried light deck planking confined to the sides of the ship deck. The 2nd and 3rd levels each were reserved for one light cross-beam forming an oblique platform, parallel to the feet of the oar crew. Obviously, in each space segment some kind of foot-stretchers existed (Fig. 4.7–4.9).

All the details mentioned match the three-dimensional system to place tholes, thwarts and foot-stretchers in such a way as needed for effective rowing. The longitudinal and vertical distances are best adapted to individuals some 1.7 m tall. It is hard to believe such a complex structure of a single-banked unit could have been constructed without planning. Actually, the contrary is true: as the thwart bearers in the A type wrecks were nailed to the forward or after edge of each 3rd

Fig. 4.6: Mainz, wreck no. 1.—Boat according to dendrochronological analysis built around A.D. 385.—Internal view of bottom and wall of the port side with installations connected to oar propulsion (Photo by Römisch-Germanisches Zentralmuseum Mainz).

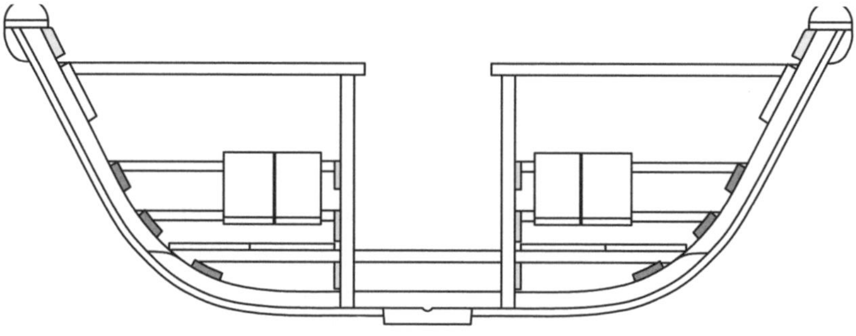

Fig. 4.7: Cross-section of Mainz, type A boats.—Reconstruction based on constructional features found in wrecks no. 1 and 5.—No scale.—Colour code corresponds with plan drawing Fig. 4.9.

floor-timber (Fig. 4.10), the propulsion systems of boats no. 1 and 5 are geometrically related to the principle for placing the floor timbers. Their arrangement was based on a concept which designed intervals of slightly less than 1 m, i.e. the ancient *interscalmium* (cf. Bockius [8]). That reminds us of the distances of the layout markings: three-times

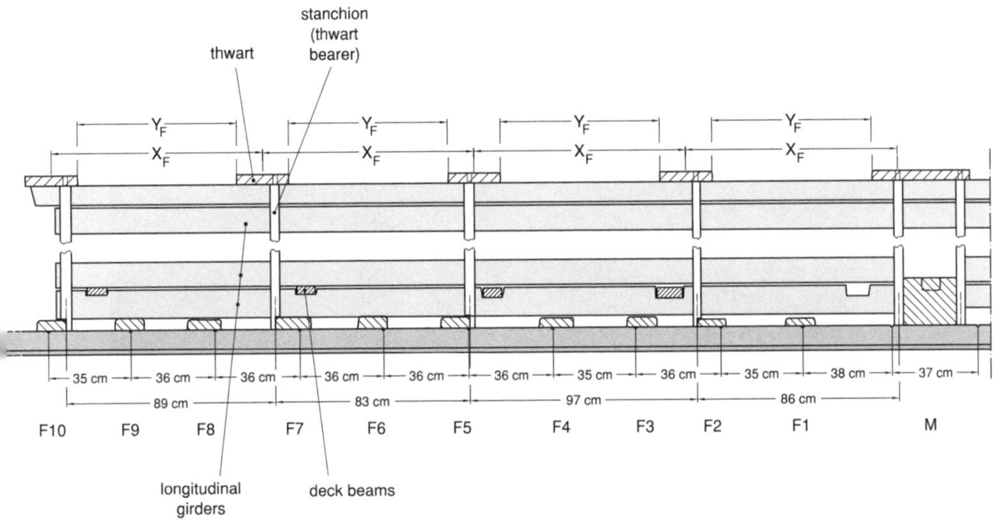

Fig. 4.8: Internal installations of the oar system of Mainz, type A boats.—Reconstruction and distances based on findings of wreck no. 5 (Drawing by the author and U. Kessel).

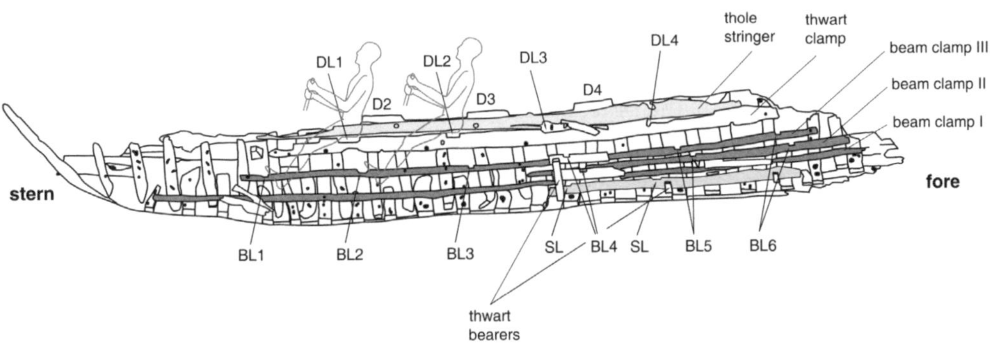

Fig. 4.9: Mainz, wreck no. 1.—Side-plan with internal view.—Longitudinal girders with recesses originally used to insert deck-beams (SL), foot-stretchers (BL) and thwarts (DL); near the keel remains of three thwart-bearers; gunwale with still three tholes (D) preserved.—cf. cross-section Fig. 4.7 with same colour code.—No scale.

the average 32.5 cm of wreck no. 1 (Fig. 4.3) results in some 97 cm, twice 46.8 cm of wreck 2 (Fig. 4.4) almost 94 cm. But it is not so easy as it seems to be: three-times the average spacing of wreck no. 5, i.e. 36.1 cm (Fig. 4.11), would come to an extraordinarily large modulus of 108.3 cm. Much too comfortable for ancient military oar systems, the *interscalmium* of boat no. 5 was indeed almost 20 cm shorter.

Fig. 4.10 [Col. Pl. XIV]: Mainz, wreck no. 5.—Fragment of thwart-bearer preserved on starboard aft of the mast-frame (Photo by the author).

Compensated by the practical execution, stanchions were nailed either to the fore or to the after edge of floors (Figs. 4.8 and 4.10). Assuming the benches were not placed with their centre on top of the stanchions originally, a uniform oarsman's room of only around 90 cm would have been achievable. In case of a more flexible handling, sequences of a bit shorter and longer spaces are conceivable.

The *interscalmium* by most scholars is regarded as a fixed standard. Nevertheless, the practical execution in our cases reflects flexibility, caused by the individual space relation of layout markings and corresponding frames. The error amounts to half the width of a floor timber; just a few cm plus-minus. In one and the same boat, small variations of the oarsman's rooms were caused, which might have been accepted because they easily could be compensated by manning with rather tall or small crew members.

Who other than naval architects might have developed such an oar system, consisting of fittings and components of accommodation, whose positioning was related to the human body? Whether such experts—let us say, a combination of designers and senior boat builders—had been able to predetermine the ship's shape, is another problem. Although

Position	to Position	Sectional Distance
F16	→ F15	34.0
F15	→ F14	no measure
F14	→ F13	no measure
F13	→ F12	35.9/36.2
F12	→ F11	35.3/35.8
F11	→ F10	35.5/36.5
F10	→ F9	34.5/35.5
F9	→ F8	35.5/36.5
F8	→ F7	35.0/36.5
F7	→ F6	35.0/36.5
F6	→ F5	35.5/36.5
F5	→ F4	35.5/37.0
F4	→ F3	35.0/36.0
F3	→ F2	35.5/36.0
F2	→ F1	34.0/35.0
F1	→ M / fore	37.5/38.5
M / fore edge	→ M / after	37.0
M / after edge	→ A1	35.5/37.0
A1	→ A2	36.5/38.5
A2	→ A3	36.0/37.5
A3	→ A4	36.0/36.5
A4	→ A5	37.0/37.5

average 36.1 cm
(foreship 35.9)
(amidship 36.8)

Fig. 4.11: Mainz, wreck no. 5, stern and plank keel: interrelated distances of setting-out marks found underneath floor-timbers.—Measurements in cm taken by the author.

there is evidence from Greek and Roman temples, whose construction details were based on sketches carved into stone, nothing is known about ancient ship design other than the substance of its products. But, as is well-known, simple methods exist to build a ship without having the opportunity to refer to lines-plans. Just to mention one technical solution: any current boat builder should be able to copy a prototype. He first has to take proper measurements, which he then

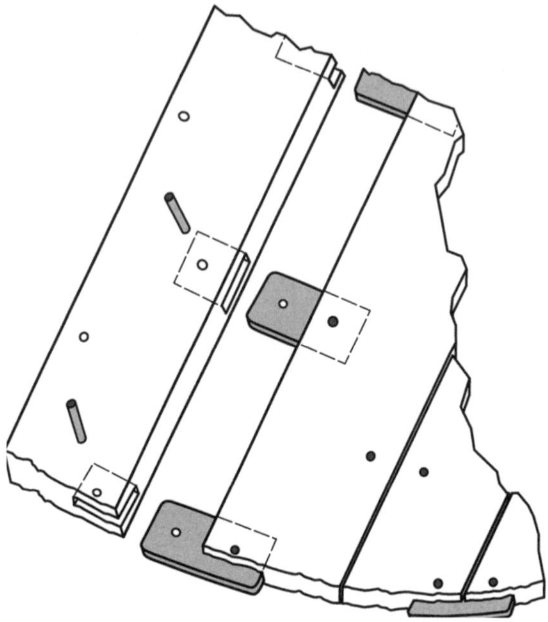

Fig. 4.12: Ancient Mediterranean edge-joint carvel planking with mortise-and-tenon construction.—No scale.

transfers onto practical boat building material. To come to the essential the point now: there are clues found in the Mainz wrecks which seem to indicate the same.

The Mainz ship finds belong to the so-called Romano-Celtic shipbuilding tradition (cf. Bockius [4], pp. 198–207), which is archaelogically defined as follows: vessels are built predominantly, if not exclusively, from oak, the planking without edge-joining; mast frames were used for rigging instead of keelsons, typical of ancient Mediterranean vessels. The most important phenomenon which distinguishes classical and provincial customs may be the edge-joining (Fig. 4.12), missing in Romano-Celtic ships; another difference consists in the joining of frames and planking: The frames had been fastened by iron nails, most of them double-clenched inboard. As in particular shown by wreck no. 5, a certain number of iron nails were plugged by wooden pegs, a procedure also well known from the Mediterranean. But the Mainz ship finds indicate still another category of timber joining which was of a temporary character only: remains of treenails or pegs (Fig. 4.13) occur in quantities of up to 70 per wreck (cf. Bockius [4], pp. 21–25,

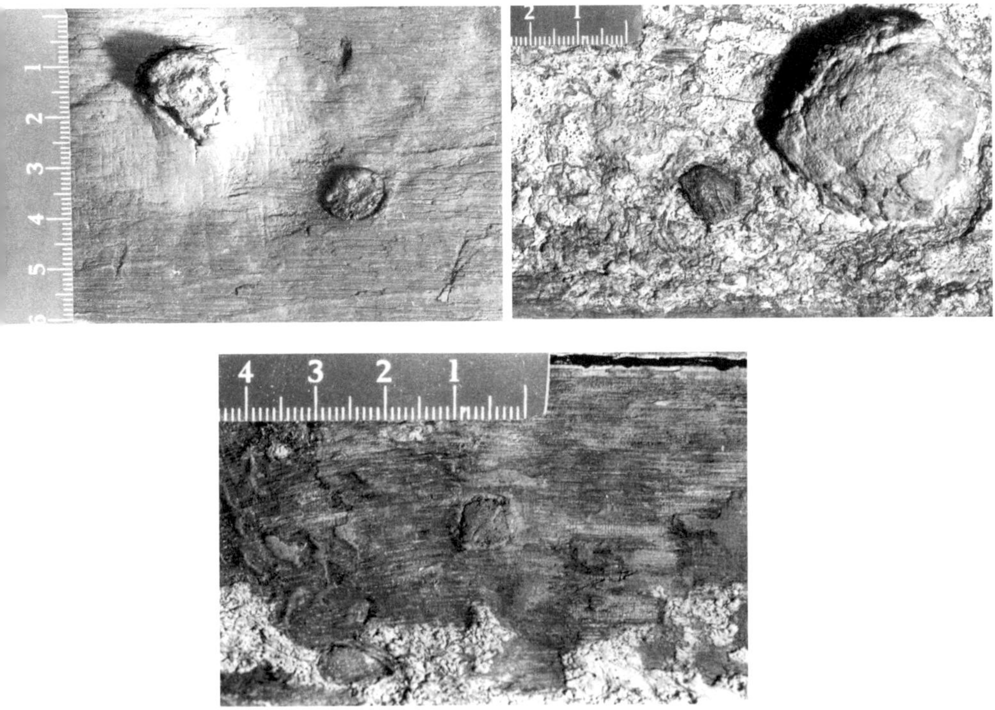

Fig. 4.13: Selection of wooden pegs or tree-nails to be found with the planking of the late Roman Mainz wrecks (Photos by the author).

56–59, 101–109, 141–165, index 5a–c, 14a–b, 17a–b, 20, 26.). Sticking in keel and planking, most are found underneath ribs, without penetrating into them. So, they evidently belong to a phase of boat building before the adjustment and fixing of the ribs. A good deal of such findings implies sets of groups distributed over the preserved parts of the vessels in transverse plane sets. The most striking scheme is revealed in wreck no. 2 (Fig. 4.14). There, each one or two pegs per strake scattered from the keel to the sheer plank on port, represent planar transverse lines. These are separated from each other by 1 to 2.8 m longitudinally according to what is defined to be a joint series. However, these sets remain incomplete—it can be assumed that such sets initially continued on the starboard side of the boat which remained unexcavated in the ground.

A different scheme, though comparable, is indicated by the outer skin of wreck no. 1 (Fig. 4.15): its sets, 2.0 to 2.7 m apart, seem to be shorter, at least in the after part of the wreck. A third planar set towards the

Fig. 4.14: Mainz, wreck no. 2.—Boat according to dendrochronological analysis built in th early 4th c. A.D.—Stern post, keel and planking projected in longitudinal construction sequence showing the distribution of pegs or tree-nails (dots) and empty bore-holes (circles) from th interior side.

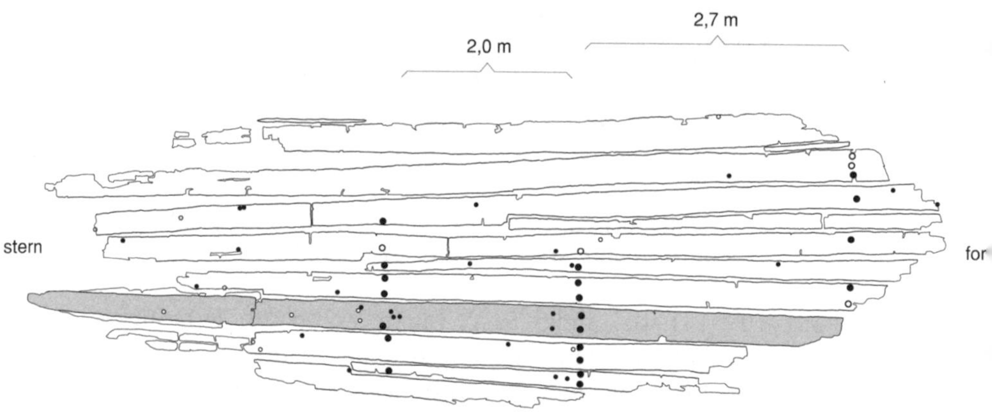

Fig. 4.15: Mainz, wreck no. 1.—Stern post, keel and planking projected in longitudinal con struction sequence, showing the distribution of pegs or tree-nails (dots) and empty bore-hole (circles) from the interior side.

midship section of the craft may have crossed the keel plank running from sheer strake to sheer strake. Obviously, objects had been connected to the planking in an early phase of boat building before frames had been inserted; whether by wooden or iron nails is a different question. In the latter case, after the removal of such transversal elements, of course nail holes had to be closed for reasons of water tightness.

There is another perspective connected with the interpretation of such findings: as the Mainz boats have been slender, sharp ending vessels, the character of the temporary installations would necessarily

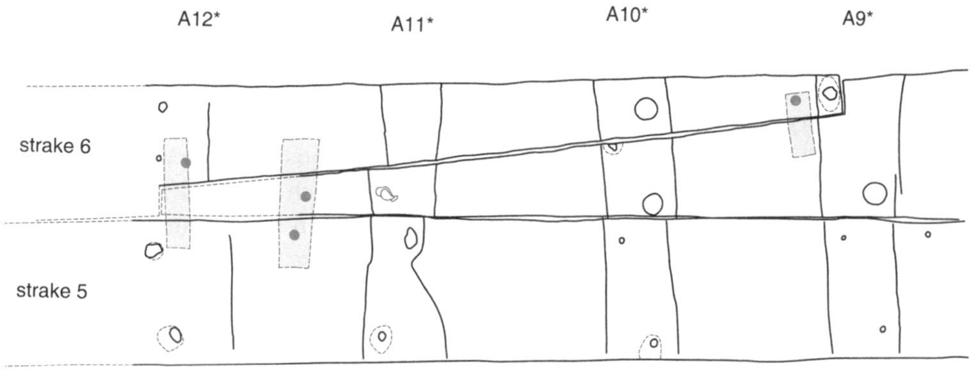

ig. 4.16: Mainz, late Roman wreck no. 4.—Constructional detail: inboard view of plank carf fastened to frames A9* to A12* by iron nails, in a prior phase of boat building temorarily secured by cleats (squared areas), the latter indicated by impressions and relicts of tree-nails.—No scale.

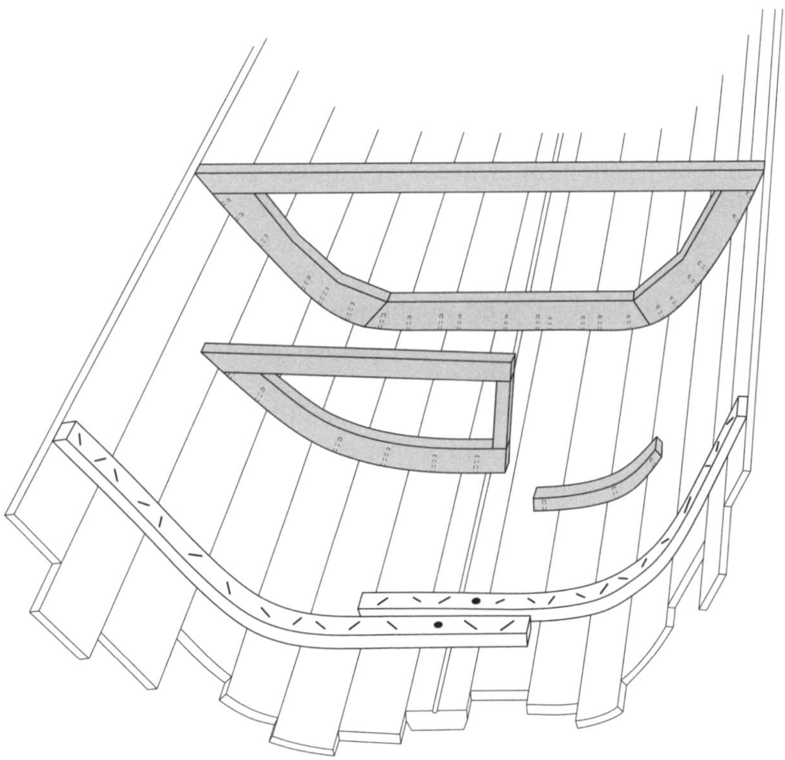

Fig. 4.17: Hypothetical reconstruction of auxiliary methods in Roman boat building to create ship shape: carvel planking by temporary installations of moulds, semi-moulds and cleats.

correspond to moulds. Actually, the transversal arrangement of the pegs in some cases accurately follows a vertical plane with an error of just 1 or 2 cm. The latter suggests the application of rather light auxiliary frames made from timbers just a few cm thick, thick enough to hold light planking nailed to the temporary construction. Whether the same is true for wrecks nos. 3, 4 and 5 which revealed a rather small number of sets of a few pegs concentrated in up to four strakes or interrupted by strakes without peg hole findings remains open. In certain cases, treenails had been used to fix the extremities of stealers and the segments of scarf-joints in position until frames were inserted. As an example, details of wreck no. 4 appear instructive, where imprints of temporary stiffeners were still visible before the conservation treatment on the inner surface of the planking (Fig. 4.16). Such traces of boat building methods caution us that a generally valid interpretation of this phenomenon cannot be established.

Nevertheless, it seems plausible that some kind of a mould marking procedure (Fig. 4.17) was used to build vessels from which hundreds, if not thousands were put into action to patrol the late Roman frontiers along Rhine and Danube during the late 3rd and 4th centuries. Beyond the interpretation of the Mainz wrecks, even literary documents and peculiar archaeological evidence of military architecture led to the conclusion that uniform types of such single-banked craft used for amphibious warfare on rivers had been the result of mass-production. How else but by mould marking and construction could such large series production be realised? And would mass-production of war material initialised by military authorities be possible without proper planning? The author tends to deny that.

The fore-mentioned marking and building procedure may have achieved also the reproduction of shapes and the verification of dimensions. Moulds or templates supplemented by rather simple manuals or lists which had to give exact instructions to position each auxiliary framing along the X-axis may have been shared among shipyards. However, it seems that ancient naval architects and boat builders had been aware of the geometrical complexity of slender, sharp ending vessels. Based on a tradition which in the Mediterranean occurred very early, at the latest starting in the Early Bronze Age, around 3000 B.C., boats and ships were built from edge-joint planking, either by lashing, later by mortise-and-tenon or sewing (for a general view: cf. Steffy [9], Ward [10], Bockius [11], pp. 192–214). Shapes were created by the determination of the main dimensions of an individual ship, by eye and experience. To what extent the result was reproduced, whether by navy dockyards only or also in the commercial sphere, and as late as from the late Roman period on, are different questions.

As a matter of fact, to create the shapes of certain Romano-Celtic boats and ships, each prototype had to be built in some way. Hypothetically, such a prototype might have been a shell-based product. In that case, the dimensions and outlines of strakes, which could be edge-fastened as usual in Mediterranean shipbuilding, their bend and twist would have defined the ship body. Whether this was possible without any auxiliary construction, seems less plausible, particularly if large ancient vessels, as 40 m long freighters or the Nemi barges, over 70 m long, were concerned. However, there are indications from individual Romano-Celtic wrecks, e.g. the Barland's Farm boat in Wales (Nayling and McGrail [2]) which point to the pre-erection of frames in the lower part of the shell. Hence, at least in this individual case some kind of a frame-based design may have taken place on the shipyard, in order to complete a fragmentary existing body step by step.

Should some kind of free ship design have existed in antiquity, rather simple solutions might be expected. We may not leave out of consideration the use of combinations of rules and standard curves, presumably templates, variably used as *sesto* etc. Or still simpler: a spline and a string, the latter to generate circular arcs. What may appear as pure theory, originates from tiny clues pointing into such a direction. As can be shown by some drawings based on a selection of frames found with the Mainz wrecks, circular arcs can be generated roughly matching the curves. As a small series of half frames selected from wreck no. 2 indicates, the radii gradually are becoming longer nearer the stem (Figs. 4.18–4.19). This is only a non-systematic spot check which proves nothing, it needs to be performed for the whole series. There was still no opportunity to check this hypothesis, but readers of the monograph on the Mainz ship finds are provided with the full graphical documentation (cf. Bockius [4], pp. 252–301, figs. 50–74) which enables them to examine the data by themselves.

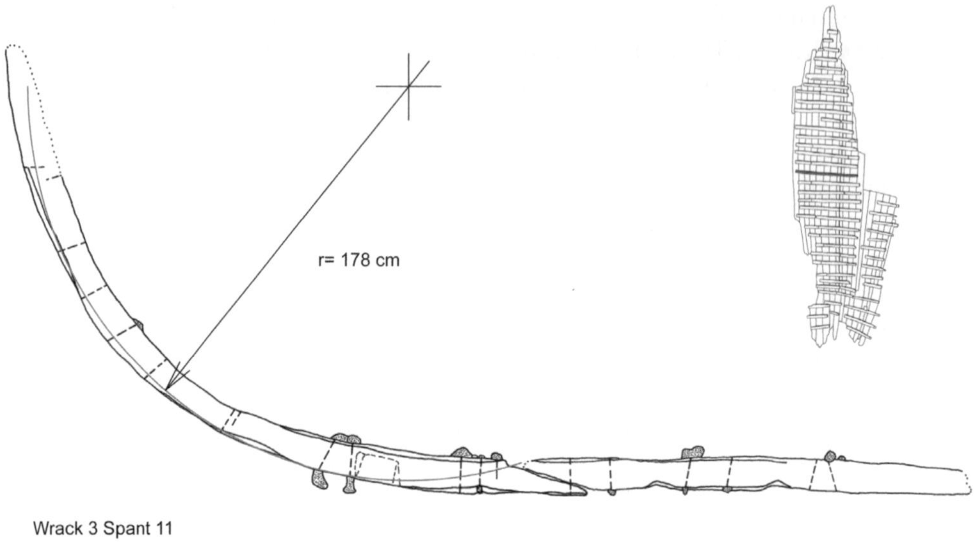

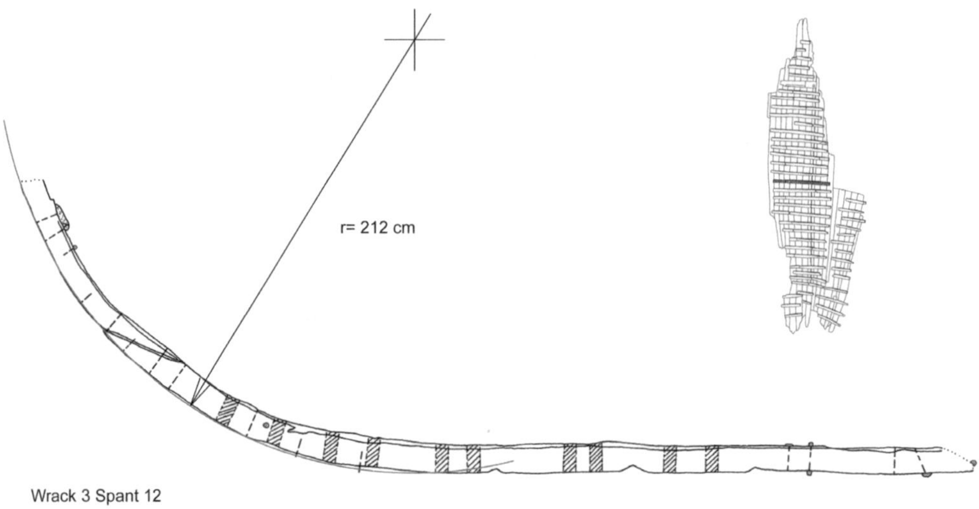

Fig. 4.18 [Col. Pl. XV]: Mainz, wreck no. 3.—Selection of frame curvatures compared with projections of circular arcs.

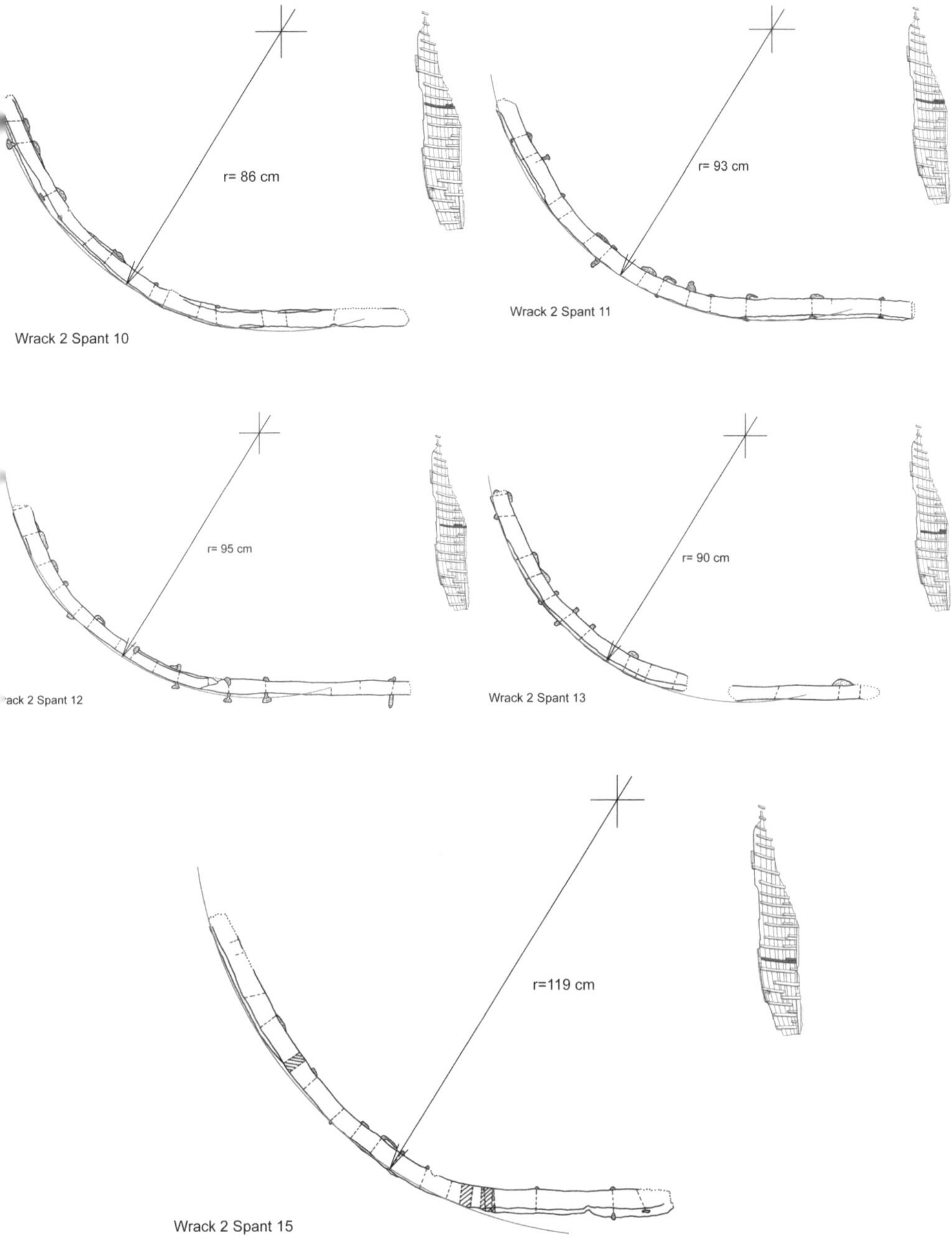

Fig. 4.19 [Col. Pl. XVI]: Mainz, wreck no. 2.—Selection of frame curvatures compared with projections of circular arcs.

References

1. Pomey, Patrice: "Principles and Methods of Construction in Ancient Naval Architecture", in: Hocker, Frederick M., Ward, Cheryl A. (eds.), The Philosophy of Shipbuilding. Conceptual Approaches to the Study of Wooden Ships, Texas A&M University Press, College Station, Texas, pp. 25–36, 2004.
2. Nayling, Nigel, McGrail, Seán: "The Barland's Farm Romano-Celtic Boat", Council for British Archaeology, Research Report no. 138, pp. 205–211, York.
3. Bockius, Ronald: "Die römerzeitlichen Schiffsfunde von Oberstimm in Bayern", Monographien des Römisch-Germanischen Zentralmuseums Mainz, No. 50, Verlag des Römisch-Germanischen Zentralmuseums, Mainz, 2002.
4. Bockius, Ronald: "Die spätrömischen Schiffswracks aus Mainz, Schiffsarchäologisch-technikgeschichtliche Untersuchung spätantiker Schiffsfunde vom nördlichen Oberrhein", Monographien des Römisch-Germanischen Zentralmuseums Mainz, vol. 67, Verlag des Römisch-Germanischen Zentralmuseums, Mainz, 2006.
5. Höckmann, Olaf: "Maatvoering on Roman Boats from Mainz, Germany", in: Instituut voor Pre- en Protohistorische Archeologie, Albert Egges van Giffen (ed.), address presented at the farewell ceremony of Dr. Maarten Derk de Weerd 1 February 2000, Amsterdam, pp. 17–26, 2000.
6. Rule, Margret, Monaghan, Jason: "A Gallo-Roman Trading Vessel from Guernsey, The Excavation and Recovery of a Third Century Shipwreck", Guernsey Museum Monograph No. 5, pp. 25–30, fig. 15, pl. 18, 1993.
7. Bockius, Ronald: "Zur Rekonstruktion des römischen Plattbodenschiffs aus Woerden", Jahrbuch des Römisch-Germanischen Zentralmuseums Mainz, vol. 43, part 2, pp. 511–530, 1996.
8. Bockius, Ronald: "Gleichmaß oder Vielfalt? Zum *interscalmium* bei Vitruv (*De architectura* I 2,21f.)", in: Studia Antiquaria. Festschrift für Niels Bantelmann zum 60. Geburtstag, Universitätsforschungen zur prähistorischen Archäologie, No. 63, Dr. Rudolf Habelt GmbH, Bonn, pp. 111–125, 2000.
9. Steffy, John Richard: "Wooden Ship Building and the Interpretation of Ship-wrecks", Texas A&M University Press, College Station, Texas, 1994.
10. Ward, Cheryl A.: "Sacred and Secular: Ancient Egyptian Ships and Boats", Archaeological Institute of America, Monographs New Series, no. 5. University Museums Publication, Philadelphia, Pennsylvania, 2000.
11. Bockius, Ronald: "Abdichten, Beschichten, Kalfatern. Schiffsversiegelung und ihre Bedeutung als Indikator für Technologietransfers zwischen den antiken Schiffbautraditionen", Jahrbuch des Römisch-Germanischen Zentralmuseums Mainz, vol. 49, pp. 189–234, 2002.

CHAPTER FIVE

ANCIENT GREEK AND ROMAN ARCHITECTS' APPROACH TO CURVATURE—THE CORINTHIAN CAPITAL, ENTASIS AND AMPHITHEATERS

Mark Wilson Jones

The architecture of antiquity is substantially characterised by conventions associated with the orders (chiefly Doric, Ionic and Corinthian) and the most prevalent building types (temples, theatres, tombs and so on). In part due to the legacy of Vitruvius, Renaissance treatises, pattern books and Beaux-Arts teaching, in part due to modernist reactions to the same, it is commonly presumed that these conventions stifled creativity and reduced design to a formulaic process. In reality there was the latitude for considerable flexibility not only in the organization of ground plans and in the choice of elevational vocabulary (witness mixed and hybrid orders), but also in all aspects of the architectural project. To understand the way ancient architects manipulated the shapes and spaces they created it is necessary to grasp certain fundamental principles that can be observed at least as far back as the classical period.

These principles revolve around geometry and proportion, but they had little to do with the kind of 'secrets' that were so often championed in the 19th and 20th centuries, including the much vaunted golden section and notions such as 'Dynamic Symmetry'. If such may be said to have existed, the secret of ancient design was its marriage of rule and variety. On the one hand it was indeed important that civic buildings and the orders with which they were dressed should belong to recognised conventions. But on the other hand diversity and novelty were equally in demand.

Whereas in the 18th and 19th centuries fairly close copies of buildings were fashionable (with versions of the Monument of Lysicrates for example being produced all over Europe and North America), in antiquity direct citation was exceptional. Notable examples are the reworkings of Mnesikles' 5th century propylaea on the Athenian acropolis in the form of the propylaea built in the Roman agora and at the sanctuary of Demeter at Eleusis, the late first century B.C. and the first half of the second century A.D. respectively. In both cases, however, the act of 'copying' was relatively flexible. Trajan's Column famously provided the basis for the design of the Column of Marcus Aurelius, but while they share the same kind of helical relief and the same 100 ft height for the columns themselves (excluding pedestals and statues), there are also extensive differences (Wilson Jones [1] chapter 8). Even in the design of the Greek Doric temple, arguably the most consistent of all ancient building types, the modular-proportional methods employed did not equate to mere mechanical reproduction. In my view the Greeks used a modular-proportional method based on a module equivalent to the width of the triglyph, a method that was to find a later echo in the writings of Vitruvius (Wilson Jones [2, 3]). Different temples can look so similar to modern eyes because we are not attuned to the fine level of variation and adjustment that permeated each exemplar.

So how did ancient architects mediate between the apparently conflicting polarities of rule and variety? Although Vitruvius does mention the need on occasion to bend the rules according to circumstance, he did not discuss this question in anything like a systematic manner. Nor was it properly resolved in the Renaissance or subsequently; indeed commentators on ancient architectural design never really succeeded in going beyond Claude Perrault's findings of the late 17th century to the effect that Roman monuments display too much variety ever to have been the product of agreed methods.

But Roman architects (and their Greek forebears) did use methods. Of course they did, otherwise it would be impossible to explain the conformity in appearance that is no less incontrovertible than the variations. This point is also confirmed by analyses of ancient design practice such as those by Coulton [4, 5], along with collected essays on this theme [6, 7]. In order to understand the way in which the methods used were able to admit variation, I propose focusing on the design of the Corinthian capital, the swelling of column shafts known as entasis, and the layout of amphitheatres. Each are pertinent to the theme of this volume by virtue of the curving shapes they involve.

The Corinthian capital

After centuries of development and experimentation on Greek territory, around the time of Augustus Roman architects privileged a single morphology, often known as the 'Normalkapitell' following in particular

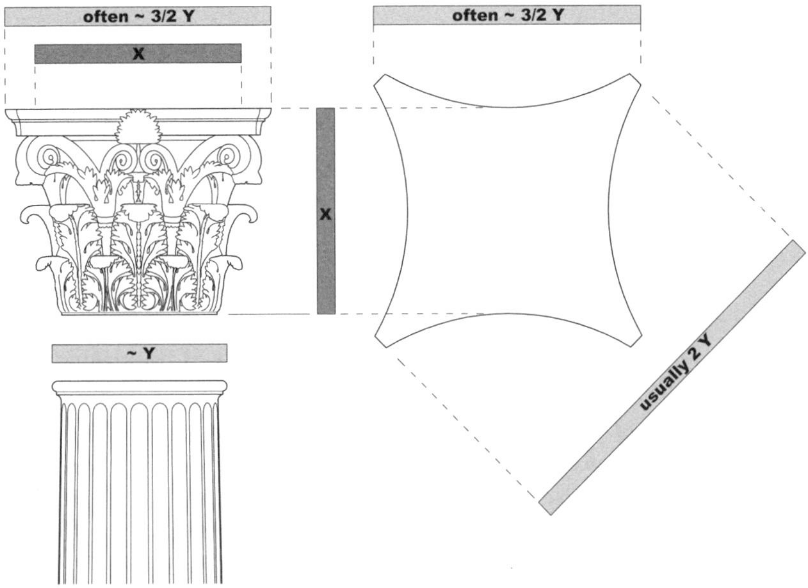

Fig. 5.1 (Col. Pl. XVII]: Standard proportions for orthodox Corinthian capitals in the Roman imperial period. The key dimensions typically correspond to the lower diameter of the shaft (Y) and the diameter of the flare of the shaft where it meets the base (X). The ratio between 'X' and 'Y' is not fixed, but is most commonly 7:6, 9:8 and 10:9. (Drawing by author.)

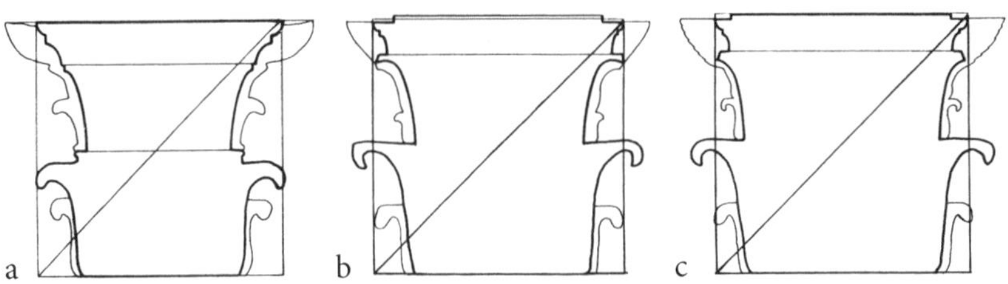

Fig. 5.2: Cross-sections of three Corinthian capitals compared, each of which conform accurately to the cross-sectional rule, as indicated by the overlaid squares.

a. Tholos by the Tiber (mid-late 2nd century B.C.), 1:30
b. Temple of Vespasian and Titus (circa 90 A.D.), 1:40
c. Temple of Hadrian (140 A.D.), 1:40.
(Drawing by author.)

the study of Heilmeyer [8]. Production became so standardized that it is possible to observe, for example, the capital type of Augustus' Temple of Mars Ultor in Rome providing the template for smaller scale examples in places such as Tunisia and France. Yet in general the conventions at work were sufficiently elastic to permit a range of variants, as can be seen upon detailed inspection of the individual constituents of different capitals: acanthus leaves, stems or cauliculi, helices, volutes, abacus and central flower. Also instructive is the question of proportions, since mathematics allows us to discriminate more clearly between that which remains the same and that which changes, while moreover moving discussion in the direction of the problem of shape definition. Satisfying proportion or mathematical harmony—a concept that embraced concordance between number, measure, ratio and shape—was of course a basic premise of ancient design as Gros [9] underlines (see also Wilson Jones [1], chapter 2).

Although it is not mentioned by Vitruvius (nor any of his Renaissance followers), the design of the Corinthian capital revolved around perhaps the single most dominant of all ancient rules for design, applying at a guess to two thirds or more of civic examples, whether Hellenistic, Roman or Byzantine. This set the total height of the capital as equal to the axial width of the abacus, measured on the orthogonal axes, excluding the flowers [Fig. 5.1]. It therefore governed the section on the main axes [Fig. 5.2], which is why I have christened it the "cross-section rule" (Wilson Jones [1], chapter 7; [10]). Following its publication over fifteen years ago this rule has yet to be called into question, so it now seems safe to assume that this was indeed a key feature of ancient procedures. (It is overlooked, however, by Greco [10a].) The very simplicity of the proportional relationship, 1:1, argues in its favour, while occasional masons' marks on the abacus confirm that the cross-sectional width was a key constraint on the carving of capitals. Significantly, at places like the quarries at Proconnesus it was established early on in the process.

While the cross-section rule may go back as early as the Tholos of Epidauros [Fig. 5.3, a], it started to dominate practice by the mid-to-late Hellenistic period. It characterises the so-called Italic Corinthian capital [Fig. 5.3, b–c] just as much as the 'Normalkapitell', although the latter is of greater interest due to its subsequent predominance over a wide geographical panorama (Heilmeyer [9]; Freyberger [11]; Wilson Jones [1]; 140–141 [10]). From the Augustan period onwards the majority of imperial Corinthian capitals maintain this same proportion,

Fig. 5.3: Plans and elevation of selected Corinthian and Composite capitals from a wide chronological range which accurately fit the cross-section rule. (Drawing by author.)

a. Tholos of Epidauros (360–340 B.C.)
b. Solunto, Sicily (late 3rd/early 2nd century B.C.)
c. House of Augustus, Rome (30s–20s B.C.). The right hand side illustrates the unfinished and the left hand the finished state.
d. Temple of unknown divinity, Sabratha (mid 2nd century A.D.)
e. Arch of Caracalla, Ostia (early 3rd century A.D.)
f. "Wind-blown" capital from Deir Sambul, Syria (5th century A.D.).

right the way down to schematic late-antique examples that in terms of style are quite another thing [Fig. 5.3, f]. Composite capitals often conformed to the same system [Fig. 5.3, e].

As is easily shown by comparative illustration [Fig. 5.3], it was evidently possible to maintain the "cross-section rule" in spite of fluctuations in style, in morphology, and in secondary proportions (for example the relative height of the tiers of acanthus leaves compared with the total height of the capital). At first sight this may seem surprising, for on account of the legacy of Renaissance treatises and pattern books we have been accustomed to conceive of proportion as a means of defining and transmitting certain forms. In other words it is widely presumed that such-and-such a proportion yields such-and-such an appearance. But objects with the same proportions do not necessarily look the same. In fact, the Romans put their faith in the cross-section rule precisely because it suited two key principles, first that it did *not* restrict the look of capitals, and second that there were other proportions which had a more immediate impact on their appearance (Wilson Jones [1], 143–148, 156).

It is interesting to reflect on the implications of the cross-section rule for the shape of the abacus, and the specific curvature given to its four sides. It happened that in monumental imperial practice another rule was common (if not quite so common as the cross-sectional rule), namely one that fixed the diagonal width of the abacus as double the lower column diameter [Fig. 5.1]. The concurrent application of both the cross-sectional and diagonal rules was fairly frequent (Wilson Jones [10]; [1], 149–151). It should be emphasized that this combination could occur in capitals of varying slenderness. This meant that the ratio of the overall width of the capital to the cross-sectional width had to vary too, which in turn induced variation in the arcs used to set out the curving sides of the abacus. A slender capital had to have an abacus with shallow curvature [Fig. 5.3, b]. A squat one had to have an abacus with pronounced curvature [Fig. 5.3, c]. Thus there existed no standard template for the abacus, and designers must have been used to adjusting centres and radii so as to suit both chosen proportional schemes and their personal taste as regards such details as the resolution of the corners of the abacus. In short, the principles at work demanded elasticity of execution.

The profile of entasis

Progressively introduced during the 6th and 5th centuries B.C., the entasis of columns shafts became the most ubiquitous of all the so-called refinements that are such a hallmark of Greek sacred architecture (Hauck [12]; Penrose [13]; Goodyear [14]; Haselberger [15]; Pakkanen [16].) Indeed columnar entasis (along with the diminution that it invariably accompanied) was the only refinement used as a matter of course in Roman times, not to mention the practice of classical architecture as it was later revived in the Renaissance.

In discussing this topic, we are fortunate in being able to side-step the debate on the subject that preoccupied architects and scholars from the 16th century to the early 20th century, a debate dominated by speculative expositions of Greek mathematics and theoretical contemplations of the merits of curves such as the parabola, the hyperbola (even pairs of hyperbolas) and conchoids, as proposed by Penrose [13], Goodyear [14] and Stevens 17]. We now know with unusual confidence how ancient architects proceeded thanks to growing knowledge of their working drawings/templates (Heisel [18]; Haselberger [19]; Inglese [19a]; see also [6, 7]). The most famous of these drawings, discovered by Haselberger [20, 15] inscribed on one of the walls of the Hellenistic Temple of Apollo at Didyma, defines the profile of one or more of the peristyle columns [Fig. 5.4]. The curve on the drawing itself is an arc of a circle, disarming in its simplicity. Yet although the horizontal scale was full size (1:1) the vertical scale was compressed by a factor of sixteen (i.e. 1:16, with a dactyl on the drawing representing a foot on the building). This ingenious device meant that when executed the arc became transformed into a shallow portion of an ellipse. Voila! (It may be noted that the geometrical principle is similar to that used by Sebastiano Serlio (Book IV, 5.4), although the ancient method yielded far greater precision because horizontal dimensions were full size (Haselberger [15, 20] and Wilson Jones [1], 130). (For methods of entasis construction proposed in early modern treatises on architecture such as Serlio's *Regole*, see below the contribution of A. Becchi, this volume pp. 279–296.)

A completely different method is shown on another drawing/template inscribed on a wall surface belonging to the scaenae frons of the theatre at Aphrodisias (Hueber [21]). This drawing defines the outline of a shaft by pairs of lines at an oblique angle to one another [Fig. 5.5], creating

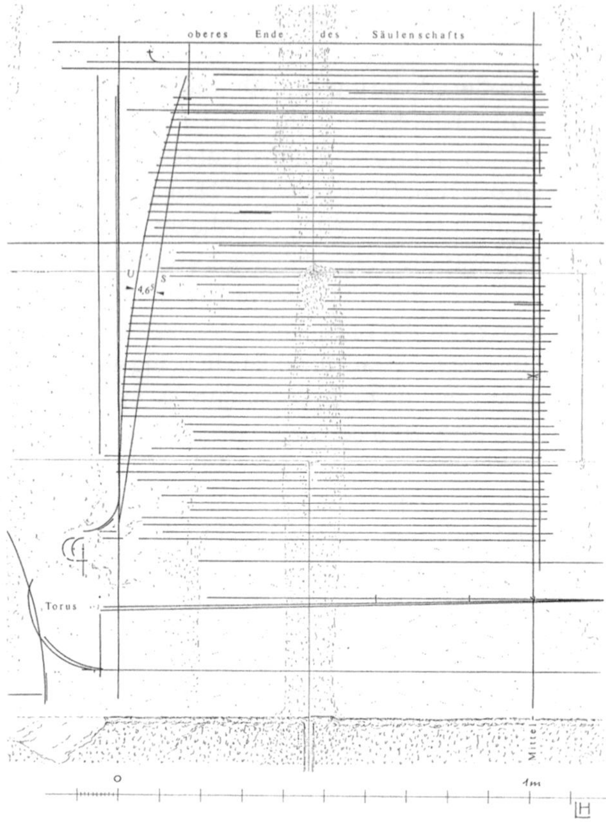

Fig. 5.4: Working drawing from the Temple of Apollo at Didyma defining entasis. The horizontal scale is full size while the vertical scale 1:16, or one dactyl to the foot. (Haselberger 1999, Fig. 1.32)

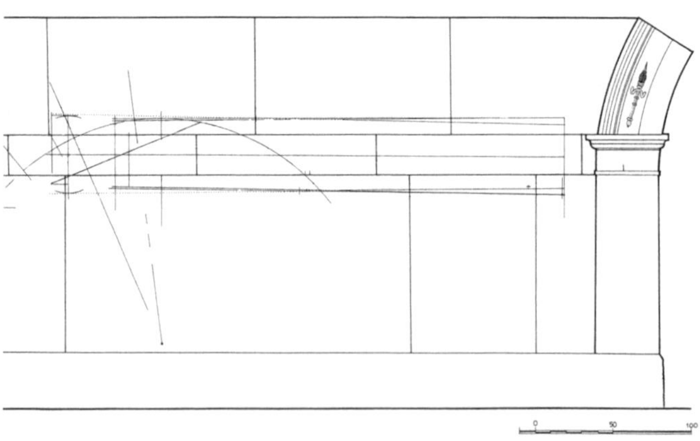

Fig. 5.5: Working full-size drawing defining entasis from the scaenae frons of the theatre at Aphrodisias. (Hueber 1998, Fig. 9.)

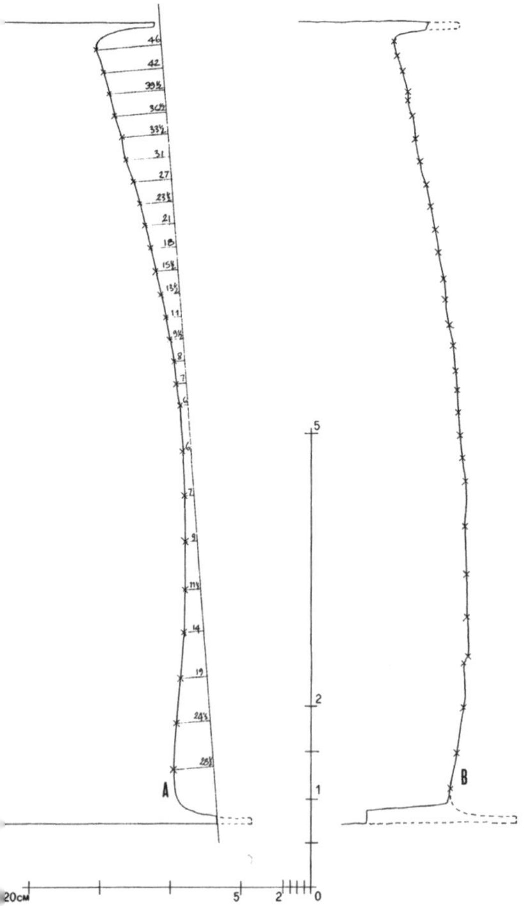

Fig. 5.6: The profile of a shaft from the Forum of Trajan represented with a scale compression of 1:16. This conforms to a variation on the Didyma method. (Drawing by author.)

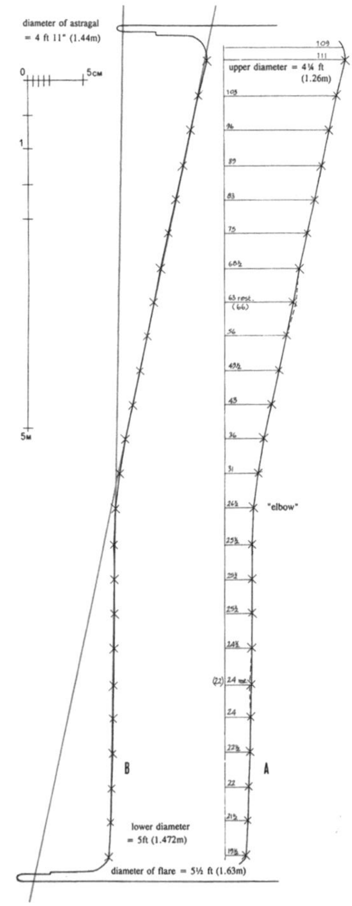

Fig. 5.7 [Col. Pl. XVIII]: Profile of a shaft from the Temple of Hadrian (Rome, c. 140 A.D.) represented with a scale compression of 1:16. (Drawing by author.)

what might be called a 'cranked' profile (Wilson Jones [1], 128). A comparable procedure was also advocated by a Renaissance authority, this time Leon Battista Alberti (Book VI, 12). The result is somewhat cruder than that obtained by the Didyma method, but it is hard to imagine anything easier to execute, although the quality of the effect depended on the smoothing into a curve of the transition between the linear sections, and so on the skill with which this was done.

It is possible to detect a wide range of Roman solutions that are consistent with one or other of these methods. One of the smoothly swelling shafts from the Portico Octavia in Rome fits a single curve akin to that specified at Didyma, except that the curve was positioned so as to generate a maximum radius at a little below 1/3 of the height of the shaft (Wilson Jones [22], 234, fig. 13.10), creating a bulging or cigar-shaped profile [Fig. 5.6]. As is apparent to the naked eye, the shafts of the Hadrianeum in Rome do not curve at all [Fig. 5.7]; they have a 'cranked' profile of a similar kind to that specified at Aphrodisias. The same seems to apply at the Round Temple by the Tiber, and possibly the Erechtheion (as can be deduced by comparing Penrose [13], Pl. 14 and Wilson Jones [22], 227–229, figs. 13.2–13.4). Here, then, are clear-cut Roman instances of each of the two main methods. (In his discussion of entasis Vitruvius cited some kind of drawing, since lost; presumably it represented one or other of these methods.)

In Roman practice entasis was in fact more commonly a mixture or composite of the two methods just described. Whether found in Rome or in places as far away as the cipollino quarries in Euboea, the surveys of shafts that I conducted in the early 1990s return variation after variation of such composite profiles. Shafts from the Forum of Trajan have a profile that divides between a curve for roughly the lower two-thirds and a straight taper for the rest (Wilson Jones [22], fig. 13.5). The same principle is found it seems at the Pantheon (in both the interior and exterior orders), where the transition takes place instead slightly lower, at about 3/5 of the height of the shaft (Wilson Jones [22], fig. 13.10, middle and right). Rather unusually my survey of an example at Euboea (Wilson Jones [22], fig. 13.23) shows that it has a relatively short (but very pronounced) curve, occupying less than half of the shaft before a straight taper takes over [Fig. 5.8]. Each of the profiles mentioned could have been defined using a compressed scale drawing/template of the kind found at Didyma, but using an arc in conjunction with a straight tangent. It is safe to conclude that the entasis of the great majority of Roman column shafts was created by adapting and combining in some way the methods attested at Aphrodisias and Didyma, that is to say using arcs in conjunction with straight tapers on a compressed scale drawing or template.

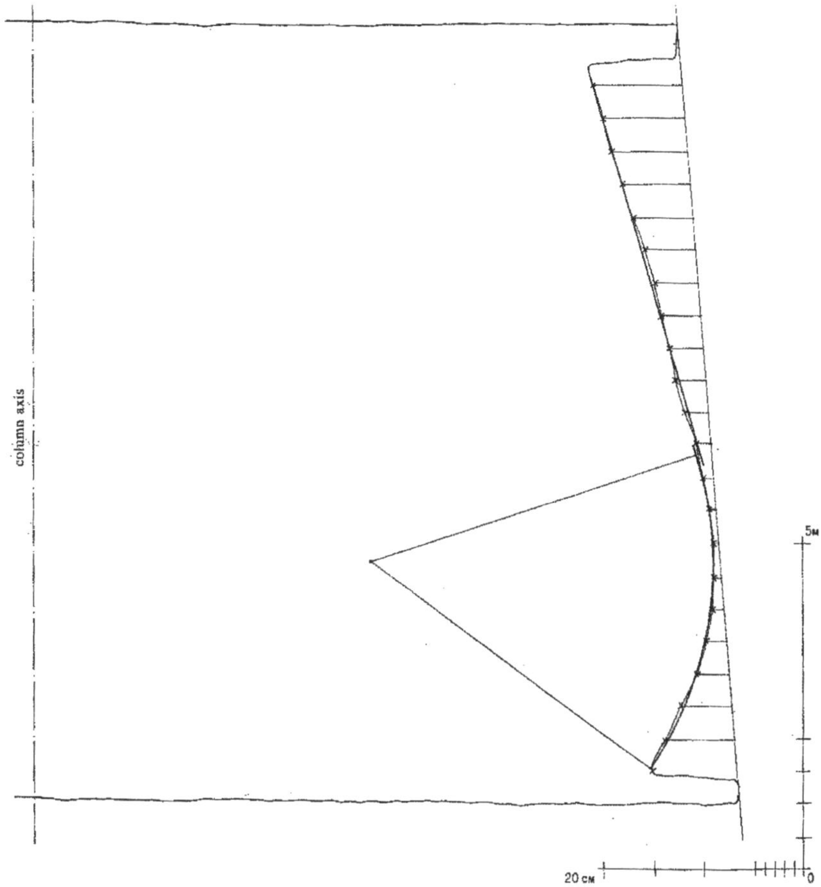

Fig. 5.8: Profile of a shaft in the cipollino quarries at Kylindri above Karystios, Evvia (ancient Euboea), Greece, represented with a scale compression of 1:16. (Drawing by author.)

Trajan's Column represents a particularly interesting case since the exterior displays a different solution to that used for the helical stair within. On the exterior an ample curve is sandwiched between two straights, pointing to a template involving a segmental arc and two tangents [Fig. 5.9]. As I have already argued (Wilson Jones [22], 231–233) the staircase represented a different problem; while its diameter had to diminish broadly in sympathy with the exterior there was no advantage in producing a visually satisfying effect, since no one could

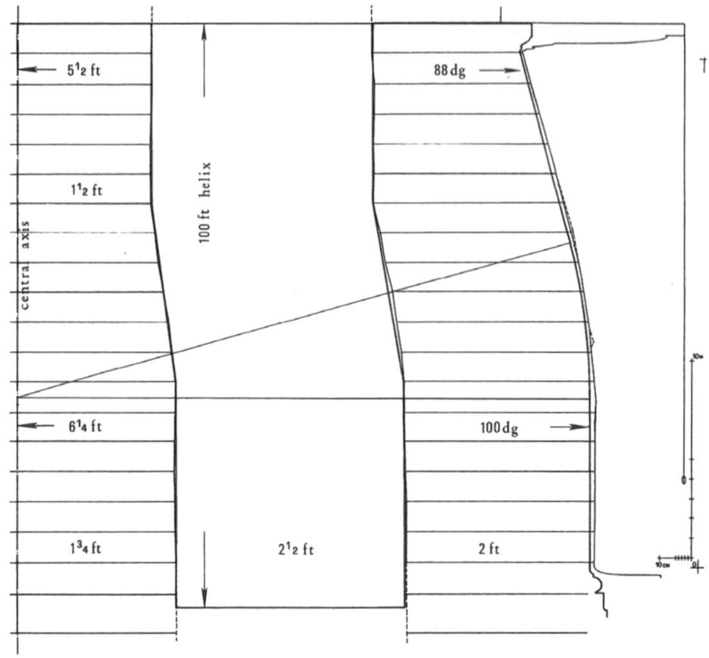

Fig. 5.9: Cross-section of the Column of Trajan, Rome, represented with a scale compression of 1:16, with hypothetical scheme overlaid. (Drawing by author.)

ever enjoy it. Accordingly the profile conforms to a simple variant of the cranked method, this time involving three straight sections: vertical-inclined-vertical.

Practical issues relating to specific cases therefore had an impact on particular designs, as would have individual architects' predilections. Other factors may have been the skills available in particular quarries or workshops, and any engrained production procedures. No doubt some teams of craftsmen were adept with one method, others with another; perhaps architects and contractors negotiated as to the precise solution to be adopted.

Although problems of measurement and analysis make it impossible to be dogmatic, it seems likely that the construction of the compressed scale profile was predicated on simple metrical premises, with round numbers of feet and/or digits being used for the arcs and their positioning. A good example is Trajan's Column. It can hardly be a chance coincidence that a hypothetical template with a compressed vertical scale of 1:16 (one digit to one foot) points to a segmental arc 100 digits in radius, an especially suitable choice given that the outline of the structure represented in this manner measures 200 digits in diameter and 100 digits tall [Fig. 5.9]. Although the height of its helical stair is exactly 100 ft tall, the height of the exterior column (base, shaft and capital combined) is not in fact the same, but about half a foot taller, seemingly because of compromises that had to be made in the course of detailed design or implementation (Wilson Jones [1] chapter 8). Simple proportions and dimensions are likely to have been predominant, yet since no two solutions are the same it is equally clear that architects eschewed slavishly sticking to a fixed formula, but worked freely around principles which were thoroughly embedded in their experience.

The setting out of amphitheatres

Can similar approaches be seen to operate in the design of whole buildings? An answer broadly in the affirmative comes from the analysis of Roman amphitheatres, a building type chosen here on account of the key role played by curvature. The defining geometrical characteristic of the amphitheatre is of course its elliptical/oval plan, one developed out of an understanding that it offered more dynamic qualities than the relatively static properties of circle, while being better adapted to processions and spectacles that involved a linear or axial component, as Golvin [23] has demonstrated and I concur (Wilson Jones [24]).

Much research has gone into determining whether the Romans set out their amphitheatres on the basis of the ellipse or the oval, and unlike our previous two case studies there is more room for scholarly disagreement. The question is of a certain fascination, since although the two shapes are very similar in general terms one approximating the other—their geometrical definition is quite different. The challenge that the ellipse presents for an architect or surveyor is its ever changing curvature, and the impossibility of using fixed arcs. Instead it has two focal points on its long axis, by definition located so that the sum of the distances between them and any point on the ellipse remains constant; this provides the key for setting one out on a building site using what Golvin [23], 388 and other French scholars term the 'méthode du jardinier' [Fig. 5.10]. By contrast the oval is a compound shape that aims at a comparable effect by means of segmental arcs of fixed radius combined so that they share

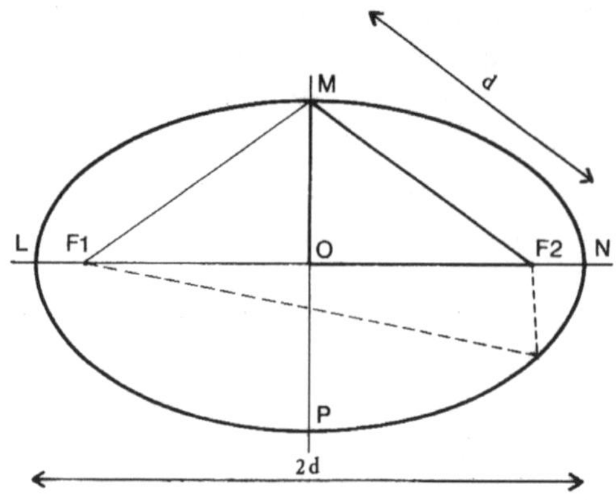

Fig. 5.10: The so-called Gardener's method for setting out an ellipse using two fixed focal points (F^1, F^2) and a fixed length of cord (F^1M + F^2M). Note that ratio FM: OM = the aspect ratio between the length and width of the arena (LN: PM). (Drawing by author.)

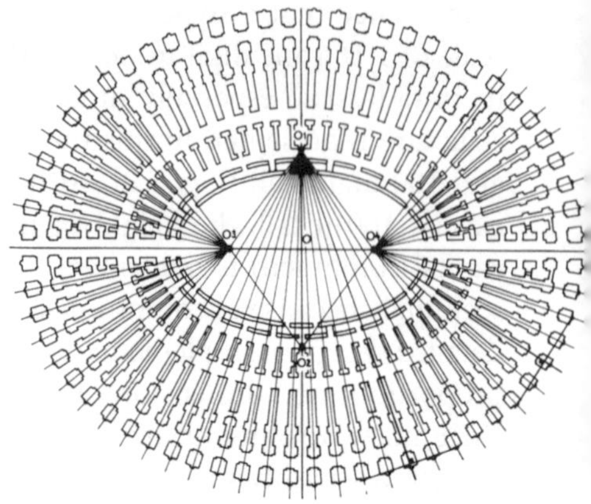

Fig. 5.11: Plan of the amphitheatre at El Jem (Thysdrus), Tunisia, with geometrical overlay. (Drawing by author, after Golvin.)

common tangents [Fig. 5.11]. The simplest oval has four arcs drawn from four focal points; more arcs and focal points may be added [Fig. 5.12], the greater their number the closer the oval resembles the ellipse. The ellipse, then, is the primary formal idea, which the oval mimics by combining portions of fixed curvature.

The ancients knew of both geometries. The means of generating an ellipse by stretching as it were a circle seems to be testified by geometrical constructions that survive from both ancient Egypt (a diagram on a pottery sherd or ostrakon illustrated in Heisel [18], 130, fig. A.18), and Hellenistic Ionia (the working drawing used to define the entasis at Didyma already discussed). This is intriguing, although this does not mean that the creators of these designs necessarily understood the mathematical definition of the ellipse beyond knowing that their methods produced the desired result. There is also the explicit reference by a surveyor working in the time of Trajan to the setting out of arenas with four arc ovals, the phrase used being "*harenas ex quattor circulis*" (Hallier [25], 74 and n. 18.). Examples of amphitheatres with both types of plan have been established beyond reasonable doubt; sometimes the ellipse was used, sometimes the oval.

While Gilbert Hallier [25] has made on the basis of traditional methods of measurement a convincing case that the ellipse was often used for provincial military amphitheatres, it is appropriate to turn to

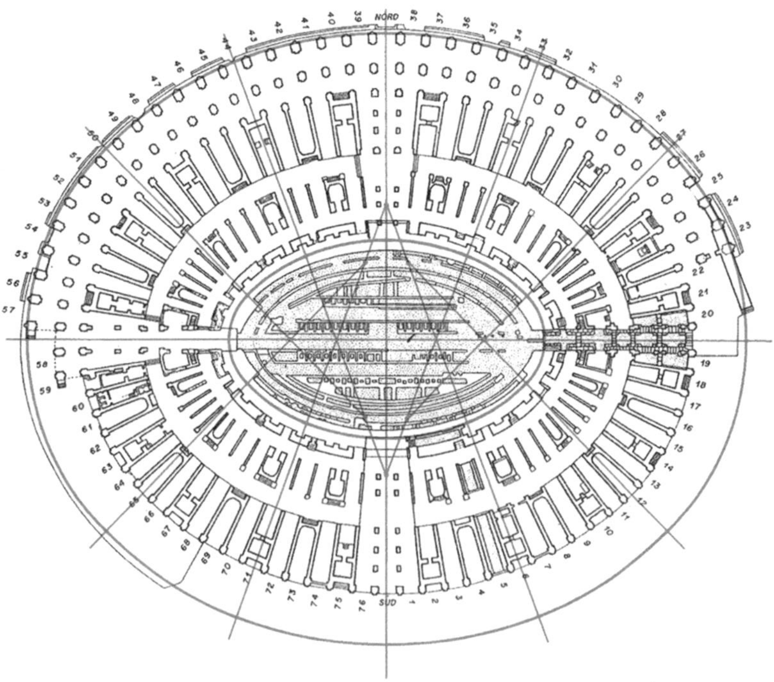

Fig. 5.12 [Col. Pl. XIX]: Plan of the Colosseum, Rome, with geometrical overlay. (after Golvin, with overlay by author.)

recently developed digital techniques that can help in identifying curving geometries. In a study that subjects the results of a digital survey to statistical analysis, Sylvie Duvernoy and Paul Rosin [26] demonstrate that in all likelihood the amphitheatre at Pompeii, the oldest extant Roman example of this building type, conforms to an elliptical plan. This concurs with my own findings, in as much that I found it impossible to trace on-site the short ends of the arena using a single arc, a fact that rules out a four-point oval.

Even with the aid of statistics it is not however possible to determine with certainty between an ellipse and an 8-point oval, given the very slight difference between these two shapes relative to the 'noise' inherent in the data (the product of tolerances both in the original construction and in surveying, as well as of earth movements, decay and damage). Yet there are other considerations which come down in favour of the ellipse in the particular case of the amphitheatre at Pompeii. One is the question of practicality. Rather than starting from a levelled ground-plane (as for most monumental civic examples), this structure relied to a significant extent on ground modelling. It is

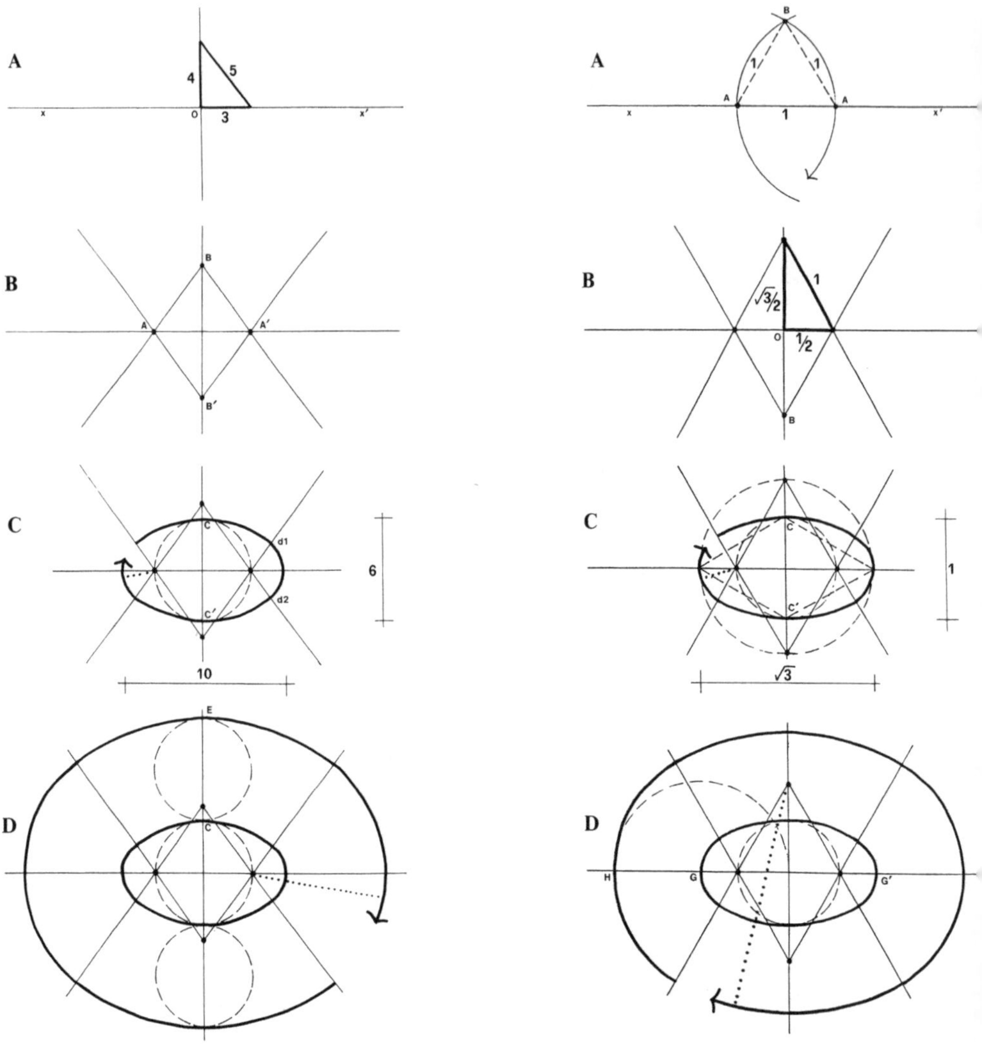

Fig. 5.13: Methods for laying our monumental civic amphitheatres using the Pythagorean-triangle-and-inscribed-circle scheme (left), and the Equilateral-triangle-and-inscribed-circle scheme (right). (Drawing by author.)

important to note that the arena lies below the level of the rest of the structure; given that the tracing of an oval calls for the focal points on the short axis to fall outside the arena, this would mean that these would be located at higher level, with all the attendant difficulties implied. On the other hand operations could have been contained within the level area of the arena itself when tracing out an ellipse using 'the gardener's method'. Another point in favour of the ellipse here is the simplicity of the scheme that can be reconstructed for the original layout, one in which the arena and two other key rings of the structure match ellipses measuring round numbers of Roman feet. The width and the length of these three elliptical rings at Pompeii match simple multiples of a module of 12 Roman feet, that is to say 10:19, 13:22 and 26:35 modules (Duvernoy and Rosin [26], 31–33).

The situation at Pompeii can be contrasted with the general tendency for monumental civic amphitheatres to be laid out as ovals. Several features of these buildings are more natural and easier to both explain and to achieve on the basis of an oval, concentric, geometry than an elliptical one. First, in many instances radial walls converge noticeably on just four points, as for example at Capua and El Jem [Fig. 5.11]. Second, the width of the cavea and the associated annular corridors tends to be relatively uniform (Hallier [25] 72 ff.; Wilson Jones [24], 398). Third, in some cases rings of curvature can be shown to accurately match segmental arcs; in the case of the Colosseum and the amphitheatre at Pozzuoli I succeeded in simulating *in situ* large portions of their arenas using arcs swung from the focal points (Wilson Jones [24], 394). Fourth, features articulating the exteriors (principally piers and openings of the repeating bays of fornices) tend to be regular in terms of size and rhythm (Wilson Jones [24], 407–8). As if to underline this point, in some well-known examples, those at Verona, Pozzuoli, Capua and Nimes, the interaxial interval of the perimeter bay is 20ft; in others, those at Italica and El Djem besides the Colosseum, the *intercolumniation* is 20ft. Fifth, arenas have an aspect ratio that exhibits a convergence on 5:3 (1.666:1) and √3:1 (1.732:1), a pattern which can be explained by the popularity of two specific oval schemes, one based on a 'focal triangle' corresponding to the 3:4:5 Pythagorean triangle, the other based on the bisected equilateral triangle [Fig. 5.13]. Both these triangles of course were well known in antiquity.

It is furthermore possible to identify the main steps in the design process that determined the precise configuration and curvature of monumental amphitheatres. In conjunction with the focal triangles just mentioned, it was common for the width of the arena to be made equal to the distance between the focal points on the major axis, an arrangement which can be visualized by means of an inscribed circle. The two most popular layouts were thus the "Pythagorean-triangle-with-inscribed-circle-scheme" (generating an arena with a 5:3 aspect ratio), and the "equilateral-triangle-with-inscribed-circle-scheme" (with a √3:1 aspect ratio), as illustrated in Fig. 5.13 (Wilson Jones [24], 401–405; Golvin [23], 298 ff., 387 ff.). It was almost instinctive for a Roman architect to define such schemes with suitable 'round' dimensions, e.g. 250 ft for the length of the arena or 500 ft for the total length; but while such instances do occur what is in fact notable is the frequency with which linear proportions and dimensions do *not* match ideal dimensions. For example, the aspect ratio of the arena at Verona is not 1.731 (√3:1, as predicted by the hypothetical equilateral triangle scheme) but 1.700, while the overall length of the building is not 500 ft but 512 ft. The explanation has to do with setting out the exterior façade so that it conformed to regular intervals like 20 ft. In theory this could be achieved by establishing the necessary circumference by multiplying the desired bay width by the number of bays, and then calculating the radii of the arcs of the oval that would produce that circumference. In practice, however, this depended on a calculation that was not straightforward, which explains why architects proceeded to some extent by trial and error. First of all an initial, or provisional, layout was proposed in terms of linear dimensions. Then the circumference would be calculated and divided by the chosen number of bays, producing an interval that was unlikely to match a whole number of feet. For this purpose, rather than use trigonometry, in all probability Roman architects resorted to an approximation which involved treating the oval as if it behaved like an ellipse, i.e. by taking the circumference as the average width multiplied by π; that is to say:-

$$C = [(L + W) \times \pi]/2$$

where L is the length of the building and W its width (Wilson Jones [24], 410; Trevisan [27], 1999, 121). While keeping the original focal triangles the same, this was then remedied by shrinking or expanding the length of radii to be used for the definitive layout. As a result the final scheme could no longer enjoy simple overall linear dimensions.

At Verona for example the starting point was the 'Inscribed-Equilateral' layout based on a focal triangle measuring 72 ft and 125 ft on the two short sides; this generated an arena length of 250 ft, while the choice of 125 ft again for the width of the cavea produced an overall

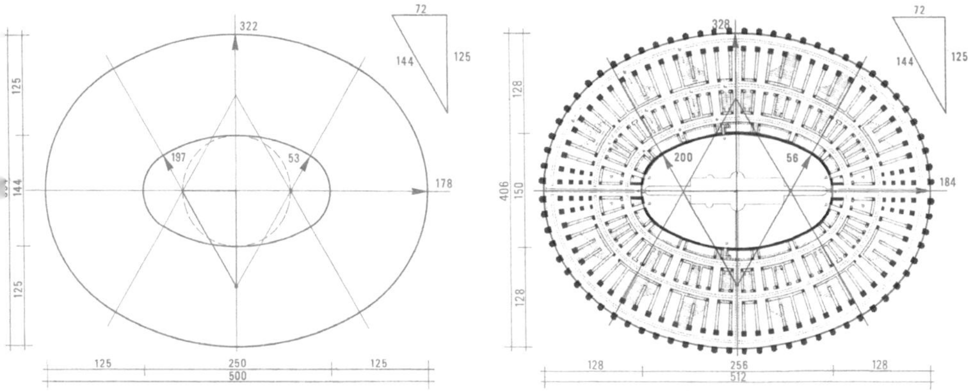

Fig. 5.14 [Col. Pl. XX]: Outline layout of the amphitheatre at Verona: a. hypothetical initial design based on the Equilateral -triangle-and-inscribed-circle scheme and simple dimensions; b: modified scheme ensuring that the centres of the perimeter piers are spaced 20 ft apart. (Drawing by author.)

length of 500 ft [Fig. 5.14, a]. (125, 250 and 500 ft relate to one another as 1:2:4.) Calculation would then have yielded an interaxial interval of 19½ ft for the perimeter arcade. While the focal points stayed the same, in order to achieve 20 ft exactly the arc lengths were increased, hence the overall length of 512 ft as opposed to the ideal 500 ft [Fig. 5.14, b]. Several other monumental amphitheatres manifest similar approaches, with the popular 'Inscribed-Equilateral' or 'Inscribed-Pythagorean' layouts repeatedly being modified so as to arrive at a desired composition for the facade, a consideration that was significant given that the façade of a large amphitheatre had a major impact on the urban landscape (Wilson Jones [24]). Such concerns were of scant relevance when laying out provincial and military amphitheatres, especially where the cost and impact of their exteriors was minimized by exploiting sloping terrain. It is therefore no surprise that in this context the ellipse appears to have been used as much, if not more, than the oval (Hallier [25]).

Apart from that at Verona, many of the best known ampitheatres fit very well with this analysis, including in my view the Colosseum [Fig. 5.12]. This interpretation is however a question of debate, witness the collection of studies by members of the University of Rome, "La Sapienza", published in 1999 in a single volume entitled *Il Colosseo. Studi e ricerche* [28]. The problem is actually not easily resolved; as at Pompeii it is only possible to rule out with confidence the 4-point oval. Best-fit analysis of the survey data from the perimeter of the Colosseum shows that it matches equally well an ellipse and an 8-point oval [Fig. 5.15].

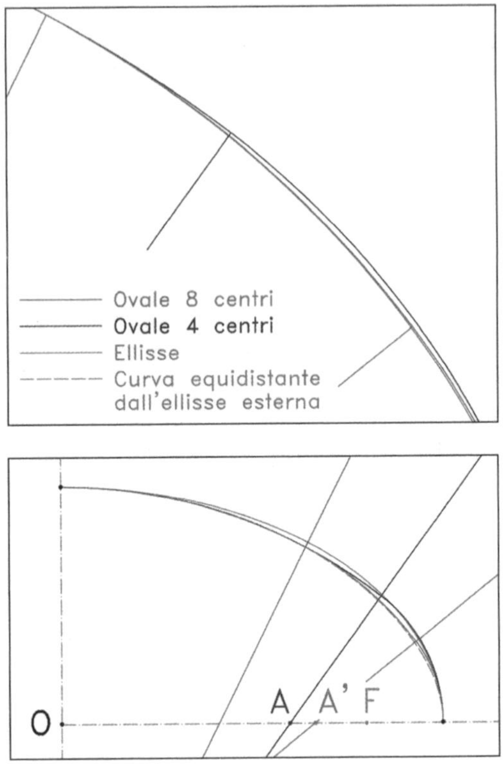

Fig. 5.15 [Col. Pl. XXI]: Diagrammatic part-plan of the Colosseum, Rome, comparing the actual outline with 4 and 8-point ovals and an ellipse. (*Disegnare idee immagini* 18–19 (1989).)

On these grounds alone there would be nothing to choose between them, but as we have seen earlier there are several other characteristics of amphitheatre design to be taken into account. Those who advocate the cause of the ellipse often do so it seems more out of personal conviction than an objective evaluation of the various lines argumentation (Michetti [29], de Rubertis [30], Birardi [31]), which, if conducted even-handedly, gives more weight to the oval (Docci [32]; Casale [33]; Sciacchitano [34]; Trevisan [27]). After all, radial walls converge on focal points corresponding to an oval construction [Fig. 5.16], the width of the cavea (and its annular corridors) is very uniform, the fornices of the perimeter are very regular in conforming to an intercolumniation of 20 ft, while the overall proportions and dimensions provide a good match with a version of "Pythagorean-triangle-with-inscribed-circle-

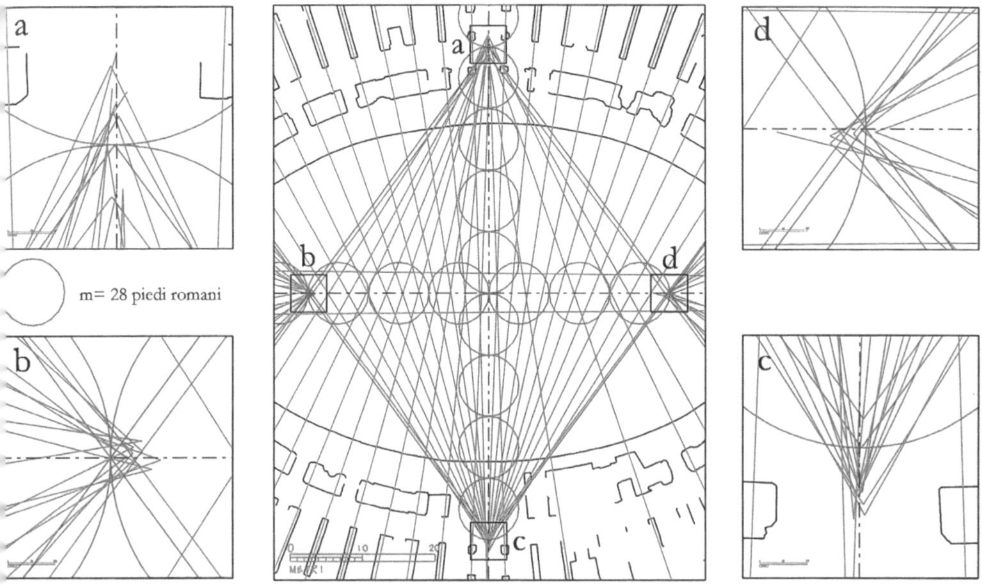

Fig. 5.16 [Col. Pl. XXII]: Diagrammatic part-plans of the Colosseum, Rome, showing the convergence of axes relating to radial walls. (*Disegnare idee immagini* 18–19 (1989).)

scheme" adapted to an 8-point system (Wilson Jones [24], 418, fig. 22c; Trevisan [27], 123–7, figs. 10, 11, 13).

Leaving to one side the disputed geometries of individual cases such as the Colosseum, it is clear that whatever methods were generally used they had to be flexible. The strategies just described had to accommodate a variety of options for several separate considerations: the width of the cavea in relation to that of the arena; the number of annular rings composing the cavea; the overall number of bays (e.g. 60, 64, 72 or 80); the bay width (e.g. 16, 18, 20 or 23 ft). Further details had to be decided, such as the configuration of the axial bays and their immediate neighbours. Given the large number of design choices to be made it is no surprise that a great many variations could emerge despite adherence to just two dominant points of departure. There were of course also a minority of exceptions that adopted alternative design strategies that did not make use of the 3:4:5 or the bisected equilateral triangle.

Amphitheatres, then, illustrate well the paradox of rule yet variety highlighted in the introduction; they were both the same and different. Certain key 'rules' guaranteeing essentially recognisable and satisfying

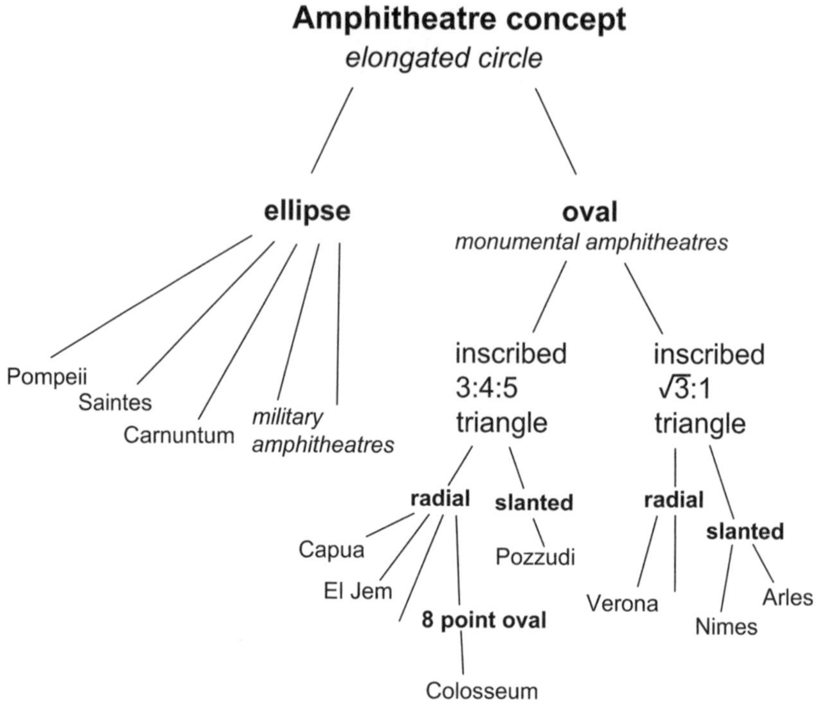

Fig. 5.17: Schematic tree diagram showing the principal flow of decisions involved in determining specific amphitheatre layouts. (Drawing by author.)

outcomes coexisted with a series of secondary options and proportions; it was the exercising of taste and choice while considering practicality too that gave rise to a variety of solutions, each one of which contributed to the individual character of any specific example. This design strategy can be represented in diagrammatic form, as illustrated in Fig. 5.17. It can be likened to a tree of possibilities, in which a single (or forked) trunk represents the dominant common premise (or premises), while the outer branches represents the discreet identities that result from the accumulated effect of choices made in the course of detailed design.

* * *

The precise outline of an amphitheatre was thus not defined by some absolute geometry; it rather had a principled basis in geometry and proportions that was capable of flexing in response to a series of considerations. This complements the conclusions already reached in regard to the shapes used for the curving abacus of Corinthian capitals and the entasis of column shafts, revealing a pattern of approach that characterises the ancient sensibility towards design. Greek and Roman architects created shapes with extreme attention to key geometrical and proportional characteristics, but in certain respects only; otherwise they valued flexibility and improvisation so as to create both the sculptural effectiveness and variety of form which is the source of so much curiosity and delight.

References

1. Wilson Jones, Mark, *Principles of Roman Architecture*, London and New Haven, 2000.
2. Wilson Jones, Mark, "Doric measure and Doric design, 2: A Modular Re-reading of the Classical Temple", *American Journal of Archaeology* 105 (2001): 675–713.
3. Wilson Jones, Mark, "Ancient Architecture and Mathematics: Methodology and the Doric Temple", *Nexus* 6 (2006): 149–170.
4. Coulton, J. J., "Towards Understanding Greek Temple Design: General Considerations", Annual of the British School at Athens 70 (1975): 59–99.
5. Coulton, J. J., *Ancient Greek Architects at Work*, London, 1977 (2nd ed. 1988).
6. Hoepfner, W. (ed.), *Bauplanung und Bautheorie der Antike*, DAI Diskussionen zur archäologischen Bauforschung 4, Darmstadt, 1984.
7. Bommelaer, J.-F. (ed.), *Le dessin d'architecture dans les sociétés antiques*, Strassbourg, 1985.
8. Heilmeyer, W.-D., *Korinthische Normalkapitelle*, Heidelberg, 1970.
9. Gros, Pierre, „Le fondements philosophiques de l'harmonie architecturale selon Vitruve." *Aesthetics. Journal of the Faculty of Letters, Tokyo University* 14 (1989): 13–22.
10. Wilson Jones, Mark, "Designing the Roman Corinthian capital", *Papers of the British School at Rome*, 59 (1991): 89–150.

10a. Greco, M., "La realizzazione scultorea dei capitelli d'ordine corinzio: i non 'finiti', procedure grafiche incise come semplificazione del processo di lavorazione del manufatto," in *Il disegno e la pietra*, ed. R. Migliari, Rome, 2000.

11. Freyberger, K., *Stadtrömische Kapitelle aus der Zeit von Domitian bis Alexander Severus: zur Arbeitsweise und Organisation stadtrömischer Werkstätten der Kaiserzeit*, Mainz, 1990.
12. Hauck, G., *Die Subjektive Perspektive und die horizontalen Curvaturen des Dorischen Styls*, Stuttgart, 1879.
13. Penrose, F. C., *Principles of Athenian architecture*, London, 1888 (1st ed. 1851).
14. Goodyear, W. H., *Greek Refinements*, New Haven, 1912.
15. Haselberger, Lothar, ed. *Essence and Appearance. Refinements in Classical Architecture: Curvature*, symposium, Philadelphia (1993), Philadelphia 1999.
16. Pakkanen, Jari, "Entasis in fourth-century B.C. Doric buildings in the Peloponnese and at Delphi", *Annual of the British School at Athens* 92 (1997): 323–344.
17. Stevens, G. P., "Entasis of Roman Columns", *Memoires of the American Academy at Rome* 4 (1924): 121–152.
18. Heisel, J., *Antike Bauzeichnungen*, Darmstadt, 1993.
19. Haselberger, Lothar, "Architectural Likenesses: models and plans of architecture in Classical antiquity", *JRA* 10 (1997): 77–94.

19a. Inglese, Carlo, *Progetti sulla pietra*, Rome, 2000.

20. Haselberger, Lothar, "Werkzeichnungen am Jüngeren Didymeion." Vorbericht, *Istanbuler Mitteilungen (DAI)* 30 (1980): 191–215.
21. Hueber, F., "Werkrisse, Vorzeichnungen und Meßmarken am Bühnengebäude des Theaters von Aphrodisias, *Antike Welt* 29 (1998): 439–445.
22. Wilson Jones, Mark, "The Practicalities of Roman Entasis", in Haselberger ed. [15]: 225–249.
23. Golvin, J.-C., *L'Amphithéâtre Romain*, Paris, 1988.
24. Wilson Jones, Mark, «Designing amphitheatres», *Mitteilungen des Deutschen Archäologischen Instituts* (Rome) 100 (1993): 391–441.
25. Hallier, Gilbert, „La géométrie des amphithéâtres militaires sur les limes du Rhin et du Danube", *Akten des. 14. Internationalen Limeskongresses in Carnuntum* (1986), Vienna, 1990: 71-82.
26. Duvernoy, S. and P. Rosin, "The Compass, the Ruler and the Computer", *Nexus: Architecture and Mathematics* 6 (2006): 21–34.
27. Trevisan, C., "Sullo schema geometrico degli anfiteatri romani: gli esempi del Colosseo e dell'Arena di Verona", *Il Colosseo. Studi e ricerche*, issue of *Disegnare idee immagini* 18–19 (1999): 117–132.
28. *Il Colosseo. Studi e ricerche*, issue of *Disegnare idee immagini* 18–19 (1999), Dipartimento del Rappresentazione e Rilievo dell'Università degli Studi 'La Sapienza' di Roma, Rome.
29. Michetti, A., "Possibili costruzioni delle ellissi del Colosseo", in *Il Colosseo* [28]: 89–98.
30. Rubertis, R. de, "Un enigma avvincente. Il tracciato planimetrico ellittico del Colosseo", in *Il Colosseo* [28]: 99–105.
31. Birardi, G. *et al.*, "Il rilievo fotogrammetrico del Colosseo", *Bolletino della Società Italiana di Topografia e Fotogrammetria*, 2 (1987): 93–119.
32. Docci, M., "La forma del Colosseo: dieci anni di ricerche. Il dialogo con I gromatici romani", in *Il Colosseo* [28]: 23–32.
33. Casale, A., "Alcune ipotesi sul progetto e sulle geometrie del Colosseo", [28]: 81–88.
34. Sciacchitano, E., "Il Colosseo. Geometria dell'impianto" *Il Colosseo* [28]: 107–116.

MIDDLE AGES

CHAPTER SIX

"TO DESIGN" AND "TO BUILD" MEDIAEVAL SHIPS (FIFTH TO FIFTEENTH CENTURIES)—THE APPLICATION OF KNOWLEDGE HELD IN COMMON WITH CIVIL ARCHITECTURE, OR IN ISOLATION?

Eric Rieth
Translation: Richard Barker

1. Introduction

As the title of this paper indicates, our intention is a consideration of the notions of geometric knowledge and know-how made use of in naval architecture in the Middle Ages. More precisely, we would wish to concentrate our remarks on one principal question corresponding to the theme of the workshop: did the knowledge and the knowledge particular to naval architecture and, more specifically, those connected to the geometric conception of the forms, have any relationships with those proper to civil architecture, or did they, on the contrary, function in an isolated fashion? The whole difficulty lies in this comparative reading. To present geometric knowledge and know-how attested in the mediaeval shipyards is, certainly, a complex enterprise more relevant to our habitual field of research. However, this is not the case when it becomes a question of taking into account the very rich territory, difficult for a non-specialist, of civil architecture. In these conditions, as an archaeologist who has oriented his studies towards the history of mediaeval naval architecture, we will limit ourselves, with prudence, to formulate a certain number of questions and to propose some provisional axes for research which, we hope, will be discussed, corrected and completed after this workshop. It is consequently a question of a study at its beginning, simply marking one stage of research in hand.

The manner in which the conception of forms can be defined within the body of frame-first carvel construction, characteristic of the mediaeval Mediterranean, will be considered in the first part. In a second part the fashion in which that geometric conception of forms is attested in the different types of written, graphic and archaeological sources will be examined. Finally, in a third part, two questions will be discussed that seem particularly important to us with respect to the theme of the workshop: the first concerns the manner in which the men of the mediaeval shipyards are designated; the second bears on the fashion in which the knowledge and know-how are acquired and transmitted. Throughout our study, we will try always to keep in mind this cross-reference between naval and civil architectures.

2. Nature of the problem

The chronological time frame envisaged, which has been suggested by the organisers of the round table, runs from the 5th to the 15th centuries and corresponds with the traditional limits of the mediaeval period, considered as the historic landmarks of 476, the end of the Roman Empire in the West, and 1492, the discovery of America by Christopher Columbus and the achievement of the Reconquest with the fall of Granada. It is clear that mediaeval chronology is very relative, and that according to the territories, contexts and the topics concerned, it will demand review. It is thus that the perspective of a "long Middle Ages" [32], to adopt the expression of the mediaevalist Jacques le Goff, in which it can be extended up to the 19th century in certain fields, does not appear at all iconoclastic today. In accord with this vision of a "long Middle Ages", our paper will sometimes make reference to facts situated outside the chronological corpus between the 5th and 15th centuries.

The history of mediaeval naval architecture has been profoundly revised during the last two decades thanks to the support of archaeological sources. It will not be to the point within the limits of this paper, to sketch, even broadly, the most recent results of researches.[1] We will confine ourselves here solely to aspects susceptible to contribute to a better perception of the geometric processes of conception of the forms of a hull.

In this perspective, we will not touch on the case of Northern Europe and "shell-first" clinker architecture. Similarly we will not evoke the case of "bottom-based" shipbuilding, or, more exactly, of the adaptation of that architecture of fluvial origin to the maritime environment,

[1] For a recent synthesis we will refer to our work [40].

of which the cog family represents the best known architectural line. Descending along the Atlantic littoral, we will not discuss the problem of the progressive abandonment of clinker architecture in the course of the second half of the 15th century and its replacement by the so-called carvel architecture, according to the testimony of written mediaeval sources. In this respect, we will mention only briefly the debate on the origin—or origins—of this architectural phenomenon, on its causes, on the possible role of the North of the Iberian peninsula as a space for transition and evolution, nor the problem related to the chronological rift between written and archaeological attestations of that carvel architecture.[2] Still less will we consider the questions, in constant evolution, bearing on these long developments.

It is clear that each of these different architectural systems, with its singularities, is translated, in terms of geometric conception of the forms of hulls, by specific knowledge and know-how. In this respect, we cannot resist the temptation to make some account of a very interesting case, which alone could be the subject of a workshop: this is the "mental template"[3] as a geometric concept which, within the corpus of Scandinavian clinker architecture, is defined as follows by Ole Crumlin-Pedersen: "...the boat-builder must have been working from a mental template that is a pre-conceived concept in his mind that defined not only the size and structure but also the details of lines and shapes of the vessel-to-be" [9, p. 231]. Could a relationship have been established between the processes of geometric conception of a hull and that (or

---

[2] A series of questions has recently been emphasised by the archaeology from the excavation and study of the wreck, from the beginning of the 7th century A.D., of Port Berteau II, Charente-Maritime, France [46]. The carvel hull of this fluvio-maritime coaster of regional origin (the Aunis-Saintonge shoreline), fifteen metres long and built on a keel, belongs to a "frame-first" or "proto-frame-first" architecture. This architecture of carvel type poses the question of its interpretation within the history of Atlantic naval architecture in the Middle Ages, a question the more delicate since this wreck is thus far the only one of this type and of that period to have been discovered along the Western coastline. Is it a question of a traditional regional architecture? Could that architecture of carvel type from the Merovingian period have co-existed with other regional forms of naval architecture? Did relationships exist between the architecture of this wreck and those from antiquity of the so-called Roman-Celtic tradition (Blackfriars 1, Guernsey, Barland's Farm)? Was the architecture of the Port Berteau coaster maintained locally after the development of clinker architecture attested in Aquitaine from the 12th century? How to understand that evidence of carvel architecture from the High Middle Ages against the first written records of ships built carvel in the second half of the 15th century?

[3] On this notion of "mental template", compare the pages dedicated to this subject by David A. Taylor [51, pp. 102–106].

those) demanded at the same period and in the same Nordic environment in civil architecture, that being a question for example of the great houses of the Trelleborg type or the wooden churches from the end of the Viking period and the beginning of the Middle Ages? Indeed, would this comparison even be pertinent?

Within the body of our paper, we will essentially adhere to the Mediterranean area. Geographically it constitutes an entity that can be divided into several sub-units, those of the Western and Eastern basins on the one hand, and those of the Northern and Southern shores on the other. In a very schematic fashion we can say that the first years of the Early Middle Ages in the Mediterranean are the setting for a double and complex modification of the architectural system of Graeco-Roman antiquity, a double modification quantifiable by processes of transition on the one side and by processes of rupture on the other and which, in the current state of archaeological research, will be situated between the 6th and 7th centuries A.D.

Transition: this is characterised, schematically, by a construction of the "shell-first" type for the bottom of the hull (the planking raised without recourse to a pre-established and pre-designed transverse carpentry frame), according to the constructive model from antiquity, and by a "frame-first" construction (with transverse carpentry erected before the planking is put in place) for the part situated in the prolongation of the bilge and running up to the upperworks of the hull. For a long time, the two wrecks for reference [53, 6, 10, 50] have been those of Pontano Longarini (Sicily) and of Yassi Ada I (Turkey). That second wreck, dated to the second half of the 7th century, is that of a Byzantine coaster associated with the Eastern part of the Mediterranean basin. Recently, new wrecks showing "transition" have been discovered and excavated, of which that of Dor D, located in the waters of the Tantura Lagoon (Israel) and dated from the middle of the 7th century [27] is most significant.[4]

Rupture: this is materialised, at the level of archaeological data, by carvel planking which no longer contains—except sometimes at the stage of vestiges of guides—traces of pre-assembly by mortises and tenons or by pegs. In that configuration, the planking, with which we find associated caulking by means of a strand of tow placed by force, comes to cover the transverse carpentry previously pre-designed and

[4] For a synthesis of the wrecks of the Tantura Lagoon, cf. [30].

established, in whole or part, on the keel. It is that transverse carpentry, conceptually "active" according to the qualification justly proposed by L. Basch [4, p. 34ff], that henceforth plays the dominant role at the heart of the whole structure of the hull.

The study of this process of architectural rupture has for a long time rested on the cases of two wrecks, that of Saint-Gervais II [24], located in the Gulf of Fos-sur-Mer, near Marseille (France), and dated to the 7th century, on the one end of the Mediterranean, and at the other, that of Serçe Liman (Turkey), dated from the first half of the 11th century [5, 48]. Latterly, other discoveries and underwater excavations have been made. This is the case of the wreck from Dor 2001/1 (Israel), dated from the end of the 5th-beginning of the 6th centuries A.D. [38], and of Tantura A (Israel), dated from the 6th century A.D. [28]. It is also the case for the 9th century wrecks from Tantura B (Israel) [29, 30] and from Bozburun (Turkey) [19, 20].

It is the path of rupture and of "frame-first" architecture that will prevail. Of the three architectural assemblies proper to the whole hull, that is the longitudinal carpentry (keel, stem, sternpost), the transverse carpentry (frames) and the planking, the "frame-first" architecture privileges the transverse carpentry. That transverse architectural logic is marked at two principal levels: that of the structure of the hull on the one hand, and that of the processes of conception on the other. Before returning in a moment to this role accorded to the transverse carpentry, we will pause for a question connected to this double process of modification, by transition and rupture, of the Graeco-Roman naval architecture. In the hypothesis where relationships will exist between naval and civil architecture during this period, how will this double process of transformation, implying in all logic profound changes in knowledge, at the level of conception, and in know-how, at the level of construction, be expressed at the level of geometric conception of the forms of terrestrial edifices? If one places it, for example, in the hypothesis of processes of rupture, does that "architectural leap" from "shell-first" conception to that of "frame-first" signify the elaboration of an autonomous geometric knowledge, solely developed within the milieu of the shipyards? And in this case, could this new geometric knowledge then have been transmitted to the milieu of the terrestrial workshops? In what form? By what intermediary? We leave to the historians of civil architecture the task of replying to these questions, for which, indeed, even the formulation does not appear evident to us.

With regard to the central position of the transverse carpentry (the frames) in the structure of the hull as a whole, J. Richard Steffy has perfectly summarised the spirit of the methods of "frame-first" conception, in considering that the shipbuilders of the Middle Ages must have visualised the the forms of the hull according to a "transversal perspective" [49, p. 419]. That, in contrast to the "longitudinal perspective"[5] proper to "shell-first" architecture of the Mediterranean shipbuilders in antiquity, corresponds, again after J. R. Steffy, to an authentic "structural philosophy". It is in this relationship with that architectural culture in which the frames and, more particularly the master section, occupy a determinant position from the point of view of the geometric variation of the forms of the hull, that a comparison has been established by certain historians of civil architecture between the master-frame of a boat and the masonry vault or the wooden trusses (*fermes*) of the carpentry of a terrestrial edifice.

Does that apparent analogy between a key element of "frame-first" naval architecture and an equally fundamental element from civil architecture, suffice to envisage a connection between the knowledge and know-how of the men of the shipyard and more particularly the master-carpenters, those who conceive the forms, and their homologues in terrestrial workshops?

To try to address that question in the most precise manner possible as a function of our documentation and at the level of advancement of our study, we propose to define the notion of the geometric conception of the forms, a notion which is at the centre of the debate, from four different points of view.

The first point of view: geometric conception of the forms and naval architecture. It is a question, very certainly, of an obvious point, but one that it is perhaps not useless to recall here. To conceive a boat, however small it may be, leads us to conceive a floating architectural structure—a first characteristic, which is to be moved—a second characteristic, in a mobile environment—a third characteristic. That structure could be qualified as a "dynamic architectural assembly" wholly relevant both to the field of architecture, in the classical sense of the term, and to that of mechanics for the functions of propulsion and

[5] We emphasise that this transverse vision of the forms cannot be totally absent from a "shell-first" longitudinal conception. In one way or another, the master-carpenter must necessarily be able to visualise the transverse form of the master-frame.

steering. To the dimension of architecture, in the strict sense, that of the technical system is therefore found indissolubly associated. In this sense, the knowledge, and more specifically the geometric know-how of the master ship-carpenters of the Middle Ages on the one hand and the logic of making use of this knowledge on the other, appear to us distant from those of the master carpenters and masons of mediaeval civil architecture, whether from the workmen in wood or in stone, from urban or rural workshops. The dimension of the technical system, which seems to us an essential and singular characteristic of naval architecture, and of its implications at the level of the technical culture of the master ship-carpenters appears essentially absent from the universe of civil architecture.[6]

Second point of view: geometric conception of the forms and evolution of the forms. The basic characteristic of the geometry of a hull is the variation of the geometric figure of the master section towards the extremities of bow and stern. Even in the case of the hull of a boat with forms only slightly evolved in relation to its "body" (its central part), as is, for example, the hull of the coaster from the first half of the 11th century from Serçe Liman, in Turkey, these fashions fore and aft are translated by a certain diminution of the length of the flat floor of the master-floor and by a relative increase in its rising. It is at this level that another basic conceptual and structural difference between the carpentry of a boat and that of a terrestrial edifice is situated, in the form of the "nave" (*nef* = *nave or ship*), or "upturned hull", to take up the images traditionally cited in works on the history of civil architecture.

In the architectural system of ships built "frame-first" proper to the mediaeval Mediterranean and in which it is inscribed, the geometric conception of the forms rests on the modification of the pre-defined shape of the master-section. More precisely, it is starting from a modification of certain values of that central figure, and not of its geometric construction, that the figure of the other transverse sections is to be pre-defined. In every case, from the hull with the simplest geometric shapes (more or less developable forms) to those with the most complex forms (with hollowing at the garboard and reverse curves in the upper

[6] The problem can be posed in different terms in the case of water- and windmills, where the structure of the building ought to be conceived in terms of the particular constraints connected with the mechanism of the mill. The notion of the technical system associated to that of architecture then seems to us very pertinent, notably in the case of watermills and their relationship with the hydraulic arrangements.

futtocks and toptimbers for example), the knowledge used by the master ship-carpenters for boats seems to us to make appeal to the preoccupations and to an intellectual logic different from those of masters of masonry and carpentry. An example that can appear almost a caricature but which is however very indicative of that difference is provided by carpentry. In the case of the carpentry of a building on land, the form of the roof-truss (*ferme*) does not evolve (without exception) along the length of the structure, while in the case of the carpentry of a boat the form of the master-section is continually modified along the axis of the keel. In this case, the constructors find themselves strongly challenged intellectually, and also materially, by two geometric logics.

Third point of view: geometric conception of the forms, and "instruments" for conception. With reference to that transverse geometric conception and its evolution of the forms of the hull, the know-how is translated by recourse to privileged "instruments" of which the principal is the master mould (*maître-gabarit*). This mobile and portable instrument reproduces in two dimensions at a scale of 1:1, that is to say at full size for the work, the geometric outline of the master section. The displacements of this instrument in the horizontal plane (reduction of the length of the flat of the floor) and vertical (increase in the rising of the floor), principally, allow the determination of the geometric figure of the frames on one part of the length of the hull, moulded starting from the outline of the master frame. One of the most ancient written mentions of this master-mould, which occurs in mediaeval Latin texts in the context of "*pro sextis*", i.e., "for the moulds", goes back to the year 1273 [17], and the words *sèsto* or *sexto*[7] occur in Venetian/Italian texts of the somewhat later Middle Ages, a period which also produces one of the most ancient archaeological attestations of the usage of this instrument of conception. Indeed, it is from the study of the wreck of Culip VI, in Spain, dated to the end of the 13th-beginning of the 14th centuries, that the employment of the method of the master-mould has been brought into evidence, limited, always, to the floor timbers, taking note of the preservation in the wreck of these elements alone of the transverse carpentry [45].

---

[7] It is interesting to note that one of the archaic senses of the masculine noun *sesto* refers to a notion of geometry, in fact for the curve of an arch. Thanks to Jens Høyrup for bringing this point to our attention.

Fourth point of view: geometric conception of the forms and the supply of timbers. With regard to "frame-first" naval architecture and the determinant role accorded to the transverse geometric conception of the frames, the curved timbers, the most difficult to procure, occupy henceforth a central place in the provisioning of shipyards. Without touching here on the questions of the impact of these new requirements for curved timber on the Mediterranean forestry environment, the practices and methods of management of the forests as well as of legislation, it is certain the "forestry problem" of curved timbers is going to become more and more important for nations with a maritime vocation. If mediaeval terrestrial carpentry does not exclude curved timbers, it is however certain that it privileged straight timbers of long span, and that in this domain, the preoccupations of those responsible for the provisioning of the shipyards and of terrestrial workshops hardly seem to coincide.

After having run through the nature of the problem in broad terms and in a manner necessarily schematic and incomplete, we will examine in the second part of our study the fashion in which the sources testify to this geometric conception of forms.

3. Sources and perception of the knowledge

We begin with the written sources, operating with a summary distinction between those that give notice of private actions (contracts, specifications (*devis*), markets, accounts) passed before a notary or emanating from authority, and those which are connected, according to the typology of the written sources established by the historian Bertrand Gille[8] to that of the "technical notebooks" (*livres de recettes techniques*) for naval architecture.

The documents of the first category provide, in the majority of cases, allusives and indirect information on the subject of geometric conception of the forms of the hull. Besides, the more technical documents from this category of sources have a content above all practical and reflect a culture of know-how. An example is furnished by the account for construction of galleys at Narbonne, in the South of France, between

[8] The notion of "*livres de recettes techniques*", translated by the term "technical notebooks", has been developed by the historian Bertrand Gille [17, pp. 515–518 in particular].

the years 1318–1320 [47]. The account mentions on several occasions the use of moulds (*gabarits*) called "*galbe,*[9] *mensura, modulus*". The context of the passage does not allow us to know if these terms are synonyms or correspond to particular types of mould. In summary, the account indicates in an explicit fashion that these moulds are associated with the creation of the frames of the galleys—"...*galbe pro madayris galearum mensurandis*" (f° 57 v)—in two principal ways: as models or patterns from which the selection of the curved standing timbers and for their possible pre-shaping in the forest or in the shipyard; as master-moulds for the determination of the moulded frames. In both cases, and it is a question there of a fundamental fact, the moulds enter into a practical geometry at 1:1 scale, that is to say, at the scale of execution of the frames.

This same construction account makes a statement, similarly, of the use by the master-carpenters of red paint to mark the wood: "...*tinctura rubea, que dicitur bol cum qua sinabantur seu tingebantur fustes per magistros*" (f° 93 v). This marking with paint is the only reference to an expression that could be termed graphical. Is it a matter of a simple sign identifying the wood to cut or to shape or a question of marks of the type of those (central marks on the keel and lateral for the bilge) attested in the floors timbers of the mediaeval wreck of Culip VI and corresponding to those mentioned and illustrated in the *Livro primeiro da Architectura Naval* (c. 1600) by the Portuguese João Baptista Lavanha [31, f° 67–69]?

Although the context of production of the yard for the construction of galleys at Narbonne may be compared with a State shipyard, the only "instruments" of conception reflecting a geometry above all practical (at the scale of execution) for conception of the forms are the moulds. Other documents seem to reveal a second family of "instruments" for the geometric conception of forms which, unlike the moulds, solely enter

[9] In his study on Tunisian nautical vocabulary, the linguist Albert Gateau [15, vol. II, p. 116] emphasises that the different forms of the French term *gabarit*, that is, *garbi* (Provençal), *garbo* (Italian, Spanish), *gálibo* (Spanish), *gàlib, gàlip* (Catalan), probably all originate in the arabic *qâleb* (singular), *qwâleb* (plural). Now, one of the questions about the Mediterranean origins of "frame-first" construction concerns the shipbuilding techniques in use during the High Middle Ages in the shipyards of the Southern shore of the Mediterranean. Would the arab-muslim expansion towards Spain and Portugal be translated into an influence on the practices of the shipyards of Al-Andalus? Thus it is that several significant terms in the vocabulary of Portuguese naval architecture, such as *almogama*, which indicates a pre-moulded quarter-frame, are of arabic origin.

into the course of construction: these are the ribbands for construction. In the specification for 1273 concerning the the construction of a horse transport (*huissier*) for the account of king Charles I of Anjou [14, pp. 266–267] mention is thus made of provision of pine wood "...*pro formis et sextis*". In our opinion, these two terms, which do not appear to be considered as synonyms by the author of the specification, could include two meanings, those of mould (*sextis*)[10] on the one hand and of ribbands or of forms (*formes*: "*formis*") on the other. According to a definition given by a treatise on galley construction of 1691, the forms are the "...pieces of pine wood to which one gives 3–1/2 pouces square and the greatest length that one can, so that they better comply, and that they take the form of the galley" [13, p. 30]. This same treatise of 1691specifies that the forms "...serve to trace out the crook timbers (*fourcats*) of the stern and bow by trial and error in position, because one cannot work them at all in advance with simple moulds like the floor timbers of the bottom".

If these *formes* or ribbands seem foreign to the world of terrestrial workshops, this is not the case for the moulds that one indeed finds, not in the hands of the master carpenters of simple houses, but principally in those of the stone-cutters. The moulds (*molde, mole*) of terrestrial workshops effectively recall the *modulus*, in the sense of pattern or model, of the master ship-carpenters of the galleys at Narbonne. Thus it is a question of a simple "instrument" intended to reproduce a geometric outline at the scale of execution. As the historian of architecture Roland Recht puts it, "...These are the sheets (*panneaux*) of wood or metal to which are given the forms of the different mouldings that will be utilised in the edifice. The stone-cutter then places them on the squared stone and the contour follows" [41, p. 87].

We now consider briefly the second category of written sources, those termed "technical notebooks" for naval architecture. These documents of Venetian origin have their historians to whom we will turn: from the "classics" such as A. Jal, R. C. Anderson and F. C. Lane to our contemporaries, such as Sergio Bellabarba, Alvise Chiggiato, Mauro Bondioli, Ulrich Alertz and David McGee, the last two specialists of this technical literature being present at the workshop. Amongst the

[10] On this question of moulds (*gabarits*), we refer to our study [42] translated for the Workshop by Richard Barker: "From words to technical practices: moulds and naval architecture in the Middle Ages".

questions that interest us more particularly, there are those bearing on the purpose of these "technical notebooks", on their author(s), on the manner in which the contents of these "books" has been gathered, on the part of oral theoretical knowledge and practical know-how fixed by writing it down. As an archaeologist working on the history of mediaeval naval architecture, we will limit ourselves to formulating some remarks, leaving the specialists on these Venetian documents to reply to our questions.

First remark: with these manuscripts of Venetian origin mostly dated to the 15th century, we remain in the field of practical geometric conception of the forms of hulls in which one of the keys is the system of reduction termed "partisone" (*partison, partixon*). As A. Chiggiato [8, p. LXIII] has very justly emphasised, this system covers a much broader sense than that of simple divisor. This veritable "geometric operating system" allows the determination of a progression starting from the values of two known extreme values, calling upon the elementary geometric procedures of which the *Libro* of Zorzi Trombetta de Modon, from the mid-15th century, provides two of the most classic examples, the one in the form of a half-moon (*meia-lua* etc.), the other in the form of an isosceles triangle.[11] Such procedures are applied to define, starting from the geometric figure of the master-section, the reduction of the floor (*partisone de fondo*, the length of the flat floor timbers to be moulded), the rising of the ends of the moulded floors (*partisone della stella*), the *trébuchement* (*partisone de ramo*, the splaying of the top of the moulded frame timbers, relative to the floor) and the *recalement* (*partisone del scorrer*, adjustment at the foot of the moulded frames), these last two operations apparently limited, at that period, to the environment of the Venetian shipyards.

Second remark: with regard to the problematic possible relations between naval and civil architectures, the attestation, in the "technical notebooks" of 15th century naval architecture, of geometric procedures of the half-moon type, notably, raises the question of a shared usage of such procedures. In the working hypothesis where the response would be positive, which remains in our opinion a matter of debate, it would signify nothing more than a geometric practice held in common. It would simply signify, it seems to us, a comparable level of geometric

[11] The semi-circular figure is called "*mezzo-redondo*", the triangular is named "*scagion*" [36, f° 45].

knowledge founded on the similar geometric practices of the workshops which, after all, does not appear extraordinary.

After the written sources, we consider the case of the graphic sources, which the "technical notebooks" of 15th century naval architecture from Venice, too, reveal. As a long-term objective, we have created from a study of the modalities of the formation of graphic expression in naval architecture, a systematic analysis of the illustrations of two of these "technical notebooks": the *Fabrica di galere* and the *Libro* of Zorzi Trombetta de Modon [43, 44]. From the different graphic collections present in these manuscripts, we will pause to consider more particularly those associating, for a specific model of boat, the outlines of the master-section and of the stem and sternpost. The *Libro* of Zorzi Trombetta thus contains thirty-one figures of this type. These would seem to mix two representations. The first will be that of the geometric construction of the master-section and of the posts, and the second, superimposed on the first, that of the moulds. These figures raise a whole series of questions. We will limit ourselves to those for the construction of these figures, and their interpretation.

One can state, first of all, that arcs of circles are absent, graphically, from these figures. The construction of the contour of the half of the master-section, the most emblematic for the geometric conception of the forms in the context of "frame-first" architecture, rests on a series of points defined on abscissae for the values of the half-width and in ordinates for those of the depth. Faced with this manner of working they come immediately to the spirit of the evidence of Baldissera Drachio, dated to the end of the 16th century [11, f° 6v, 33] who describes how, in the shipyard, to draw out full scale, the *corba di mezaria* in making a lath (*una cantinella sotile* Drachio writes) pass through the points marking the different half-widths. That lath defines the contour of the master-section from the flat of the floor to the straight line of the deck beam, passing through the point of the bilge (*escoue*)—the *poselese della paraschuxula* or *poselese de fondo* of the Venetian texts.

The testimony of Drachio reveals an application, in the shipyard, of geometric construction forms which, according to the documents, does not appear to call upon any arcs of circles. That practical geometry appears completely comparable to that utilised in the mould lofts of the royal arsenals of the modern period where the contour of the frame stations were traced out at full scale starting from tables of numeric data called offsets (*devis de gabarits*). Those correspond to a strict numeric transcription of the plans established by the master constructors. It is very evident that by the 18th century the graphic conception of the transverse sections (frame stations) made great use of arcs of circles. Now, a study[12] of the figures of the 15th century Venetian "technical notebooks" carried out by our friend René Burlet, a specialist in the history of the architecture of galleys of the modern era, has shown that another construction resting on the outline of two tangent arcs of circles (the one of short radius for the lower part, the other of large radius for the upper part) could be substituted, without any particular difficulty, for the geometric construction of the master-frames of the galleys on the basis of values defined by abscissae and ordinates. The same is true for the outlines, always more complex, of the stems and sternposts of galleys, again by means of two tangent arcs of circles.

Faced with this double possible reading, the one direct, the other experimentally reconstructed, of the graphic documents of the Venetian "technical notebooks", the question is posed of the interpretation of these figures. Richard Barker [3, p. 119] has perfectly summarised the problem. The recourse to the outlines defined by coordinates could be the simplified result from a method of designing with arcs of circles; it could also correspond to an initial step related to the definition of tangent points between components arcs of a geometric method; it could result, in the end, in the existence of different methods of design according to the nature, State or private, of the shipyards and of the types of vessel. We leave the debate open while emphasising that this reveals, it seems to us, a problem for the geometric conception of forms that appears specific to the field of naval architecture.

After the written and graphic sources, we examine the case of the archaeological sources. Within the collection of mediaeval wrecks excavated in the Mediterranean, whose number is much reduced in comparison with wrecks from antiquity, those that have been the object of study oriented towards the analysis of the processes of geometric conception of the forms of the hull are very rare. The most significant concern the 9th century wreck of Bozburun, in Turkey, under study by the American archaeologist Matthew Harpster [18][13] that of the 11th century from Serçe Liman, in Turkey, by the great pioneer of the analysis of wreck, J. Richard Steffy [48] and that from the end of

[12] With thanks to René Burlet for having replied with great attention to our questions on this unpublished study.

[13] With thanks to Dr Matthew Harpster for having provided us with the text of his paper.

the 13th-beginning of the 14th centuries from Culip VI, in Spain [45], these three being the wrecks of coasters of more or less similar size.

In the three wrecks built carvel "frame-first", the conception of the whole of the forms of the hull calls upon the same basic principles, that is, the predetermination of the geometric figure of the master-section and of that of balance frames limiting the moulded length, and of the intermediate sections. It is starting from the master-section and by the modification of these values—the length of its flat floor and the height of its rising—that the other moulded sections are geometrically predetermined.

If the principles are similar, there exist a certain number of particularities between the two wrecks from the Early Middle Ages from Bozburun and Serçe Liman and that of the central Middle Ages from Culip VI. The first two are included within the nautical context of the Eastern Mediterranean, as much at the level of construction as that of fitting out, while the third is attached to the nautical environment of the Western Mediterranean. If the three wrecks possess flat floors and full and capacious forms, in relation to their function of transport, it is certain that the master-section of the coasters from Bozburun and Serçe Liman appears more like that of a "box" midship section[14] than that of Culip VI. Besides, the geometric conception of the forms of the wrecks from Bozburun and Serçe Liman, as far as it has been reconstructed, seems essentially to rest on a series of segments of straight horizontal and vertical lines in the forms of the hull, geometrically simple and without curves, resulting in a form that is almost developable. The only arc of a circle, existing over a very reduced extent, which does not seem to intervene directly in the geometric conception of the forms, is located at the level of the bilge, and is thus perhaps already equivalent to the fairing arc required in the later methods of *trébuchement*. In the case of the wreck of Culip VI, the absence of conservation of the futtocks does not permit us to propose a reconstruction of the processes of geometric conception of the master-section.

One of the important questions will be to know whether, in the history of Mediterranean architecture, the geometric model of the

[14] The form of the master floor timber of the wrecks of Bozburun and of Serçe Liman is very close to those of the wrecks of the 10th century from Bataiguier and from Agay A, located along the Provençal coastline. We note that these two wrecks are called Saracen because of the known origin of their cargo (ceramics) and not as a function of that of the ships, which remains unknown. Cf. [23, 25, 26].

coasters from Bozburun and Serçe Liman, a model forming part of the same family, could represent that attested in the "technical notebooks" from the 15th century. A second question, also historically important, will be that of situating, in terms of the constitution and transmission of geometric knowledge, the role of the shipyards of the Eastern and Western Mediterranean on the one hand, and of the North and South shores on the other.

Throughout the second part of our study there has appeared, more and more clearly it seems to us, a specific set of problems for the geometric conception of forms in naval architecture. The two questions touched on in the third part, those of the manner in which men of the mediaeval shipyards are designated and that of the manner in which the knowledge and know-how are acquired and transmitted, bring forward another clarification of this cross reference between naval and civil architectures.

4. Some questions

The designation of the personnel of the shipyards: for Western France, the accounts of the Clos des Galées, in Rouen (1293–1418) furnish precise indications, within the corpus, we recall, of a shipyard controlled by the royal power and equivalent to an arsenal in charge of the construction, maintenance and repair and fitting out of sailing and oared ships, built in carvel and in clinker [37]. We emphasise that although situated in Normandy, traditional territory for clinker naval architecture, this arsenal called upon Mediterranean specialists, come to build within the walls of the Clos carvel galleys endowed with "frame-first" naval architecture. At the head of this State establishment in Rouen was to be found the master and warden of the Clos de Galées. Master of the works and repairs of ships, whose role is that of a kind of intendant, in the modern sense of the word, a senior administrative and financial officer of the Clos. Under the direction this very powerful personage the following technical personnel worked:

1. A "master carpenter of the works on the king's ships" whose function, comparable to the king's master-carpenters of the modern era, is at the same time to conceive the vessels, galleys, barges...and to their construction.

2. The master carpenters (*maistres d'aesse*), that is to say, according to the terminology borrowed from the Mediterranean world, masters of the axe, responsible in the shipyard for the actual construction of the vessels according to the architectural project determined by the king's master carpenter.
3. The carpenters, specialist workmen, working on the contruction of the ships under the direction of the above.

Are that same hierarchy and the same designations found in the corpus of the private shipyards of the North of France? Examination of the Statutes of the Brotherhood of Ship-carpenters of Abbeville in Picardy [52] for the years 1488–9 shows, apart from apprentices, two other levels of workmen, the "masters of the craft of ship-carpentry" and the "ship-carpenters", just termed "workmen" (*d'ouvriers*)". Taking account of the limited scale of the private shipyards, the hierarchy is simplified logically in relation to that of the "State" shipyards of the type of the Clos de Galées in Rouen. But, fundamentally, more or less no difference seems to exist between the two technical environments at the level of what the men are called.

If we turn towards the Mediterranean shore, the accounts of a great shipyard at Capelles, near Narbonne, intended for the construction of five galleys for the account of Philippe V, the Long, king of France and of pope John XXII, provide data for comparison [47]. This important shipyard, though temporary (1318–1320), presents the characteristics of a State organisation marked, notably, by the presence of a representative of the king and pope, for the purposes of control and of overseeing progress.

At the head of the shipyard there was a "great master of galleys", (*maior magister galearum*) (f° 10), whose function is in all probability analogous to that of the master constructor of the galleys of the modern era. Under this specialist responsible for the whole programme of construction of the five galleys, came the "masters of the axe" (*magistri de Aysia*), whose activities are not limited to the work in the shipyard, but equally to the cutting of trees in the forest, as the account specifies it: "…*erant magistri de Aysia et scayraverunt arbores*" (f° 31v).

In taking just the case of the North and South of mediaeval France, the documentation allows us to sketch the content of knowledge and know-how that are hidden behind the terms designating the different players in the shipyards. One can observe that this hierarchy of knowledge, and its terminology, will be maintained in France up to the last decades of the 17th century in the environment of State construction (the arsenals), and rather later still in that of the private shipyards. In the arsenals, the conception of warships is entrusted to the "master carpenters of the king maintained in the ports". At the period of the first navy of king Louis XIV, which was achieved in 1690 with the death of Seignelay, Secretary of State for the Navy, these master carpenters mostly possessed an elementary and practical geometric know-how. One fact is very significant in this respect: the absence of a knowledge of geometric design. The official correspondence between the intendments of the ports and the court states makes statements on several occasions explicitly deploring the gap between the advanced knowledge existing in the environment of civil architecture, that of the great edifices both State and private, and that, much more summary, of the specialists in naval architecture of both the State and private shipyards.[15] One fact is very revealing for this disequilibrium between the states of knowledge in France: in a letter dated 1679, the intendant of the port of Toulon remarks that he has been obliged to call upon "an architect and master-carpenter from Paris" to make "the drawings of vessels", that is to

[15] A revealing example of what was at stake in this rift is provided by a letter of 16 May 1679 from the son of the old Intendent of the Arsenal of Toulon: "....if the master carpenters do not learn to thus make the drawings in the manner in which they are made for buildings on land, not only will they never be able to build anything with certainty, but they will not even be able to conform with the rules and proportions that will be given to them in future", Archives Nationales, série B3-Marine, 32e registre, letters received, f° 324, cited by Lemineur, Jean-Claude [34, p. 220].

Translator's note. A similar complaint appears in Philibert de l'Orme's "Premier tome de l'architecture" as early as 1567, but explicitly applied to masons: "Briefly by the means of the said tracts one can make everything, and the master mason can render himself capable of conducting all sorts of works: if he knows well how to understand measures and proportions, with the practice of them. But justly, I can complain here that today I do not see many workmen take the trouble to study and know what concerns their office, rather amusing themselves with a lot of mundaine and frivolous things which are not of their vocation. So that if they continue in it, it will be difficult for the kings, princes great lords and others who would build, to be well served by the said workmen (I say many of them), whom I wish to advise fraternally, to admonish and beg that they should consent to recognise and study and learn what is required and necessary to their art and office. For the knowledge of which I have written for them from my heart, but with the greatest labour, the present work of architecture" (p. 57v). The translator however suspects that this is only half a story, and wishes to know how anything had ever been built in France before these complaints, in building or shipbuilding).

say the plans, and to initiate the sons of the two master-carpenters of Toulon, Coulomb and Chapelle, into geometric draughting.[16]

In the last decades of the 17th century, all the sources join together to show that, almost without exception,[17] the knowledge of geometric drawing is only attested in the corpus of civil architecture. The proof is in the fact that in France it was necessary to wait until the year 1697 to see in a published work the first definition of the propriety of projection in plans applied to the establishment of plans of ship. In his *Théorie de la construction des vaisseaux*, Hoste recalls, doubtless to increase the value of his book in the face of criticisms made by a certain number of those responsible for the navy of Louis XIV, the rift that seems to exist at the level of geometric design between terrestrial and naval architecture in France at the end of the 17th century: "It is important that the shipbuilders should make plans for the vessels, like the architects do for buildings; if one trusts to the eye to rule the least things, the whole work becomes defective". He then enunciates on the propriety of plans in these term: "The plan of a vessel is nothing other than the projection of the various parts of the vessel, made on one of its plans by perpendicular lines on the same plan. As the plans of a vessel ought to furnish an exact figure of all the parts of the vessel, three different projections are made. The first is made in the plan of the master floor [the transverse cross-section]. The second is made on a vertical plan perpendicular to that, which cuts the vessel into two equal parts [longitudinal vertical section]. The third is made on the horizontal plan, which cuts the vessel from bow to stern" [22, p. 147ff].

[16] The letter dater 17 November addressed by Arnoul, Intendent of the Arsenal of Toulon to the king is very clear: ".... There is the man named Chaumont, who is an architect and master carpenter from Paris, and who is a very wise youth and very assiduous beyond the things that concern his profession: I have employed him for two years for all the drawings of vessels that I have had the honour to send to you, such that he has become very capable in this matter, so that I can say that he is even necessary in the plan that you have of making general proportions for shipbuilding, for instructing the carpenters, and for them to learn to work by rule, and it is already a year since I made the sons of M.Coulomb and M.Chapelle work under him...", Archives Nationales, série B3-Marine, 34e registre, letters received, f° 499, cited by Lemineur, Jean-Claude [34, p. 221].

[17] One of the French exceptions is that of Fournier who, in his "Hydrographie" published in 1643, provides examples of the geometric method of construction of a master-frame. It is certain that geometry forms part of the culture of Fournier, a theoretician of naval architecture, but in no way a practitioner of shipbuilding. The difficulty is to relate this geometric knowledge of Fournier to the reality of the shipyards of the period of Richelieu. On Fournier's method: cf. Barker, Richard [2, p. 45ff in particular].

With the first decades of the 18th century, the situation evolves with a first change of denomination. The king's master carpenter, who conceived the ships, is henceforth called the king's master-constructor (*maître-constructeur du roi*). In this new configuration, the title of master carpenter is now reserved solely for those who actually construct. Theoretical knowledge is thus associated with the first, practical know-how with the second.

A second important modification comes with the Ordinance of 25 March 1765, according to the old master-constructors the title of "engineer-constructor". That appears very revealing of the cultural step separating the two architectural worlds. The reference to the word engineer marks well, it seems to us, the fact that in naval architecture, the conception of a ship, including the geometric conception of its forms that only represents one aspect, depends upon not only the field of knowledge of architecture, as in civil architecture, but also that of mechanics, which, in itself, is particular to naval architecture.

In this new professional context, the "permanent master carpenters" (*maîtres-charpentiers entretenus*), supervise in the navy yards the construction of ships for the king "on the plan and under the orders of the engineers".[18] As to the carpenters, these are the workmen who fashion the pieces with the axe and adze, assemble them.... With good reason and with full respect for the workmen in the shipyards, the authors of the *Encyclopédie Méthodique Marine* state: "It needs plenty of intelligence in those of the carpenters who are called foremen (*chef de pièce* or *chefs de brigade*), who lead the others... They have a *natural practical geometry*[19] which never fails to become the subject of admiration of knowledgeable persons capable of reflection".[20]

If we consider that between the 14th and 15th centuries, and the end of the 17th century, the knowledge of the king's master carpenters, without speaking of those of the private shipyards, is impoverished in terms, notably, of geometric knowledge, one can consider, more or less under the title of working hypothesis, that the master-carpenters of the time of king Philippe V, the Long, were holders of a technical culture comparable to that of their successors from the time of the first Navy of king Louis XIV.

---

[18] Article *charpentier (maître)* [12].
[19] Our emphasis.
[20] Article *charpentier* [12].

The second question that we wish to discuss bears on the training of the men and, more particularly, of the master-carpenters of the Middle Ages in charge of the architectural conception. Again, we make appeal to the example of France. Following the usages current from the end of the 13th-beginning of the 14th centuries, the corporations of craftsmen developed without however becoming generalised. It is only in the course of the 16th century, with the Royal Ordinance of September 1555 on the Admiralty that the apprencticeship and admission to Master controlled by the corporations was extended to the majority of the ports of the kingdom, with few exceptions. As, for example, the first article of chapter IX of the Ordinance for the Navy of August 1681 records it, "the port of la Rochelle is of the number of those where these crafts [carpenters, caulkers, augermen of ships] are not Freemen at all".

Examination of the statutes of the corporations allows us to know with precision the modalities of the acquisition and of the transmission of knowledge within these bodies, above all, of the private shipyards. For want of being able to present the statutes of the mediaeval period, for lack of sources, those of the modern era allow us to evaluate the training leading to the status of Master ship-carpenter. A revealing example of these statutes is provided by the "Regulation in the form of statute that the masters of the crafts of ship-carpenter, caulker and augerman of the town of Honfleur ought to observe", dated to 1704 [54].

The training is based on essentially practical teaching "by gesture and word".[21] Each technical action, from the simplest to the most complex, is imitated, repeated, controlled, corrected.... until total mastery is achieved. At the end of two years' apprenticeship, the apprentices pass a proof, consisting, notably, of the realisation of a *masterpiece*, of which the model can be chosen from amongst several options. Three essential aspects of this training are to be emphasised. Firstly, none of the different types of *masterpiece* refer to knowledge relevant to the geometric forms of the hull. That part of the training appears to reduce to its "writing" in the statutes of the corporation. However, that does not necessarily mean that it is absent from the apprenticeship. The documentation is simply silent on this subject. Secondly, the role of oral acquisition of knowledge seems to be important. Twice (articles 13 and 15), the statutes of the "Regulation..." of Honfleur that the

[21] This expression is borrowed from the title of the famous book by the pre-historian André Leroi-Gourhan, "Le Geste et la parole" [35].

apprentice has chosen to "make" the *masterpiece* or of "speaking its proportion". Thirdly, according to a classic strategy for preservation of family privilege, the statutes of the corporations, and those of Honfleur do not escape the rule, tend to favour the children of a master-carpenter and in that way to protect a cultural and economic patrimony in maintaining it within a narrow social group. Article 22 of the statute thus stipulates that "The sons of masters will not be subject to undergo any questioning, nor to make any *masterpiece*, but to make eighteen months apprenticeship only". That "closed" knowledge has two major consequences: on the one hand it leads to the creation of dynasties of master-carpenters and, on the other, it favours the transmission, by reproduction, of knowledge and connected know-how, through the generations, and to the elaboration of familial architectural traditions. One question is posed: would these traditions allow the introduction of architectural innovations, or, on the contrary, tend to curb them?

In the domain of the training in knowledge, the corporations in the Middle Ages occupy a central position and, from this point of view, the master ship-carpenters like the masters of masonry and of carpentry followed a similar course. Outside the corporations, the Middle Ages saw the development of brotherhoods for the crafts in which the social function was very important. We consider the case of of the brotherhood of carpenters of Abbeville, in the Somme, which we have already cited. Certainly, it is a question of a brotherhood localised in the North of France and which does not necessarily translate the reality of Mediterranean brotherhoods. We have taken it up for documentary reasons.

The lodge of this brotherhood is situated in the church of Saint George in Abbeville. The statutes of 1488–9 mention the various obligations to which the members of the brotherhood are subject, "...masters and companions of the craft of ship-carpenters of this said town of Abbeville" [52]. According to the traditional modalities of the brotherhoods, it is indicated that the apprentices "...wishing to learn the said craft of carpentry" ought to lodge a certain sum of money for the mess and also provide drink for the said "*carpentiers*". More important for our subject is article 8. This stipulates that "...no masters of the said craft of ship-carpenter can take house-carpenters to work with them, nor other workmen than of the said craft of ship-carpenter while there are companions of the said craft of ship-carpenters who have no work...in this said town". In other words, house-carpenters could only work in the shipyards in the sole case where there were no more ship-carpenters available in the territory of Abbeville. That measure of protection seems to translate into the existence of relationships between the crafts of naval architecture and civil architecture at the level of the carpenters, of those who, according to the words of the master-constructor of the king's ships Blaise Ollivier, employ "…the axe, the adze, the hand-saw, several chisels, a wooden mallet, a line, a square, a bevel, a plummet, a compass and augers of all sizes" [39, p. 94]. Effectively, these tools are bearers of a know-how and of a "natural practical geometry", according to the expression already cited, drawn from the *Encyclopédie Méthodique Marine*, forming a sort of technical language shared by the carpenters of boats and of houses.

A recent example of this common technical culture is provided by the reconstruction at Rochefort of the 26-gun frigate *Hermione*, built in the same port in 1779. This is being undertaken by a company specialising in the restoration of historic monuments but without experience in shipbuilding, which had been chosen from amongst several others of whom two were specialists in shipbuilding. The shipyard, begun in 1997, is proceeding as fast as funding permits. One remark of the head of the shipyard shows well the possibilities of a gangway, at the level of know-how and of practice, between naval and civil architectures: "We are above all carpenters. Certain members of the crew have built boats, others not. We have the experience of very complex and monumental carpentry, such as the dome of the Hôpital du Val-de-Grâce in Paris, or that of the Manufacture des Rames [oars] at Abbeville. For this shipyard, we have twelve months for draughts. Then, on the *Hermione*, we are not in deep water. From the moment where we have the designs, we can make everything that is demanded of us" [16, p. 99].

The words of the head of the shipyard are clear. From when the draughts have been designed, that is to say that the architectural project revealing a geometric conception of knowledge has been materialised at full scale on the floor of the shipyard, the carpenters are capable of translating, on the basis of their know-how and their knowledge of practical geometry, the outlines in two dimensions on the floor into pieces of carpentry in three dimensions. Naval and civil architecture are then brought together in the same process, intended to transform raw material into wrought pieces.

5. Conclusion

At the end of this study, it is important to recall that the sources that we have available in the corpus of mediaeval Mediterranean naval architecture remain very few, partial, dispersed. In these conditions, there is a great risk of over-interpreting partial data. We have tried not to give in to the temptation. Have we succeeded?

With respect to our documentation, it seems to us that the geometric conception of hulls and the knowledge that is made use of, possess their own logic and their own history. At this level of what one can call a geometry of knowledge, different from a "natural practical geometry" or know-how, naval architecture and civil architecture seem to us to constitute singular cultural spaces. One of the reasons for this singularity has been brought up by Maurice Aymard in referring to a passage of a letter sent by V. Fausto to G. B. Ramusio in September 1530. Aymard writes: "Fausto...contrasts...to the already great difficulty of the architecture of building on land that, infinitely greater, of *marine architecture*, that is to say the simplicity of straight lines, for which the rules are relatively easy to determine...to the complexity of curved lines, and furthermore all different one from another, that happens in the construction of vessels: *curvis (lineis) atque iis subinde variantibus extruenda omnia*" [1, p. 413].

At the level of know-how and of "natural practical geometry", nothing, it seems to us, opposes the pathways between civil and naval. The testimony of the head of the shipyard for *L'Hermione* is very significant in this respect.

In the state of advancement of our research, to go beyond those remarks would be premature. There is no doubt that the territory that is open to historians of the "two" architectures, naval and civil, is vast and promising.

Acknowledgments

Many thanks to Richard Barker for his translation and commentary.

References

1. Aymard, Maurice: "L'arsenal de Venise: science, expérience et technique dans la construction navale au XVIe siècle", in Cultura, scienze e techniche nella Venezia del Cinquenceto, Atti del Convegno Internazionale di Studi, Giovan Battista Benedetti e il suo Tempo, Venice, pp. 407–418, 1987.

2. Barker, Richard A.: "Whole-moulding: a preliminary study of early English and other sources", in Nowacki, Horst and Valleriani, Matteo (eds.), Shipbuilding. Practice and Ship Design Methods From the Renaissance to the 18th Century, Max-Planck-Institut für Wissenschaftgeschichte, Preprint 245, Berlin, pp. 33–65, 2003.
3. Barker, Richard A.: "English Shipbuilding in the Sixteenth Century: Evidence for the Processes of Conception and Construction", Rieth, Eric (ed.), "Concevoir et construire les navires. De la trière au picoteux", Technologie, Idéologie, Pratiques, Editions Erès, Ramonville Saint-Agne, vol. XIII, 1, pp. 109–126, 1998.
4. Basch, Lucien: "Ancient wrecks and the archaeology of shipwrecks", The International Journal of Nautical Archaeology, vol. 1, pp. 1–58, 1972.
5. Bass, George F., Matthews, Sheila D., Steffy, J. Richard and Doorninck, Frederick H. Van: "Serçe Limani. An Eleventh-Century Shipwreck. Volume I. The Ship and Its Anchorage, Crew and Passengers", Texas A & M University Press, College Station, 2004.
6. Bass, George F., Doorninck, Frederick H. van (eds): "Yassi Ada. Volume I. A Seventh Century Byzantine Shipwreck", Texas A & M University Press, College Station, 1982.
7. Bonfiglio Dosio, G. (ed.): "Ragioni antique spettanti all'arte del mare et fabriche de vasselli: manuscrito nautico del sec. XV", Venice, 1987.
8. Chiggiato, Alvise: "Le *Ragioni antique* dell'architecura navale", in Bonfiglio Dosio, G. (ed.), "Ragioni antique spettanti all'arte del mare et fabriche de vasselli: manuscrito nautico del sec. XV", Venice, pp. LV–LXXXIX, 1987.
9. Crumlin-Pedersen, Ole, Olsen Ole (eds.): "The Skuldelev Ships 1", Ships and Boats of the North, volume 4.1, Viking Ship Museum, Centre for Maritime Archaeology of the National Museum of Denmark, Roskilde, 2002.
10. Doorninck, Frederick H. Van: "The hull remains", in Bass, George F., Doorninck, Frederick H. van (eds), "Yassi Ada. Volume I. A Seventh Century Byzantine Shipwreck", Texas A & M University Press, College Station, pp. 32–64, 1982.
11. Drachio, Baldessara Q.: "Visione", ca. 1594, Contarini, ms 19, arsenal, b.1, Venetian State Archives, Venice.
12. "Encyclopédie Méthodique Marine", Paris, 3 vols, 1 vol. of plates, 1783–1787.
13. Fennis, Jan (ed.): "Un manuel de construction des galères. 1691", APA-Holland University Press, Amsterdam & Maarssen, 1983.
14. Fourquin, Noël: "Un devis de construction navale de c. 1273", in Villain-Gandossi, Chistiane, Rieth, Eric (eds.), "Pour une histoire du fait maritime. Sources et champ de recherche", Editions du CTHS, Paris, pp. 263–278, 2001.
15. Gateau, Albert: "Atlas et glossaire nautiques tunisiens", 2 vols, Dar El-Machreq Editeurs, Beyrouth, 1966.
16. Gaubert, Yves: "Bateaux traditionnels français. Reconstitutions et répliques", Editions du Chasse-Marée/Ar Men, Douarnenez, 1998.
17. Gille, Bertrand (ed.): "Histoire des Techniques. Technique et civilisations, technique et sciences", Encyclopédie de la Pléiade, Éditions Gallimard, Paris, 1978.
18. Harpster, Matthew B.: "A 9th century AD vessel from Bozburun, Turkey: new Directions of Research", (to appear) in Tzalas, Harry (ed.), Tropis IX, Ninth International Symposium on Ship Construction in Antiquity, Athens.
19. Harpster, Matthew B.: "Dowels as a means of edge-to-edge joinery in the Bozburun vessel", The International Journal of Nautical Archaeology, vol. 34, pp. 88–94, 2005.
20. Harpster, Matthew B.: "A preliminary report on the 9th century hull found near Bozburun, Turkey", in Tzalas, Harry (ed.), Tropis VII, Seventh International Symposium on Ship Construction, Athens, vol. 1, pp. 409–418, 2002.
21. Hocker, Frederick M., Ward, Cheryl A. (eds.): "The Philosophy of Shipbuilding. Conceptual Approaches to the Study of Wooden Ships", Texas A & M University Press, College Station, 2004.

22. R. P. Hoste: "Théorie de la construction des vaisseaux", Lyon, 1697.
23. Jézégou, Marie-Pierre, Joncheray, Anne et Jean-Pierre Joncheray: "Les épaves sarrasines d'Agay et de Cannes", Archeologia, n° 337, pp. 32–39, 1997.
24. Jézégou, Marie-Pierre: "Eléments de construction sur couples observés sur une épave du haut Moyen Age découverte à Fos-sur-Mer (Bouches-du-Rhône)", in VI Congresso Internacional de Arqueologia Submarina. Cartagena 1982, Madrid, pp. 351–356, 1985.
25. Joncheray, Jean-Pierre et *alii*, "L'épave sarrasine (Xe siècle ap. J.-C.) de Bataiguier. Rapport d'évaluation de 1993", Cahiers d'Archéologie Subaquatique, n° XVI, pp. 213–222, 2007.
26. Joncheray, Jean-Pierre: "L'épave sarrasine Agay A. Campagne 1996", Cahiers d'Archéologie Subaquatique, n° XVI, pp. 223–248, 2007.
27. Kahanov, Yaacov: "Dor D wreck, Tantura Lagoon, Israel", in Beltrame Carlo (ed.), Boats, Ships and Shipyards. Proceedings of the Ninth International Symposium on Boat and Ship Archaeology, Venice 2000, Oxbow Books, Oxford, pp. 49–56, 2003.
28. Kahanov, Yaacov: "The Byzantine Shipwreck (Tantura A) in the Tantura Lagoon, Israel. Hull Construction Report", in Tzalas, Harry (ed.), Tropis VI, Sixth International Symposium on Ship Construction in Antiquity, Athens, pp. 265–271, 2001.
29. Kahanov, Yaacov: "The Tantura B, Tantura Lagoon, Israel: Preliminary Hull Construction Report", in Litwin, Jerzy (ed.), Down the River to the Sea. Proceedings of the VIIIth ISBSA, Gdansk, 1997, Gdansk, pp. 151–154, 2000.
30. Kahanov, Yaacov, Royal, Jeffrey, Hall, Jerom: "The Tantura wrecks and ancient Mediterranean Shipbuilding", in Hocker, Frederick M., Ward, Cheryl A. (eds.), "The Philosophy of Shipbuilding. Conceptual Approaches to the Study of Wooden Ships", Texas A & M University Press, College Station, pp. 113–127, 2004.
31. Lavanha, João Baptista: "O Livro primeiro da Architectura Naval", ca. 1608–1616, Gama Pimentel Barata, João, Barker, Richard, Contente Domingues, Francisco (eds.), Academia de Marinha, Lisbon, 1996.
32. Le Goff, Jacques: "Pour un long Moyen Age", Un autre Moyen Age, Editions Gallimard, Paris, pp. 447–452, 1999.
33. Lehmann, Louis Th.: "Baldissera Quinto, Drachio. La Visione del Drachio", Amsterdam, 1992.
34. Lemineur, Jean-Claude: "Les vaisseaux du roi Soleil", Editions Oméga, Nice, 1996.
35. Leroi-Gourhan, André: "Le Geste et la parole", Editions Albin Michel, Paris, 2 vols, 1964–1965.
36. "Libro de Zorzi Trombetta de Modon", ca 1445, British Library, London, Cotton ms, Titus A.XXVI.
37. Merlin-Chazelas, Anne: "Documents relatifs au Clos des Galées de Rouen et aux armées de mer du roi de France de 1293 à 1418", Bibliothèque Nationale, Collection de documents inédits sur l'histoire de France, Paris, vol. 1, 1977, vol. 2, 1978.
38. Mor, Hadas, Kahanov, Yaacov: "The Dor 2001/1 Shipwreck", The International Journal of Nautical Archaeology, vol. 35, 2, pp. 274–289, 2006.
39. Ollivier, Blaise: "Traité de construction contenant par ordre alphabétique la description des vaisseaux de tout rang, galères, frégates, 1736", (Vincennes, Service historique de la Marine, ms SH 310), Editions Oméga, Nice, 1992.
40. Pomey, Patrice, Rieth, Eric: "L'archéologie navale", Editions Errance, Paris, 2005.
41. Recht, Roland: "Glossaire des termes d'architecture navale", Dossiers Histoire et Archéologie, n° 47, pp. 84–90, 1980.
42. Rieth, Eric: "Des mots aux pratiques techniques: gabarits et architecture navale au Moyen Age", Chronique d'Histoire Maritime, n° 56, pp. 13–34, 2004.

43. Rieth, Eric: "Les illustrations d'un "livre de recettes techniques" d'architecture navale du milieu du XVe siècle: le Libro de Zorzi Trombetta de Modon", in Villain-Gandossi, Christiane, Rieth, Eric (eds.), "Pour une histoire du fait maritime. Sources et champ de recherche", Editions du CTHS, Paris, pp. 81–104, 2001.
44. Rieth, Eric: "A propos des relations entre le texte et les illustrations d'un "livre de recettes techniques": la Fabrica di galere (début du XVe siècle)", in "Utilis est lapis in structura. Mélanges offerts à Léon Pressouyre", Editions du CTHS, Paris, pp. 381–393, 2000.
45. Rieth, Eric: "L'arquitectura naval", in Palou, Hug, Rieth, Eric, Izaguirre, M. *et al.*, "Excavacions arqueològiques subaquàtiques a Cala Culip. 2. Culip VI", Monografies del CASC, 1, Girona, pp. 115–117 and 137–201, 1998.
46. Rieth, Eric, Carrierre-Desbois, Catherine, Serna, Virginie: "L'épave de Port Berteau II (Charente-Maritime). Un caboteur fluvio-maritime du haut Moyen Age et son contexte nautique", Documents d'archéologie française, 86, Editions de la Maison des Sciences de l'Homme, Paris, 2001.
47. Sosson, Jean-Pierre: "Un compte inédit de construction de galères à Narbonne (1318–1320)", Bulletin de l'Institut historique Belge de Rome, fasc. 32, pp. 57–318, 1962.
48. Steffy, J Richard: "Construction and Analysis of the Vessel", in Bass, George F., Matthews, Sheila D., Steffy, J. Richard and Doorninck, Frederick H. van, "Serçe Limani. An Eleventh-Century Shipwreck. Volume I. The Ship and Its Anchorage, Crew and Passengers", Texas A & M University Press, College Station, pp. 153–170, 2004.
49. Steffy, J. Richard: "Ancient scantlings: the projection and control of Mediterranean hull shapes", in Tzalas, Harry (ed.), Tropis III, Third International Symposium on Ship Construction in Antiquity, Athens, pp. 417–428, 1995.
50. Steffy, J. R.: "Reconstructing the hull", in Bass, George F., Doorninck, Frederick H. van (eds), "Yassi Ada. Volume I. A Seventh Century Byzantine Shipwreck", Texas A & M University Press, College Station, pp. 65–86, 1982.
51. Taylor, David A: "Boat building in Winterton, Trinity Bay, Newfoundland", Musée national de l'Homme, Centre canadien d'études sur la culture traditionnelle, n° 41, Ottawa, 1982.
52. Thierry, A.: "Recueil des monuments inédits de l'histoire du Tiers Etat. Première série. Région du nord. Tome quatrième contenant les pièces relatives à l'histoire municipale d'Abbeville et à celle des villes, bourgs et villages de la Basse Picardie", Paris, pp. 318–320, 1870.
53. Throckmorton, Peter and Joan: "The Roman wreck at Pantano Longarini", The International Journal of Nautical Archaeology, vol. 2, pp. 243–266, 1973.
54. Vintras, A.: "La construction navale à Honfleur (de la fin du XVIe siècle au début du XIXe siècle)", in Vintras, A., Banse, Jehan, Decomble, G., Abbat, Pierre, "Le corporatisme ancien de la construction navale en France", Académie de Marine, Paris, pp. 121–148, 1939.
55. Vintras, A. Banse, Jehan, Decomble, G., Abbat, Pierre: "Le corporatisme ancien de la construction navale en France", Académie de Marine, Paris, 1939.

CHAPTER SEVEN

BOAT AND BOAT HOUSE. THE CONCEPTIONAL ORIGINS OF CLINKER BOATS AND BOAT-SHAPED HALLS OF THE FOURTH TO ELEVENTH CENTURIES IN SCANDINAVIA

Ole Crumlin-Pedersen

1. The boat. The product of an artisan or an architect?

During the greater part of the history of mankind, ships and boats were created by 'simple craftsmen' who were artists in their field, capable of constructing complex three-dimensional hulls of a high visual and functional quality. These early boat and ship builders and master shipwrights were the true creators of most of the basic boat shapes known to us today. The naval architects would work from moulds or drawings to define or modify the final three-dimensional hull form and plan the building process accordingly. Their ability to do so was used as a criterion to distinguish them from the craftsmen building ships in their 'primitive' ways, without using drawings.

Modern man is inclined to interpret the original construction process for large buildings and other complex structures of the past in terms of present-day architects and engineers, considering calculations, drawings and moulds a prerequisite for the original builders. However, for several complex structures of the past this was evidently not the case. A good example is the construction of ships during the Renaissance, built on the basis of the Dutch shell-first method using clamps as vital elements in creating the desired result, such as the Swedish man-of-war *Vasa* and several large Dutch East Indiamen. They were built without any drawings at all. Such a process was indeed 'the art of shipbuilding' since it involved the trained eye of the ship builder and a sculptural talent for shapes, symmetry and individual curves that resulted in a complex and fully-functional hull form. At a later stage, this 'art' term has been disregarded in favour of the more academic naval architectural approach to the construction of ships.

Therefore, I suggest going further back in the history of boat and ship building in an attempt to identify the initial shape-creating processes forming the conceptional origin of some traditional boat types.

The Nordic boat and its 'standardised' shape

The layout and shape of boats vary strongly in different parts of the world as a result of several individual factors. There is rich variation in the way boats are built, in their basic concepts as well as in their shapes, proportions and structural details.

Among the large number of finds of boats and ships from Northern Europe, dating to the 3rd to 13th centuries, one group stands out as a class of its own: the double-ended, clinker-built vessels with overlapped plank strakes fastened to a gently curved keel, stem and stern. These vessels primarily come from Scandinavia and the coastal areas of the Baltic and North Seas and they are members of what is commonly called the 'clinker-built Nordic boat type'.

These boats, including Viking ships, comprise a group of complex three-dimensional structures that have been built on a well-defined basic concept, applied with creativity to form a multiplicity of shapes and sizes of vessels, guided by the boat builder's ability to combine functional requirements with aesthetic qualities—the artisan working without any involvement of an external architect.

Boats of this basic design built over a millennium ago, such as those found with the Gokstad ship of c. A.D. 900 from eastern Norway, are strikingly identical to a 12th-century boat found in Denmark and to the oselver boats being built today on the basis of living traditions in western Norway (Fig. 7.1). They share identical basic features, having a double-pointed, elegant shape with rising lines towards the curved stems, as well as slender frames and beams which are symmetrical to the centre plane and placed with a regular, wide spacing along the length of the hull [3].

To Scandinavians this set of features defines what is considered a 'proper boat shape' for a traditional boat. This is not strange, since the general pattern of the Nordic clinker-built boat during the centuries has been used in the construction of a wide range of shapes and sizes of vessels. For more than a millennium, clinker-built ships and boats could meet all the requirements for relatively safe transportation of warriors, merchants or emigrants on their voyages in the stormy North Sea and North Atlantic as well as for communication in home waters and on the large European rivers.

Fig. 7.1: The continuity in the Nordic clinker tradition is very apparent when comparing reconstructions of the small boat from Gokstad, ca. 900, from south-eastern Norway (1) and the Danish boat from Gislinge on Sjælland from ca. 1140 (2) with a traditional oselver boat built ca. 1970 in south-western Norway (3). (Photos © Viking Ship Museum, Roskilde).

Design by rules-of-thumb and building-by-eye

How did the builders deal with the naval architectural aspects of these vessels? Evidently extensive expertise was needed to design the larger ships. The requirements for the space and buoyancy needed to carry the weight of the cargo or the number of warriors should be met without losing the advantage of a light-weight structure and a supple structural lay-out. At the same time the lines of the hull should give minimal resistance, and the distribution of weights etc. should ensure equilibrium between the forces acting upon the submerged part of the hull and the aerodynamic forces on the sail and the sides of the hull above water. Even today this is almost too complex a system to be designed by computer prior to construction, especially when taking into consideration that for most of these vessels special measures were taken to ensure the suppleness of the structure. For these ships of the past, a high degree of specialised knowledge based on long experience combined with specific tricks-of-the-trade was needed.

Within traditional clinker building communities, there is no indication of the use of drawings or advanced calculations for the design and construction of these ancient vessels or for their descendants until modern times, when some influence from the larger shipyards can be traced. Instead, *rules-of-thumb* and *building-by-eye* were the key words to the process. Today these terms are disregarded and held up to ridicule, but they are, nevertheless, the clue to the successful original construction of highly admired and beautiful vessels such as the Oseberg and Gokstad ships or the swift and powerful Viking longships (Fig. 7.2).

The experience from the Skuldelev ships

The thorough study of the individual elements in ship finds has made it possible to trace the sequence of actual construction, as for instance in the case of the coaster Skuldelev 3 from Roskilde Fjord [5: 235] (Fig. 7.3). After having laid the keel, the stem and stern were cut to their final shape. They were secured to each end of the keel to take up the ends of the strakes of the planking. The bottom part of the hull was first constructed as a free-standing plank shell without any help from moulds or frames. By adjusting the outline and edge bevels of each plank in a couple of stages before fastening it to the previous one, the precise shape of the shell could be built up freely, guided only by the eye and a few direct control measurements to ensure symmetry. In order to check the shape in relation to other vessels built previously, further control measurements could be taken at a few cardinal points. A final adjustment of the lines could be achieved by means of slight pressure on the plank, using props from underneath or the weight of stones from above.

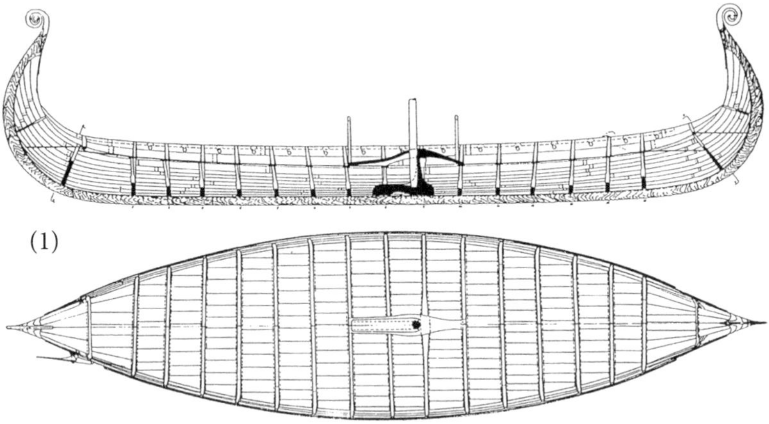

(1)

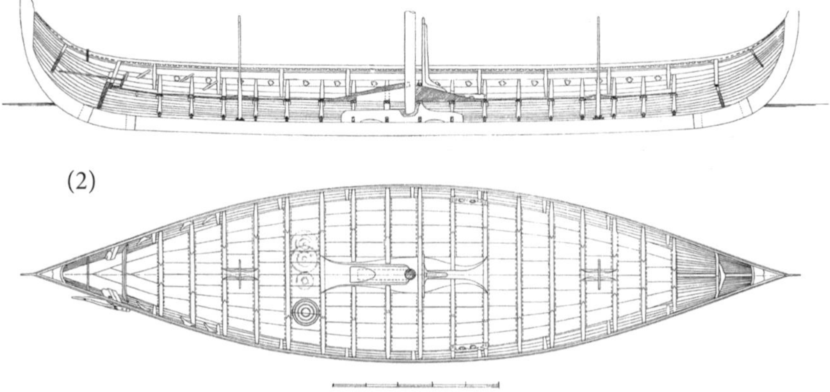

(2)

(3)

Fig. 7.2: Reconstruction plans of the Norwegian Oseberg (1) and Gokstad (2) ships (originals built ca. AD 820 and 895 respectively), and the reconstructed Skuldelev 2 longship (3), [Col. Pl. XXIII], originally built 1042 in Ireland. (© Viking Ships, Oslo, and Viking Ship Museum, Roskilde).

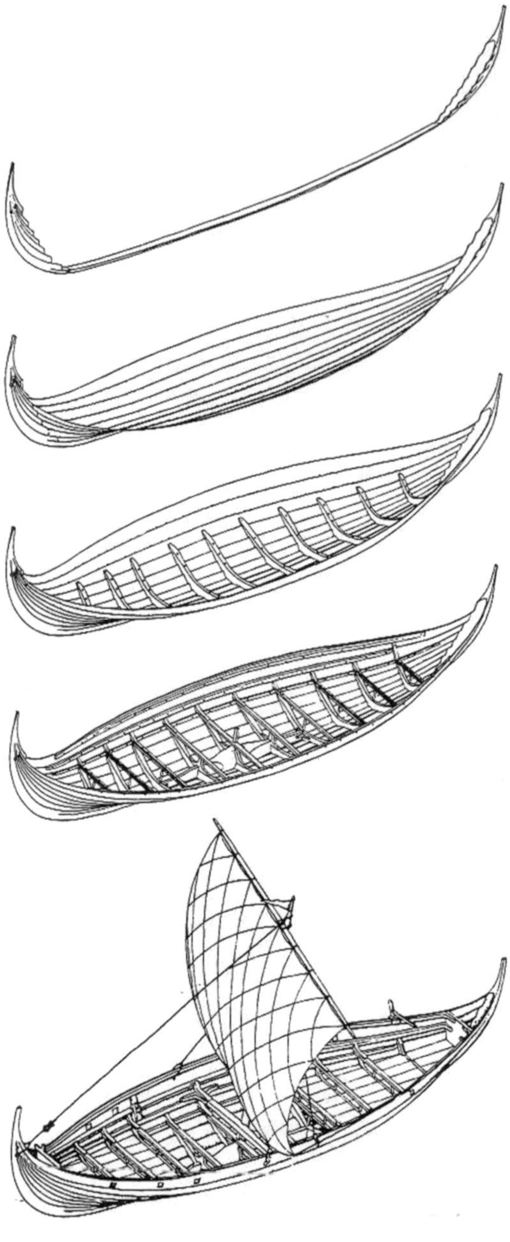

Fig. 7.3: Skuldelev 3. Stages in the construction of the hull as deduced from a study of the overlapping elements of the planking and internal timbers. (© Viking Ship Museum, Roskilde)

Fig. 7.4: Stem found in a bog on the island of Eigg in Scotland (by permission of the National Museum, Edinburgh).

In building a reconstruction of Skuldelev 3 in 1982–84 using authentic tools and techniques, the large, fully preserved and elegantly carved original stem presented a problem: how did the boat builder define its shape without using drawings? A similar stem with all its steps and lines was found in a Scottish bog and as it lacked holes for fastenings, it was clearly ready-made by a boat builder for later use (Fig. 7.4). This meant that the boat builder, already at the outset, was in control of all the lines of the planking in the boat since these should match the lines of the stem. How could this be explained for a building tradition based on rules-of-thumb rather than on drawings?

An analysis of the shape of the original stem of Skuldelev 3 demonstrated that its outline was defined by three segments of circles with radii of 1.5 m, 3.0 m and 6.0 m (Fig. 7.5). Since the original length of the keel was 9.0 m, it is very likely that the shape of the stem was originally determined on simple proportions of the keel length—just like some of the rules-of-thumb systems known in recent traditional clinker boat building. Consequently, the outline of the stem could be traced out on a plane side of the log with no other tools than a string and a piece of chalk [1; 5: 237]. This example illustrates some of the principles in building ships and boats according to simple rules-of-thumb based on, for instance, proportions of the keel length for the scantlings.

This 'design plan' for the shape of the stem actually seems to have been an old one that had been developed at an earlier stage in the history of this specific ship type. This is indicated by the fact that the stem had steps for seven strakes each side, whereas the shell of the ship had actually been built using eight strakes on each side. An extra strake had been inserted at the bilge in a slightly irregular way on each side in order to reach the desired width of the planking before mounting the side strakes, evidently due to lack of sufficiently broad planks. This observation is in accordance with other ship finds of the 10th–12th centuries, showing a rapid decrease in the access to large oak trees for ordinary shipbuilding activities [1].

The overall size and proportions of a vessel were determined by the requirements for handling the sail and with an eye to functional units such as the modulus of a rower in action or the size of standard cargo units. The shape was controlled by the trained eye to achieve the desired sculptural hull form, possibly using a few control measurements to compare with boats built previously. Considering the importance of the stems for the construction process as well as for the overall appearance

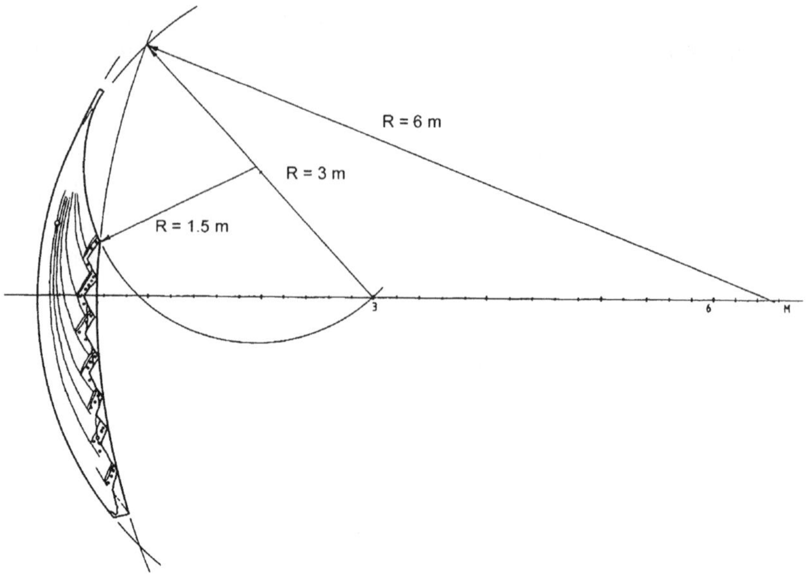

Fig. 7.5: The outline of the original Skuldelev 3 stem is defined by a simple system of circles with interrelated centres. (© Viking Ship Museum, Roskilde).

Fig. 7.6 [Col. Pl. XXIV]: The *Roar Ege* reconstruction of Skuldelev 3 during launching 1984. (Photo © Viking Ship Museum, Roskilde).

Fig. 7.7 [Col. Pl. XXV]: The shipbuilding scene on the Bayeux tapestry (by permission from the Musée de la Tapisserie de Bayeux).

of a new boat, it is no wonder that the Viking-Age term for the master boat builder was 'stem-smith' ('smith' being used as a general term for a craftsman at the time).

The floor-timbers were not inserted into the hull until the bottom planks had been mounted and fastened to each other. In the same way the following planks were mounted forming the sides of the ship, and when the shape of the planking had been approved by the master craftsman, the hull was strengthened before launching with upper frames and beams (Fig. 7.6).

An interesting example of one of the later stages of this building process is depicted in the 11th-century Bayeux Tapestry, in a scene showing orders being given to the master ship builder for the construction of ships for Duke William's invasion of England in A.D. 1066. The boat builders are working hard to complete the vessels (Fig. 7.7), using various tools for trimming, drilling and hammering, whereas one of the boat builders is shown standing in a position from where he can control the fairing of the uppermost plank and give instructions to adjust it in contact with the craftsman onboard the ship with whom he is clearly communicating [2: 188–89].

Fig. 7.8: The reconstruction of the longship Skuldelev 2 in the Irish Sea 2007. (Photo W. Karrasch, © Viking Ship Museum, Roskilde).

The concept of the Nordic boat

Together with the 'boat-shaped' lines and the symmetrical frame system described previously, the shell-first principle is a fundamental feature of the clinker tradition, to be found in all vessels of the Nordic type over the centuries. In fact, this basic set of features had such permanence over more than a millennium that it was as if a proper master plan had been imposed, followed by everyone working within this tradition regardless of the size and proportions of the vessels built or of the woodworking technologies applied.

The clinker boat builder was able to construct his vessels in accordance with this general concept, combined with his full control of the three-dimensional lines of the individual planks, resulting in a complex and highly functional hull shape and structural lay-out. These features were combined in ships operating under strained conditions in the open sea with cargoes of men or goods, such as the famous Viking ships which outclassed ships of other contemporary Northern European building traditions. In this case it seems fair to use the term 'the art of shipbuilding' since the result was calling on a range of creative skills from the master boat builder. His work was evaluated on functional as well as aesthetic criteria, and the success of the resulting vessel was a matter of life and death for everyone onboard (Fig. 7.8).

Fig. 7.9: Stages in the making of an expanded logboat in Satakunta, Finland. (National Museum of Finland 1935).

The expanded boat

The search for the origins of the fundamental concept for the Nordic type leads to the production process of the expanded boats, exemplified in the expanded logboats of south-western Finland in the 1930s [9] (Fig. 7.9).

Here, local boats for use on the numerous lakes and rivers were made from large aspen logs that were carefully carved into 'boat elements' of a shape almost like a pea-pod with thin sides and pointed ends. By gently heating a hollowed log over a fire, the wood was softened so

Fig. 7.10: The 5th-century Björke boat as excavated in Gästrikland, Sweden, in 1950, and as reconstructed 2006 at the Viking Ship Museum, Roskilde. (Photos © Gävle Museum and the Viking Ship Museum, Roskilde).

that the sides could be pressed out. At a certain stage, the ends of the boat started rising, and in the course of a single process of successful expansion, the log was directly converted from its cylindrical shape to a proper double-ended and gently curved, slender boat shell. In order to maintain this shape, symmetrical frames or thwarts were needed, distributed regularly along the length of the boat. Thus this process 'as if by a touch of magic' established the characteristics that were to become the conceptual basis for the Nordic clinker boat type.

In recent years a considerable number of finds of such expanded boats have been found in Scandinavia, along the southern border of the Baltic and the North Sea, and from south-eastern England, all dating to within the first millennium A.D. [4]. Some of these have had extra strakes added above the expanded bottom element and an appearance very much like the plank boats built from a central backbone of keel, stem and stern in normal clinker fashion (Fig. 7.10). Also, in other parts of the world the expansion technique has been used, and it may have spread around the globe at a very early stage before it was taken into the Nordic world as a basis for 'the proper boat shape'.

Nordic traditions merging with foreign elements

The idea of using separate timber elements to serve as a backbone of keel and stems and of relatively uniform planks for the shell of a hull was not an original part of early Nordic boat-building technology. However, these features were characteristic of ships and boats of the Celts and Romans [7]. During the Roman period, warriors from the North served in the Imperial army and its naval units on the Danube and the Rhine, and in the North Sea (Fig. 7.11). Here they would have observed the advantage of building boats from separate elements for keel and stems, and from sawn planking, rather than by carving out the elements of the boat shell from logs, as in the 4th-century B.C. Hjortspring boat and the expanded boats.

As a result of this inspiration, a new generation of boats of the North, like the 4th-century A.D. Nydam boats (Fig. 7.12), was built with the backbone of the keel and stems shaped much like in Roman ships, and with the planking built up of large planks that were not sawn but split from log halves and cut to shape. These modifications, however, did not affect the fundamental lines and interior structure established with the expanded boats. Thus the conceptual basis for all the clinker-built planked ships and boats to follow during the next millennium had evidently been conceived in combining the expanded boat of the North with aspects of Celtic and Roman wood-working technology.

Fig. 7.11: Reconstruction of the Roman 3rd-century A.D. Mainz B warship type for patrolling the Rhine. (© Museum für Antike Schiffahrt, Mainz).

Fig. 7.12: Reconstruction drawing of the 4th-century A.D. Nydam oak ship from South Jutland. (From Sophus Müller: *Danmarks Oldtid*, 1897).

2. The boat houses

In contrast to the ships and boats described here, from which considerable parts of their original wooden structure have been preserved, the wooden parts of Scandinavian houses of the same periods of the Iron and Middle Ages have seldom survived. When located during excavations, these houses are usually identified by the postholes dug for their main structural elements, leaving the archaeologists with a series of questions as to the original structure of the buildings and often with a confusing pattern of several generations of buildings on the same site.

Therefore many archaeologists will go no further than to record the ground plan of an excavated building site, and classify it in relation to other ground plans and the distribution of possible finds of domestic objects, and the like, indicating the interior lay-out of the building. In contrast, several interested architects have reconstructed these buildings completely, graphically or as full-scale reconstructions. Some of the results of these efforts are more fanciful than realistic, such as when the postholes of large halls are taken to be for poles to carry the weight of ships turned upside down to serve as a roof [6].

Recent examples of fishermen's sheds made from old boats turned upside down as a vault over stored fishing gear, etc., have been used to support the idea of boats serving as roofs. Such sheds may have been common in fishing hamlets before corrugated iron sheets became readily available as a better suited roof cover material. However, an old boat would not last long as a roof, unless covered by tarpaulins or roofing felt, since the decaying overlaps of the planking would be facing the wrong way and thus would give easy access for water dripping into the shed through the seams.

Therefore, clinker-built boats and ships would not have been suitable as roof elements in permanent houses or halls at any time. The suggestion that the Gothic arch or the elaborate roof structures of medieval churches were inspired or derived from the frames of ships does not take into account the principles applied in medieval shipbuilding in Scandinavia. The shell construction principle of clinker boats was entirely different from the 'skeleton principle' of building roofs on all sorts of buildings where an erected set of rafters would be covered with planking or other materials.

It is a fact, however, that several of the ground plans of the larger buildings of Iron- and Viking-Age Scandinavia have curved sides. This feature is best known from the large Viking fortresses, such as the Trelleborg site with 32 large houses (Fig. 7.13) and the huge Aggersborg site with as many as 48 houses arranged in a strict geometrical plan [11]. These impressive military camps only existed for a short period during the later 10th century, before being abandoned. However, buildings of a similar shape have been excavated elsewhere in Scandinavia, showing that this ground plan had been in use throughout the 3rd to 12th centuries for large halls as well as for more moderately-sized houses.

The origin of the concept of houses of this shape is likely to be found in boat houses. In western Norway, sheds for the storage of boats and ships had to be solidly built, with external walls of sod or stone to withstand the pressure of strong winds during winter storms [8; 10] (Fig. 7.14). For this reason the sites of these boat houses are easily recognised in the landscape, even today with all remains of wooden structures long gone. Dating to between the 3rd to 12th centuries, some of the excavated sites show that these buildings were positioned on a piece of sloping land close to the water, with an opening in the end facing the water. They have a shape and length that allowed a boat of up to 30–35 m to be stored with space along both sides for repairs and maintenance. A house to fit this function with a minimal requirement for timbers for the roof would quite naturally get an oval shape with curved sides that would also reduce the effects of wind stress on the roof.

Since wind-protecting stone or sod walls were evidently not needed in most of eastern Norway, as well as in Sweden and Denmark, only few boat houses have been identified in these areas. Few archaeologists have searched for them, since they would lie in places outside the area of farm houses. In a few cases such sites have been found at locations that were obviously suited for a boat house, but here the interpretation for this purpose has been denied since traces of a fireplace were found in the central areas of the buildings. The use of a fireplace would obviously not be possible while the ship was stored in the house. However, such a building would be empty during the sailing season and would offer itself during that period as an ideal hall for all sorts of activities. Consequently, it would have been an obvious choice to take inspiration from this constructional concept when the custom of building chieftains' halls became common during the Late Roman period and the centuries leading up to the Viking Age.

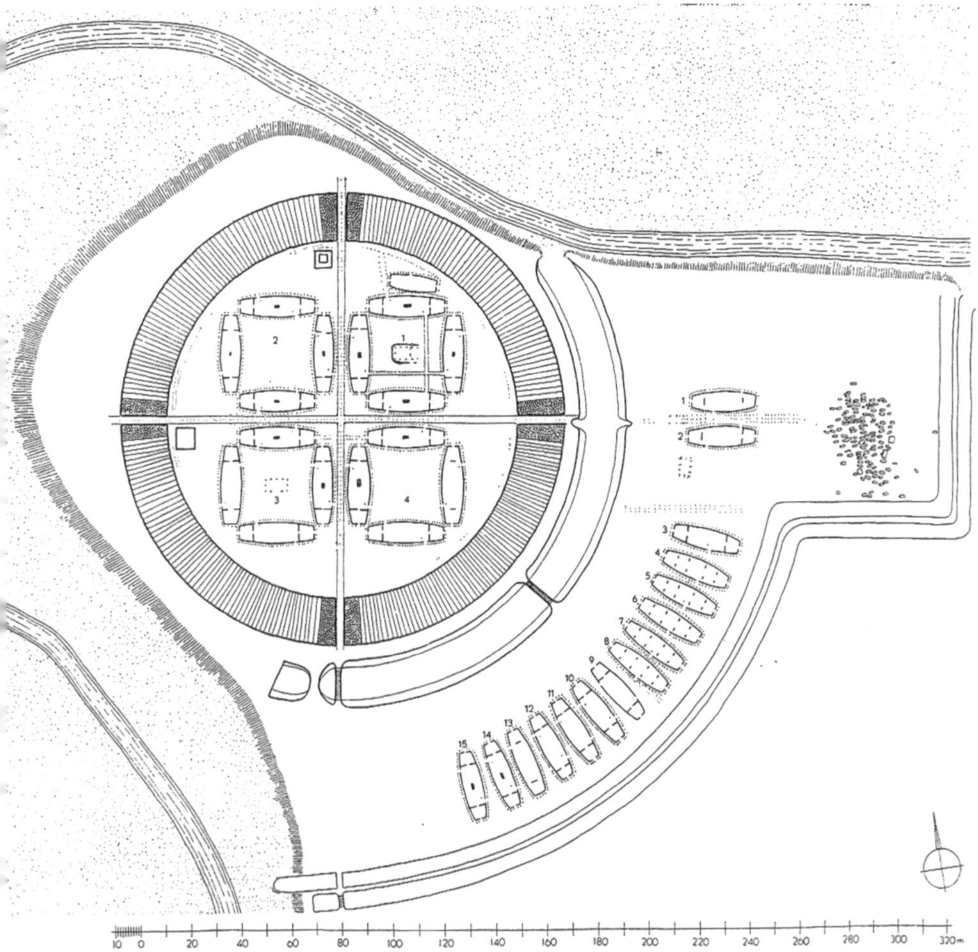

Fig. 7.13: The 10th-century Viking fortress Trelleborg on Sjælland, Denmark. (From [11]).

Conclusion

As shell-first technology was the one and only method used in Iron-Age and medieval shipbuilding within the Nordic cultural area, there are no links traceable from here to the various ways of constructing the skeleton-based roofs of houses or vaults of churches during the same period, although the same craftsmen may have been involved for both jobs.

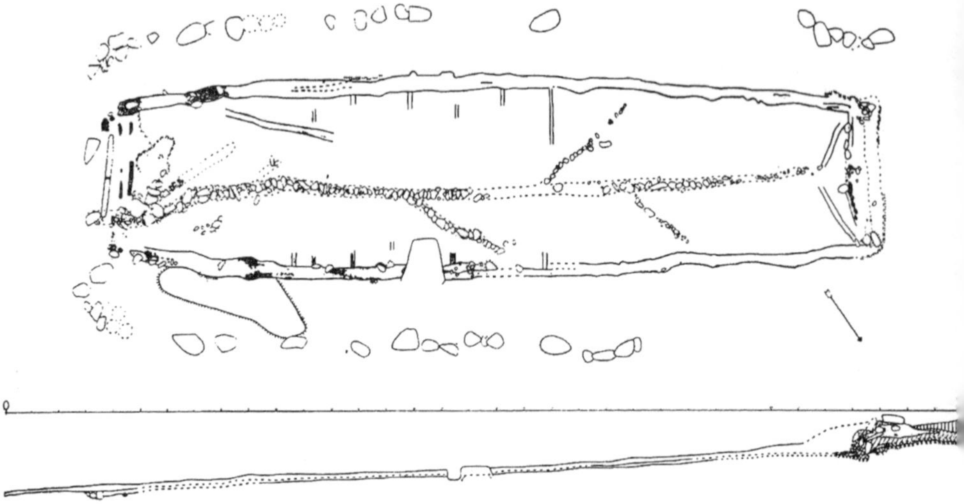

Fig. 7.14: Plan of a 27 m-long boat house from c. A.D. 500 at Bjelland, Stord, Norway. (From [10]).

The camps of the Trelleborg type demonstrate the application of advanced geometry in the planning of military installations during the Viking Age; however, there is not any indication that the shape of the hulls of ships would have been pre-determined in drawings or by the use of moulds or master-frames. This was a matter for the experienced master ship builder, the stem smith, to keep everything under control as the shell of the hull was gradually built up like a large three-dimensional sculpture matching the basic 'boat-shape' concept and the functional requirements for the new vessel. This also included the application of modules for the space needed for each rower and for the main functions of handling the cargo as well as the sail.

Thus, in order to understand and appreciate the sophisticated watercraft built by early shipwrights and boat builders of Scandinavia, one must see the construction process as based on a set of principles derived from the expanded logboat and applied skilfully to comply with the complex physical conditions of a sailing vessel at sea. This process encompassed talents that today are described separately under the headings of aesthetics and functionality—the arts and sciences—but which were integrated parts of the Viking ship builders' mental toolbox, continuously being updated through new challenges and practical experience.

References

1. Crumlin-Pedersen, Ole, 1986: Aspects of Wood Technology in Medieval Shipbuilding. In O. Crumlin-Pedersen & M. Vinner (eds) *Sailing into the Past*, 138–148. The Viking Ship Museum, Roskilde.
2. Crumlin-Pedersen, Ole, 1997: *Viking-Age Ships and Shipbuilding in Hedeby/Haithabu and Schleswig*. Ships & Boats of the North 2. Schlesvig & Roskilde.
3. Crumlin-Pedersen, Ole, 2004: Nordic clinker construction. In F. M. Hocker & C. A. Ward (eds) *The philosophy of shipbuilding. Conceptual approaches to the study of wooden ships*, 37–63. College Station.
4. Crumlin-Pedersen, Ole, 2006: Den nordiske klinkbåds grundform—en totusindårig tradition og dens rødder. In T. Arisholm, K. Paasche & T. L. Wahl (eds) *Klink og seil—Festskrift til Arne Emil Christensen*, 33–55. Norsk Sjøfartsmuseum & Kulturhistorisk Museum, Oslo.
5. Crumlin-Pedersen, Ole, *et al.* 2002: *The Skuldelev Ships I. Topography, Archaeology, History, Conservation and Display*. Ships and Boats of the North 4.1. The Viking Ship Museum, Roskilde.
6. Maiorano, Giuseppe, 2004: Viking-age ships as roofing structures in ship-shaped houses and their contribution to the origin of the Gothic Architecture. In G. Guđmundsson (ed.) *Current Issues in Nordic Archaeology*, 79–84. Proceedings of the 21st Conference of Nordic Archaeologists, 6–9 September 2001, Akureyri, Iceland. Society of Icelandic Archaeologists. Reykjavik.
7. Mees, Allard, *et al.*, 2002: *Römerzeitliche Schiffsfunde in der Datenbank "Navis I"*. Kataloge vor- und frühgeschichtlicher Altertümer, Band 29. Römisch-germanischen Zentralmuseum, Mainz.
8. Myhre, Bjørn, 1997: Boathouses and naval organization. In A. Nørgård Jørgensen & B. L. Clausen (eds) *Military Aspects of Scandinavian Society in a European Perspective, A.D. 1–1300*. PNM Studies in Archaeology & History 2: 169–183. The National Museum, Copenhagen.
9. Nikkilä, E., 1947: En satakundensisk äsping och dess eurasiska motsvarigheter. *Folk-Liv* XI, Stockholm.
10. Rolfsen, Perry, 1974: *Båtnaust på Jærkysten*. Stavanger Museums skrifter 8. Stavanger.
11. Schmidt, Holger, 1999: *Vikingetidens byggeskik i Danmark*. Moesgård Museum, Jysk Arkæologisk Selskab, Århus.

CHAPTER EIGHT

GOTHIC AND RENAISSANCE DESIGN STRATEGIES IN STONECUTTING

Enrique Rabasa Díaz and José Calvo López

1. Introduction

The training of Medieval and Renaissance artisans was confined to the boundaries of each craft; it took place within guilds or, as in the case of Spain, within families or through contracts of apprenticeship. The craftsmen involved with either architectural or boat construction were not given prior, theoretical training; as a result, they did not have a common reference frame. The construction of vaults seems to have been carried out through intuitive reflections based on experience [36]. Enrique Nuere, a leading expert in the history of Spanish carpentry, deals with the connections between shipwrights and building carpenters in these terms:

> Several authors have posited a direct relation between boat construction and building carpentry. Though it would not have been impossible for a carpenter to know both trades, they have nothing more in common than wood and tools used. These techniques, based on formulas learnt over many years, are so specialised and different from one another, that the procedures used in one of these fields cannot be applied to the other. [24, p. 43]

At first sight, the difference between shipbuilding and stone stereotomy is even greater. We know very little about the geometrical knowledge of Arab stonemasons and bricklayers in Spain. Their creations did not consist of a wide range of vaults; of all their constructions, the most interesting ones are almost entirely limited to a peculiar kind of vault with intercrossing arches, though these do show very accurate geometrical solutions.

Regarding the evolution of Christian stonework, we must take into account that at the beginning of the 15th century many Spanish stonemasons in charge of designing vaults and dressing their elements, such as Juan Gil de Hontañón and Andrés de Vandelvira, were illiterate.

However, their sons, Rodrigo Gil de Hontañón and Alonso de Vandelvira, were able to write treatises on their work [14, 52]. The master masons who were able to write do not seem to have come into contact with the scientists of the period. Furthermore, their way of working, which involved working as both architects and contractors whereby they were continually resolving material problems on construction sites, did not favour such relations.[1]

Only through Renaissance architects, such as Juan de Herrera, would a prior, theoretical training have been possible. By that time, however, the most important stereotomical procedures had already been developed. Moreover, Juan de Herrera does not seem to have been interested in stonecutting, which was left to his surveyors or auxiliaries at El Escorial.

Nevertheless, there are some geographical coincidences between stereotomy and navigation, which suggest the possibility of an exchange of ideas between both fields. Although it is not possible to reach any definite conclusions, it is worthwhile to remark these concurrences.

[1] There is, however, some indirect and scant evidence of contacts between men of science, particularly from the University of Salamanca, and stonemasons or figurative artists in command of architectural works. Both Fernán Pérez de Oliva and Juan Martínez Guijarro or Siliceus, professors of the University of Salamanca, supervised the construction of the library at Salamanca University [30, p. 69].

Significantly, Pérez de Oliva's books on cosmography have been connected with the interest in spherical developments of the Seville school of stonecutting [33, 34]. Later on, Siliceus taught Prince Philip, later King Philip II of Spain, and is said to have fostered the interest of the King in mathematics and philosophy, although this is still under discussion [40, pp. 108–121; 23, pp. 529–530]. When he left his duties as instructor of the prince, Siliceus was made Bishop of Cartagena. In the cathedral of this bishopric, which is located in Murcia for complex historical reasons, there is a quite singular ribbed sail vault in the second story of the bell-tower, built during Jerónimo Quijano's term as master mason. Contrary to the usual practice in Spanish stonecutting of the period, and even the work of Quijano's school in El Salvador in Caravaca, the ribs are not drawn on vertical planes, but rather on inclined diametral planes. Siliceus, who was bishop during the period in which the vault was built and who was in contact with Pérez de la Oliva as mentioned, could have inspired this departure from standard practice. Besides, Siliceus had published in Paris an *Ars Arithmetica*, which, in some editions was corrected by Orontius Finaeus, another well-known cosmographer. At the same time, Siliceus held Quijano in high esteem since he chose Quijano to represent the Archbishopric of Toledo at the appraisal of the choir stalls of Toledo cathedral [5, pp. 32–33, 197–210]. Another professor at Salamanca University, Pedro Ciruelo, *Darocensis*, was quoted by Rodrigo Gil de Hontañón as the author of a rule on vault curvature. It is not clear, however, whether the rule put forward by Ciruelo was a constructive rule or a geometrical rule that others applied to construction [14, f. 22 r.; 45, 149–150].

2. Common areas between stereotomy and navigation

The Gothic architecture of the Crown of Aragon, which included Catalonia, the Balearic Islands, Valencia and, in some periods, parts of the South of France, Sicily and Naples, has a number of particular characteristics. These involve both the general form of buildings, which feature spacious naves with short and wide, quadripartite vaults, and chapels between the buttresses, as well as technical details such as the decorative simplicity of the ribs and the panels of the rib vaults [9, 57].

A boat's hull is built on a series of transversal frames supporting longitudinal planking. Between the 13th and 15th centuries, and especially in the 14th, the cover of some large naves was constructed through what we term diaphragm arches. This technique uses wooden joists to form a horizontal floor or sloping roof resting on great stone arches. It seems to derive from the Syrian constructions of stone slabs resting on parallel walls above arches, which Viollet-le-Duc believed to have been brought to Europe during the crusades, fostering the development of western medieval stonecutting [9, pp. 58–63; 53; 55, pp. 23–42; 58, pp. 110–128]. The diaphragm-arch system was suitable for use in places where the forests were becoming depleted and wood was necessary for shipbuilding. This form of construction was used in the Atarazanas, or in the shipyards of Barcelona and Valencia.

In the 15th century, the characteristics of Catalan stonecutting are well defined. Rib vaults, as has been mentioned, are simple and not very ornate, and their panels usually made of well-carved ashlars, rather than irregular, rubble stones as in other places. However, brickwork panels were also frequently used and, as far as we know, the earliest infillings with Catalan vaults date back to this time.

In Palma de Majorca, an outstanding Gothic architect and sculptor named Guillem Sagrera built the city market, or *Lonja*, between 1426 and 1447 [2]. The master mason Pere Compte undertook a similar construction in the building of the *Lonja* in Valencia, using vaults with panels made of whole slabs of stone. Joseph Gelabert's manuscript [13, pp. 252–283] explains the dressing and construction of this type of panel. Enrique Rabasa recently completed the translation of this manuscript from Catalan to Spanish, to be published along with a critical edition.

As far as we know, Gelabert's treatise is the only original source that deals with the creation of panels in Gothic vaults. The passages that deal with rib vaults as though they were usual at this time are particularly striking. The severy is described as being made of whole pieces of stone, which were occasionally split into two, like those in the *Lonja*. From a stereotomic point of view, Gelabert uses a simple method, first dressing a stone into the shape of a straight voussoir, and later giving it the required warping. The result, however, is more similar to a classic, well-dressed ashlar than to a course of rubble stones for a medieval panel.

Pere Compte had been apprenticed to Francesc Baldomar, who, in 1439, built Valencia's Santo Domingo convent, including a chapel with unribbed vaults. However, these vaults take the form of conventional rib vaults, mimicking Gothic ribs through bonding ashlars (in the way that Renaissance vaults are built) [12; 54; 56]. There are also examples of such *groined* stone vaults in Assier (France) and Puglia (Italy) [31, pp. 212–213; 29].

This line of progression from the Gothic rib vault towards the Renaissance voussoir vault was interrupted. However, it is clear that experimentation throughout the Crown of Aragon until the middle of the 15th century is exceptional, in accordance with the importance of those territories in the history of boat building, sea trade and cartography. The 14th and 15th centuries were the heyday of Majorcan cartography, in particular regarding nautical maps, which were generally drawn by Jews.

As a result of America's discovery, the importance of Seville's port on the Guadalquivir River grew enormously. In the 16th century, Juan de Herrera prepared the plans of the *Casa de Contratación* in Seville. Later on, the upper story of this trade hall was covered with sail vaults, featuring many different kinds of quartering and decoration. All this required a good mastery of surface developments and spherical geometry [25; 35]. At the same time, rib vaults were still being built and many of them featured a perfectly spherical central area. This research line in spherical developments coincides with the creation of more complex cartography in Seville than in Majorca, particularly in the *Casa de Contratación*, since it should represent the shape of the Earth. Juan de Herrera was associated with the scientific studies that took place there and was assigned the task of creating an Academy of Mathematics in Madrid by Felipe II. The Monastery at El Escorial, built by Juan de Herrera, housed the best Spanish library of the era, where many works on navigation, cartography and cosmology have been preserved to the present day. Though no clear-cut relation between stereotomy and cartography in Seville and Cádiz has been proven, it is evident that the two sciences underwent an extraordinary development in the same area [33; 34].

The procedures for dressing voussoirs in spherical vaults probably came from France, though it should be noted that the first ashlar sail vault to be constructed was that erected in the Cathedral of Murcia in 1525, another coastal location [5, pp. 79–92; 6].

We will just briefly mention here other, less technical and more metaphorical affinities, such as that of helical lines. The *caracol de Mallorca*, a kind of spiral staircase with a small well in the centre was given its name because the first example is found in Majorca's *Lonja*. Another *caracol de Mallorca*, also by Guillem Sagrera, was built in Naples' *Castelnuovo*, and displays a grooved soffit [9, p. 370; 2, pp. 158–198; 54; 48; 1; 56, pp. 147–148, 152–154]; there are similar exemplars in Valencia's *Lonja*, the Vélez Chapel in the cathedral of Murcia and the cathedral of Cuenca [57, p. 164; 26, p. 161; 7]. The grooves are related to the Gothic columns in the Lonjas of Majorca and Valencia as well as many other places in Valencia and Murcia, and the famous plinth of one of the Castilnuovo towers. The Majorcan spiral staircase model was to be repeated all over Spain, occasionally accompanied by a spherical vault with an unusual finish, displaying just one spiral course (conventional spherical vaults have circular courses) [39].

Another spiral staircase of note is the *Vis de Saint Gille*, from the Saint Gilles Abbey near Avignon [31, pp. 89–90, 143–146; 32; 46; 50]. There are some examples in Egypt that seem older than the French one; the earliest Spanish example is in Barcelona's Cathedral, and dates back to the 14th century [49]. A similar model is the staircase with steps that are horizontally curved on the intrados; we are not sure whether this type exists outside Spain. There are examples of both types in the church of Santa María de la Coronada in Medina Sidonia near Cadiz, together with a spiral vault. Thus, stone spirals appeared in abundance during the heyday of these Mediterranean countries.

We have mentioned two areas where naval activity and the development of stereotomy coincide. However, we have not yet talked about Cantabria and the Basque country, an important boat building area. This was the homeland of many important seafarers and the birthplace of many stonemasons who worked all over the peninsula; the Trasmiera region is especially renowned as a cradle of many important stonemasons. In the 15th century the quality of stone construction in Spain was very poor, and though the Mudéjar artisans were able to work with this material, they preferred brick. At that time a number of German, Flemish and Breton stonemasons arrived in Spain and revitalised Spanish construction. Later, in the 16th century, they were joined by a few French craftsmen. The next generation of master masons in charge of important Spanish works throughout the 15th and 16th centuries was formed by Spaniards who learnt stonecutting techniques from these foreigners. Nearly all of them came from the Basque Country and Cantabria, two areas the Mudéjar tradition had not reached. European carpenters, however, did not come to Spain, probably because there was already a significant tradition of using interlaced frameworks (*armaduras de lazo*), which are similar in form to the Muslim kind, although all their elements serve a structural purpose [24].

Thus, in 16th-century Spain, Gothic and Renaissance construction overlapped. Although the Renaissance was new, Gothic had not died out, since it is not "late Gothic", and was used in an outstanding fashion throughout the 16th century and continued well into the 17th century. We shall now look in detail at the design strategies that characterise each form of construction.

3. Surface Intersections

Choisy [8, p. 704] has given a clear explanation of the difference in conception between the Gothic and Renaissance vault forms:

> In Blois or Saint Germain there are only gothic vaults, built "in the French style" using, like in the Middle Ages, forms that are easily built... Only in the times of Philibert de L'Orme, the problem is reversed: the builder chooses arbitrarily a more or less complex vault form, and adapts to this form a more or less costly quartering. In this way, the geometrical whimsy of modern stereotomy starts.

Thus, in both cases, surfaces are curved and their intersections form groins, but the strategy is radically different. In the case of the classic vault, the surfaces have geometric forms: spheres, cylinders, and cones, and the lines generated by their penetrations were defined afterwards, resulting in flat or warped curves of varying complexity. In contrast, the Gothic vault is begun by establishing as simply as possible, like arcs of a circle, the intersections which formed the vault's ribs. The panels were constructed over the ribbed lattice formed by the arcs.

As a result, the Gothic builder thought of the vault as an articulated network of elements—the ribs—meeting at particular points: the *tas de charge* or springing, and the keystones or junctions. The voussoirs of the ribs were dressed by means of templates with the form of the straight section or outline of the ribs; the same templates were also used to dress the keystones and the *tas de charge*. Thus, the information required to carry on the carving process is just the general layout and the elevation and template of each rib.

4. Control by Section of the Gothic vault

The templates used in Gothic construction are sections, that is, outlines of the beds of the surfaces of the voussoirs that will be hidden. In contrast, the templates used in the dressing of Renaissance ashlars are templates for the bed joints and visible surfaces such as the intrados. In fact, the attention paid to the intrados and the concern for the development of the surfaces was to play a fundamental role in the development of modern stereotomy.

In general terms, we have seen that a prior conception of the forms and contemplation of the intersections was characteristic of the Renaissance. When the Gothic builder used templates, a seemingly similar process occurred, although on a smaller scale. We shall see that the shape of Gothic mouldings and their intersections are defined by sections. The *tas de charge* and the plinth of the pillars are controlled by their horizontal sections; the keystones join the rib sections. [36, pp. 424–426]. And this eventually became standard procedure amongst Gothic constructors. In Spain, German stonemasons such as Juan and Simón of Cologne probably reinforced this tendency in their creations, notably at the *tas de charge*. Later, it became usual to decorate windows and doors with mouldings whose intersections were defined in this way.

In accordance with this control of the form through cross-sections, the building presents itself as a series of different levels. At one of the lower stretches of the most recent tower wall of the León Cathedral, we can see (figure 8.1) a protruding angle, which does not project at 45°, but at a different angle; that is, the plane of symmetry of the protrusion is oblique to the wall surface. The reason for this can be seen high up (figure 8.2), where, under the spire, there is a cross vault over

Fig. 8.1: Protruding angle in the lower stretch in a tower in León cathedral.

Fig. 8.2: Buttress supporting an octogonal-plan rib vault in a tower in León cathedral.

an octagonal layout. What we see below is the corner of the buttress required by this vault.

In the wall shown in figure 8.3, which belongs to the Salamanca Cathedral, we find a corner moulding that protrudes from the wall surface. The logic behind this is found up above (figure 8.4), where there is a pinnacle with a characteristic play on rotated squares.

Consequently, we can identify a formal approach in Gothic design: the intersection of parts based on their cross-sections or outlines. Another, even more characteristic design tendency is the connection of linear elements.

5. The Gothic design process

The Gothic vault is a volumetric, spatial object. It is controlled, however, by a two-dimensional layout: arcs of a circle on vertical planes. The

Fig. 8.3: Corner moulding protruding from a buttress in Salamanca cathedral.

Fig. 8.4: Pinnacle over a buttress in Salamanca cathedral.

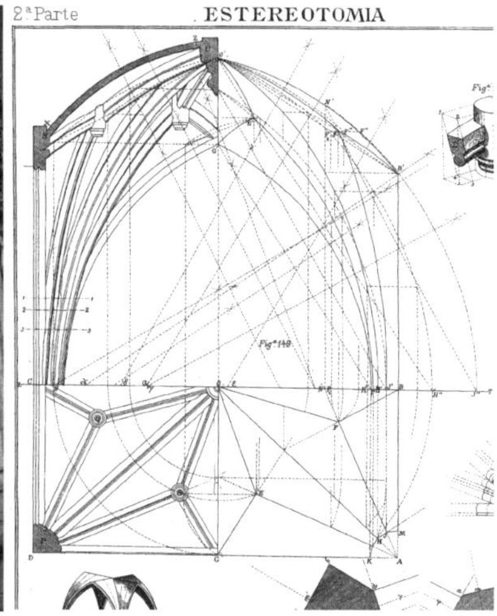

Fig. 8.5: Cross-section of a rib vault. Antonio Rovira y Rabassa, *Estereotomía de la piedra*, part II, 1899.

verticality of the components, including the decoration, and the reference of these same members to the horizontal projection—the layout of the ground plan—control the tracing of the main lines. There is no abstract, homogenous space, as we are accustomed to think of, but rather a network of material references. For this reason, it makes sense to say *on a plumb line* (*a plomo*), instead of vertical, and *level* (*a nivel*), instead of horizontal [52, ff. 4v, 7v, 8r, 23r, 43v]. The literature of construction techniques continued to use this terminology for many years.

The full-size tracings required to construct the vault are the plan (schematic) and the elevations of each of the ribs. These elevations are just circular arcs that should be adequately traced in order to join two given points. This is all that is needed to give shape to the components. The medieval constructor did not know, nor have any need for a vault section such as those in treatises from the 19th century (see figure 8.5).

The ribs are arches formed of voussoirs. The stonemason could dress one of these voussoirs taking into account only the curvature and outline or section of the voussoir. The formal problems, which we now call stereotomical, would be focused on the *tas de charge* and keystones, that is, those places where the ribs join or intersect each other.

As we have seen, these single components are shaped from outlines or sections. We are now going to detail the procedures. The *tas de charge*

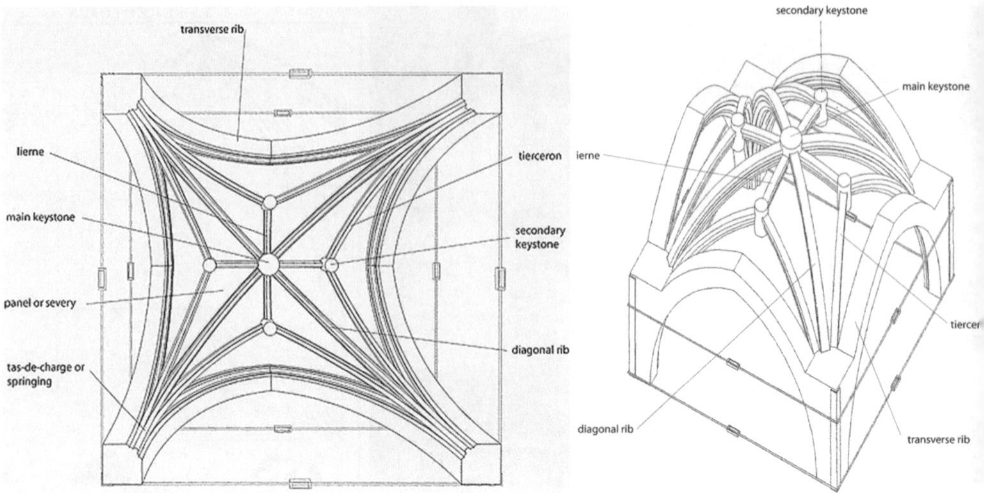

Fig. 8.6: The elements of a tierceron vault.

can be very different in appearance. In a *tas de charge*, there may be a perfect continuity between the pillar ribs and those of the vault, or there may be a point where they intersect each other in a disorganised way, when the ribs of the pillar and those of the vault do not correspond with each other. The intersecting of the ribs may also be deliberate, as in many German vaults, and in those constructed in Spain by Juan and Simón of Cologne. The stones are all carved in the same way, as explained in the mid-19th century by the English archaeologist Robert Willis [53]. The components of the *tas de charge* are separated by horizontal beds. The horizontal planes of the upper and lower beds are dressed first, and the templates for the ribs are situated over them at a suitable distance from the axis. As there are different horizontal sections, it is clear that the distancing of the template will increase with height. This distancing is a necessary piece of information, but it is easily obtained from the elevation of the corresponding rib, that is, through tracing its guiding line. Taking these horizontal sections over the upper and lower beds as a reference, the stonemason dresses the exterior surface. A certain amount of experience is needed to do this, as well as an observation of the fact that certain mouldings are in one bed but not the other, and so should disappear as they proceed (figure 8.7).

A *tas de charge* by Simón of Cologne in the Burgos Cathedral (figure 8.8) confirms this system. The ribs are separated before crossing each other. During restoration work, Enrique Rabasa was able to put

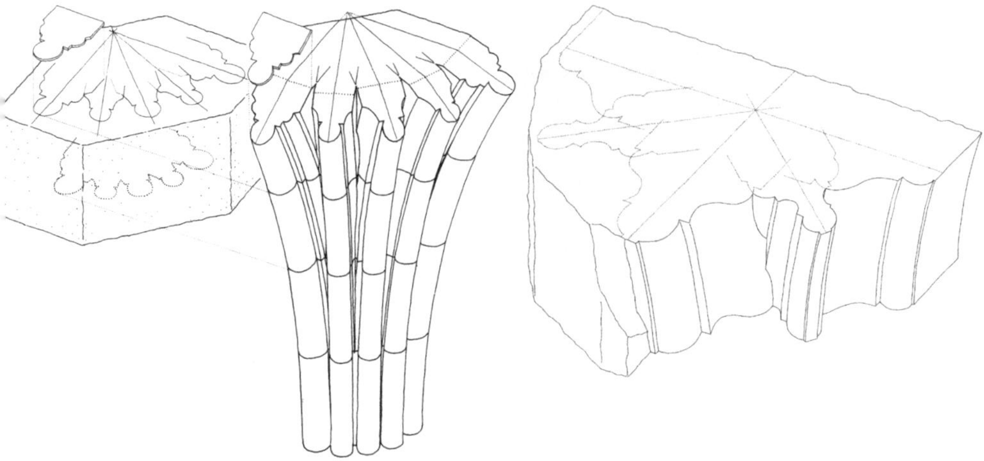

Fig. 8.7: Dressing the voussoirs of a *tas-de-charge*.

his hand between the ribs and observe that the hidden half of each rib was perfectly carved. This makes it clear that the *tas-de-charge* was not dressed in place [36, p. 424].

The keystones are shaped around a vertical cylinder (figure 8.9). When a cylinder or revolving surface hides the connection between ribs, it is usually vertical. In some cases, the end disc or boss carved with sculptural decoration is tilted in order to follow the slope of this area of the vault [37, pp. 106–112]. From this central cylinder protrude the fittings that are to meet the ribs, which reach the vertical axis of the cylinder from various spatial directions. The keystones are dressed from an operation surface—as it was termed by Willis [53]—, the horizontal surface of the extrados. The extrados is first dressed, and over it the directions that the ribs will take are traced on the layout, as well as other references, such as the circle marking the keystone cylinder. The connections that will meet the ribs are then dressed, and finished on a sloping surface, which is the bed that links with the voussoirs. The inclination of this surface can also be deducted from the schematic elevations of the ribs. The templates are then placed on the inclined surfaces. At this moment, there are sufficient references to continue carving downwards [36, pp. 424–426].

Consequently, the way to establish the directions of the ribs that project from the keystone is similar to the convention used to indicate the position of a star in the firmament; it involves the direction of its horizontal projection, or azimuth, and the inclination with respect to

Fig. 8.8: The Condestable chapel in Burgos cathedral; general view and detail of mouldings.

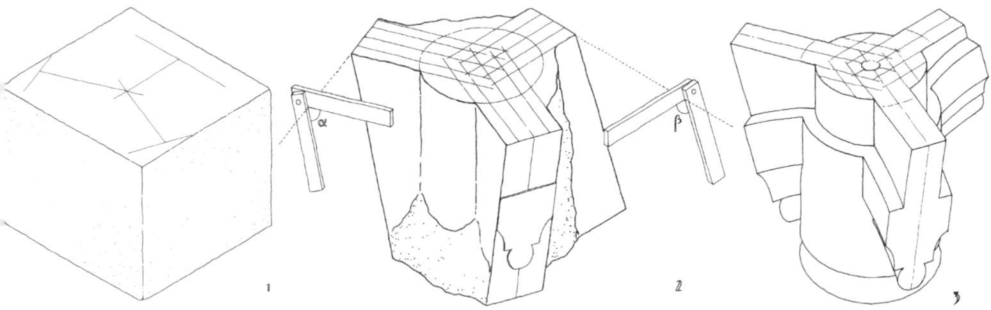

Fig. 8.9: Dressing the secondary boss of a tierceron vault.

the horizontal plane, or altitude.[2] Thus, the general design of the vault is essentially just the layout plan and the height of the keystones. The layout, the lattice, is relatively arbitrary—in Spain there is a tremendous diversity of patterns—and the height of the junctions in this network are also to a certain extent decided at discretion. If we add the design of the outline of the ribs to these initial decisions, there are no more general variables.

The rules for the layout of the Gothic rib vault are few and flexible. The horizontal projection is a network; by elevating its junctions vertically—what we call keystones—to a suitable height, this network adapts itself to the volumetric form desired. In this spatial elevation, the segments that link the junctions become ribs, and these ribs, in

[2] In a forthcoming article in *Revista de Expresión Gráfica Arquitectónica*, 2008, Miguel Ángel Alonso and José Calvo attempt to reconstruct a lost tierceron vault in the church of Santa Catalina in Valencia starting from a surviving secondary boss. After scanning the keystone with a 3D laser scanner, the key information for the reconstruction is given by the angles between the horizontal projections of a lierne and two tiercerons.

Using this information, Alonso and Calvo discard a number of hypotheses, such as a triangular vault or nine-keystone and thirteen-keystone tierceron vaults, and finally put forward the hypothesis of a rhombic plan tierceron vault. Tracing two parallels to the lierne, they deduce the direction of the wall arches; in turn they use the wall arches to draw the diagonal ribs. Once the layout of the vault is established, the curvatures of wall ribs, liernes, diagonal ribs and tiercerons are calculated using the methods explained by Vandelvira [52, ff. 123 v., 124 v.] which are based mainly in rotations around vertical axes. The most striking trait of Vandelvira's method is the shape of the diagonal ribs. The diagonals of a rhombus have different lengths, but the keystones of both diagonal ribs must be at the same height, since they should meet at the main boss of the vault with the lierne ribs. To solve this problem, Vandelvira traces the diagonal ribs as elliptical arches. The hypothesis of a rhombus-plan vault is quite likely, since there are vaults of this kind in the church of Santa Catalina, due to the irregular shape of the plot. This virtual experiment and others, such as Tomlow's [51], show clearly the systematic and yet flexible nature of Gothic geometrical rules.

principle, are merely circumferential arches over vertical surfaces. The goal of this type of inverse projection (from the layout plan to physical space) can differ greatly. In the most simple cases, it is a matter of situating the keystones so that the longitudinal or cross section of the vault—the *rampante*—acquires a more or less round shape, or that the most important peaks are of a reasonable height. In the second half of the 16th century, this was done with such ease that the alteration of the outline of the ribs became a game or a boastful demonstration of skill.[3] Eventually, it was possible to adapt the network to any previously conceived form. Thus star vaults are grouped together to form a single vault, or the network is even adapted to a classic form, as we will see further on.

As the ribs are arches on vertical planes, their meeting points, intersections of vertical planes, are vertical axes. The keystones are therefore set out vertically—the exceptions prove the rule—and the rib corresponds vertically to the layout plan, which appears materially in the process, as we shall see.

Full-size drawings of the tracing of the ribs—the simple elevations we are speaking of—have not been conserved in Spain, and there are very few extant drawings of horizontal projections. This has probably occurred because the tracings were incised on the planks of the scaffolding [14, f. 25r]. However, we do have some drawings that reproduce the process. In the drawing of a tierceron vault in the sketchbook of Hernán Ruiz [44, f. 46v] and in many others, the ribs are represented by their intrados curve. By contrast, other examples, such as the rib vault in the *Architecture* of Philibert De L'Orme [19, f. 108v], a preserved appraisal for the vaults of the parish church of Priego (Cuenca) [41], the manuscript of Gelabert [13, pp. 252–283] and some German drawings depict the entire height of the rib, including the extrados.

It makes sense to deal only with the intrados, as this line is a necessary and adequate guide for the carving of the components. It is as well the common guiding line for the arch and its centering. In a vault constructed in the Centro de los Oficios (Centre of the Crafts) at León (figure 8.10), however, the building team, directed by Enrique Rabasa, took a decision that made also necessary the tracing of the parallel line

[3] On this topic, Professor José Carlos Palacios called our attention to certain Castilian vaults that have the central keystone situated lower than the secondary ones, contradicting the habitual order displayed in the outlines.

Fig. 8.10 [Col. Pl. XXVI]: Building a rib vault in the Centro de los Oficios in León.

of the extrados. In designing the outline of the ribs, a greater height was given to the diagonal ribs than to the tiercerons and the lierne rib (and also different from the height of the four openings). As a consequence, it was necessary to take care that the various extrados surfaces of the ribs that lead to the same keystone were level with each other, with the aim of allowing the proper support to the surface of the panel.[4] Paying attention to this means simply counting on the elevation of the intrados of the rib being a few centimetres higher or lower. However,

[4] This practice is different from the usual German solution, in which the intrados surfaces of all ribs are coincident (see Jos Tomlow's contribution to this volume). Although there are Spanish examples of continuous-intrados lattices of ribs during the 15th century, such as the Llotja in Valencia, and the solution is common practice in Renaissance vaults, there are many 15th-century vaults in which the lower surfaces of the ribs are not coincident. Even Gelabert [13] gives a solution to control the tracing of ribs of different widths so that the extrados surfaces are coincident, at the expense of the intrados surfaces.

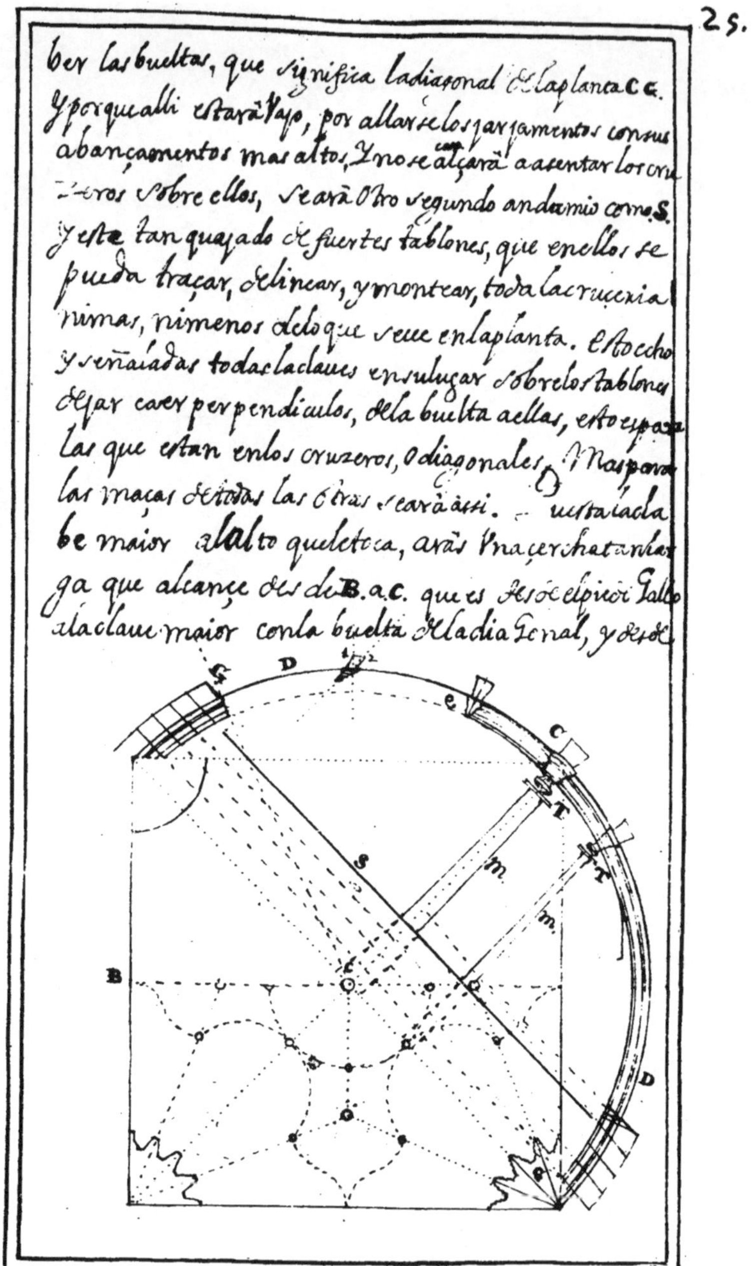

ber las bueltas, que significa la diagonal de la planta C C.
y porque alli estará bajo, por allarse los jarjamentos con sus
abançamentos mas altos, y no se alçará a asentar los cru
zeros sobre ellos, se ará otro segundo andamio como S.
y este tan quajado de fuertes tablones, que en ellos se
pueda traçar, delinear, y montear, toda la cruceria
nimas, ni menos de lo que se vee en la planta. esto echo
y señaladas todas las claues en su lugar sobre los tablones
dejar caer perpendiculos, de la buelta a ellas, esto es para
las que estan en los cruzeros, o diagonales, y mas para
las maças de todas las otras se ará assi. vestaiada
be maior al alto que le toca, arás una cercha tan lar
ga que alcançe des de B. a C. que es desde el pie de tal
a la claue maior con la buelta de la diagonal, y desde

Fig. 8.11: Rib vault in Simón García, "Compendio de arquitectura y simetría de los templos…", 1681, most probably taken from a 16th-century manuscript by Rodrigo Gil de Hontañón.

if the entire height of the rib is drawn, there will not be any possibility of confusion. This explains why this custom was followed in some of the representations mentioned. When we speak of the height and the extrados, naturally, we do not consider the projecting member where the panel abuts against the rib.

In order to mount the vault of León, the team followed the text attributed to Rodrigo Gil of Hontañón. This manuscript contains a description that says—and this is one of the clearest parts of a text that, in general, is not straight-forward—that it is necessary to establish a horizontal platform "set fast with planks", at that point where the *tas de charge* ends and the vault begins (figure 8.11). Over this platform, he adds, a horizontal projection of the ribs is traced. Where a keystone must be positioned, a prop of the right height is situated, over which a boss of a keystone is placed, supported by a bearing block.

Between these supports, trusses are set up to sustain the centerings of the ribs, and finally, the gaps between them are filled with a severy [14, f. 25r]. This is how the León vault was built, by establishing the platform over a collapsible frame. In real vaults, this role is fulfilled by a scaffolding erected on the floor or resting on the initial construction.

Thus, the assembly of the vault is conceived with the same principles as the design and dressing, with the primacy of the layout plan and the vertical correspondence of the elements above it. We could say, in short, that the design is the spatial projection of a flat network, that the dressing of the *tas de charge* and keystones requires just a schematic knowledge of the layout plan and of some arches over vertical surfaces, and that the collocation is seen as the materialisation of the layout and the elements that are situated on a plumb line over it.

The system is flexible and easy to conceive. In fact, it allows many variations such as the use of ribs that are circular in horizontal projection. These elements are often supported (by the panels), rather than being supporting elements themselves, as Rodrigo Gil would say. We can also find keystones that do not have a vertical axis. In some cases, the form has been forced in order to diminish a verticality that could clash with the overall inclination of the area. In many other cases, however, the oblique keystone is not actually a junction in the network, nor a meeting of ribs, but rather a decorative widening of the rib. As an exception and extreme demonstration of skill, some of the vaults in Prague have managed to disobey completely the rule of verticality, displaying convoluted ribs, whereby their outline twists as they advance longitudinally.

6. The Design of Renaissance Vaults

In Italy, Renaissance forms such as spheres and cylinders were produced using brickwork. In France and Spain, however, the same surfaces are used as the intrados of stone vaults with a complex quartering lattice. This requires a spatial conception of the form of each stone and of the development of the surface, and a graphical determination of the dimensions of the voussoirs in order to apply them to the carving process. In the 18th century, this culminated in the coining of a new term to denominate the knowledge and procedures used: *stereotomy* [20; 10].

In Gothic vaults the decoration is linked to the ribbed structure; it is the moulding of the ribs and the keystones. In contrast, in Renaissance vaults the decoration is a relief that is developed over a surface and which may be separate from it and its quartering.

The new Renaissance conception can be illustrated by the design of components for spherical vaults, domes and sail vaults. Normally, they are formed of round courses and each one is divided into voussoirs. According to Spanish treatises of the 16th and 17th centuries [52, f. 60 v.; 17, f. 87 v.; 13, p. 90], the habitual process for the carving of each of these voussoirs consists of first dressing the intrados face and then the joints and beds. The intrados face is spherical. The joints that separate the voussoirs of the same course are flat and the beds that separate the courses are conical. Thus, joints and beds are always surfaces that can be understood as having been created by the movement of the sphere's radius. Therefore, once the intrados face has been dressed, the beds and joints can be carved and checked or controlled ensuring that they remain orthogonal to the face [36, p. 429].

As we can see in figure 8.12, first the concave part of the intrados is dressed, which is a spherical surface. This work is checked using a curved ruler, which should be placed on the surface perpendicularly. This is possible and straightforward as the curvature of the sphere is the same in all directions. Afterwards, the perimeter is marked with a flexible template. The rest of the surfaces are then carved and the work checked with a *baivel* or bevel, which is a special square with two arms, one straight and the other curved. In this process, we have used a flexible template resting on the spherical surface.

In theory, this is not correct since it is not possible to develop a spherical surface. Nevertheless, this approximate procedure introduces an error that is certainly small, at least for a manual stone-carving process. The question is how to obtain a flat template that is an

Fig. 8.12: Dressing a voussoir for a hemispherical vault.

Fig. 8.13: Centering for the reconstruction of a vault in the Pazo de Antequeira in Rois, under the direction of Santiago Huerta.

approximate development of the sphere. There are two characteristic ways of developing an approximate sphere: by cutting it through meridians to obtain lunes, or by cutting it through two parallels to obtain truncated cones. This is what is done: each course is assimilated to the trunk of a cone, and the graphical development of the truncated cone gives the template for the intrados.

In other types of vault, such as the groined vault and the cloister vault, or in conical vaults, it is possible, with even more ease, to develop the surface of the intrados [26, pp. 25–32, 250–253]. When Renaissance builders became familiar with surface developments, they were ready to apply the decoration in relief on the development. From this moment on, the decorative form could be independent from the constructive form. Once the system has been mastered, the architect may take the liberty of creating a formal, decorative play, such as placing the circular courses of sail vaults on vertical planes. In these cases, the voussoirs are dressed in the same way, though the result is surprising. Evidently, the development of the sphere, which is one of the first problems that had to be resolved in Renaissance stonework, is also a problem of great importance for a seafarer used to handling nautical maps.

7. Centerings

It is actually the centerings that could most closely resemble a ship's hull, although unfortunately we do not have factual information on how they were made in the Middle Ages and during the Renaissance, at least in Spain.

We can only suppose that Gothic centerings were essentially a wooden framework that supported the ribs, and that the panels in some cases did not require centerings, for example, when the distance between the ribs was very short; or in areas with a steep slope, so that each course could rest on the previous one.

However, Renaissance vaults usually required a continuous wooden lining of the same shape as the vault. The centering would be a wooden vault. In some cases, this excessive expense would be reduced through the use of courses in the lower part that project forwards in horizontal beds, as in the Gothic *tas de charge*. This is what we see in the vaults of the basilica of the Monastery at El Escorial. There is a special course, with a very high extrados, which almost certainly corresponds to the transition between the *tas de charge* and the ordinary sloping beds. [37, p. 167; 4; 3; 21; 22].

Nevertheless, there are always zones that require a well-constructed lining. One may be inclined to think that this lining is held up with trusses of a roughly triangular structure, though it is more probable that the form of the structure has been achieved many times, usually through simple partitions supported by piles. This was used in a vault near Santiago de Compostela, when it was rebuilt by Santiago Huerta (figure 8.13).

The 1996 exhibition *Ars Lignea* focused on churches of the Basque Country with wooden ceilings below the roof. Many of them are wooden reproductions of the form of Gothic and Renaissance stone vaults. In the foreword to the catalogue of the exhibition [47], Enrique Nuere repeats his opinion that the influence of shipwrights should not be sought in these vaults. Rather it should be considered that in order to construct a vault it is necessary to prepare a centering, which is like a wooden vault. This leads to the idea that the carpenter could undertake the construction of the vault itself. Nuere affirms that the centering is a provisional skin, but, in the case of Gothic vaults, this is not so clear. He adds that we do not know the reasons for the use of wood in these constructions. It is true that wood is cheaper, though there may also have been other reasons, such as the structural effectiveness or the lighter weight of wood.

In Atlantic Europe, there are excellent examples of the construction of wooden vaults, such as the transept of the Ely Cathedral. This cathedral, by the way, has been known through the centuries as the "Ship of the Fens". In the Americas, probably due to seismic movements, there are many such constructions, such as the church of the Lima Cathedral and a number of later ones in Chile. In Castelnuovo in Naples, which has already been mentioned, the chapel of San Francisco de Paula is also made of wood.

We also know that in the 17th century, the Jesuit shipwright Philippe Lemaire, who built ships in Belgium and Brazil, gave wooden vaulting to the Jesuit churches in Cordoba (Argentina) and that other similar ones were constructed on the American continent [15, p. 44]. Philibert de L'Orme's *Nouvelles inventions* include a wooden ribwork for a cross vault, which was reprinted in *L'Architecture* [18, f. 27 r.; 19, f. 111 r.]. Unfortunately, our field is that of vaults made from stonework, and we are unable to supply more information of interest on this topic.

8. Form and Construction

In terms of the relation between form and construction, in short we could say that Gothic construction uses a flexible system of elements. This system can be adapted to different circumstances. The rib-and-panel system, however, determines the form of the vault, at least conceptually. By contrast, in the case of Renaissance construction, the form takes precedence. It is the first thing to be established and the construction (the quartering) should be adapted to it. In boat building, the construction of the form (the hull) is determined first; this makes it similar to Renaissance construction.

We have seen that the Gothic system was also capable of adapting itself to previously conceived forms. This was the case, for example, when Rodrigo Gil de Hontañon used the same pattern for differently proportioned areas, and even used the same one four times in the same vault, in the transept of the Salamanca Cathedral. The Saint Eustache church in Paris is also a good example of the adaptation of the lattice of the layout plan to a previously conceived space. In this nave of the 16th and 17th centuries, therefore, the Gothic system of construction was being used, though the general form is classic (a barrel vault with lunettes). Francesc Baldomar's groined vaults, which were mentioned above, were the other way around: the form is Gothic, whilst the construction is Renaissance.

In addition, the conception of the forms based on outlines or sections and the construction of surfaces resting on lineal elements are points of connection between shipbuilding and cross vaults.

References

1. Almagro Vidal, Ana: "La presenza della Corona d'Aragona nel Regno de Napoli: l'influenza spagnola nella formazione del linguaggio architettonico", in Palazzo Novelli a Carinola, pp. 47–64, Kappa, Rome, 2003.
2. Alomar Esteve, Gabriel: "Guillem Sagrera y la arquitectura gótica del siglo XV", Blume-Colegio de Arquitectos de Cataluña y Baleares, Barcelona, 1970.
3. Alonso Rodríguez, Miguel Ángel and López Mozo, Ana: "Levantamiento de la cúpula de la iglesia del Monasterio de San Lorenzo del Escorial", in *IX Congreso de Expresión Gráfica Arquitectónica*, pp. 303–308, Universidad de A Coruña, A Coruña, 2002.
4. Alonso Rodríguez, Miguel Ángel: "Sobre las cúpulas de las torres de la basílica del Escorial", in El Monasterio del Escorial y la arquitectura, pp. 487–500, Instituto Escurialense de Investigaciones Artísticas e Históricas, El Escorial, 2002.

5. Calvo López, José, Alonso Rodríguez, Miguel Ángel, Rabasa Díaz, Enrique, and López Mozo, Ana: "Cantería renacentista en la catedral de Murcia", Colegio de Arquitectos de Murcia, Murcia, 2005.
6. Calvo López, José: "Jacopo Torni l'indaco vecchio and the emergence of Spanish classical stereotomy", in Teoria e Pratica del costruire: saperi, strumenti, modeli, pp. 505–516, Università di Bologna, Bologna / Fondazione Flaminia, Ravenna, 2005.
7. Calvo López, José and De Nichilo, Eliana: "Stereotomia, modelli e declinazioni locali dell'arte del costruire in pietra da taglio tra Spagna e Regno di Napoli nel XV secolo", in Teoria e Pratica del costruire: saperi, strumenti, modeli, pp. 517–526, Università di Bologna, Bologna / Fondazione Flaminia, Ravenna, 2005.
8. Choisy, Auguste: "Histoire de l'Architecture", Gauthier-Villars, Paris, 1899.
9. Cirici, Alexandre: "Arquitectura gótica catalana", Lumen, Barcelona, 1968.
10. Frézier, Amédée-François, "La théorie et la pratique de la coupe des pierres et des bois pour la construction des voutes et autres parties des bâtiments civils et militaires ou traité de stéréotomie a l'usage de l'architecture", Jean Daniel Doulsseker, Strasbourg/L. H. Guerin, Paris, 1737–1739.
11. García, Simón: "Compendio de Arquitectura y simetría de los templos", manuscript 1681. Facsimile with transcription, Colegio de Arquitectos de Castilla y León, Valladolid, 1991.
12. Garín Ortiz de Taranco, Felipe María: "Una posible escuela hispanolevantina de crucerías anervadas", in Homenaje al profesor Cayetano de Mergelina, pp. 431–439, Universidad de Murcia, Murcia, 1962.
13. Gelabert, Joseph. "De l'art de picapedrer", manuscript 1653. Facsimile, Diputación de Baleares, Palma de Mallorca, 1977.
14. Gil de Hontañón, Rodrigo. "Manuscrito de arquitectura", c. 1550; included in [11]
15. Gómez Martínez, Javier: "Ante lapidem lignum fuit. Algo más que bóvedas de crucería", in Ars Lignea. Las iglesias de madera en el País Vasco, pp. 25–45, Electa, Madrid, 1996.
16. Gómez Martínez, Javier: "El gótico español de la Edad Moderna. Bóvedas de Crucería", Universidad de Valladolid, Valladolid, 1998.
17. Guardia, Alonso de, "Manuscrito de arquitectura y cantería. c. 1600", included in a copy of Battista Pittoni, *Imprese di diversi principi, duchi, signori…*, Book II, Venecia, 1566, preserved in Madrid, Biblioteca Nacional, ER/4196.
18. L'Orme, Philibert de: "Nouvelles inventions pour bien bastir a petits frais", Federic Morel, Paris, 1561.
19. L'Orme, Philibert de: "Le premier tome de l'Architecture", Federic Morel, Paris, 1567.
20. La Rue, Jean-Baptiste de: "Traité de la coupe des pierres où par méthode facile et abrégée l'on peut aisément se perfectionner en cette science", Imprimerie Royale, Paris, 1728.
21. López Mozo, Ana: "Las cúpulas de las torres de la iglesia del Monasterio del Escorial", in El Monasterio del Escorial y la arquitectura, pp. 510–519, Instituto Escurialense de Investigaciones Artísticas e Históricas, El Escorial, 2002.
22. López Mozo, Ana: "Extradosed vaults in the Monastery of El Escorial: The domes at the church towers", in Proceedings of the First International Congress on Construction History, pp. 1321–1326, Instituto Juan de Herrera, Madrid, 2003.
23. Marías, Fernando: "El largo siglo XVI", Taurus, Madrid, 1989.
24. Nuere Matauco, Enrique: "La carpintería de armar española", 1990, reprinted by Munilla-Lería, Madrid, 2000.
25. Palacios Gonzalo, José Carlos: "La estereotomía de la esfera", Arquitectura, no. 267, pp. 54–65, 1987.

26. Palacios Gonzalo, José Carlos, "Trazas y cortes de cantería en el Renacimiento Español", 1990, reprinted by Munilla-Lería, Madrid, 2003.
27. Palacios Gonzalo, José Carlos: "Spanish ribbed vaults in the 15th and 16th centuries", in Proceedings of the First International Congress on Construction History, pp. 1547–1558, Instituto Juan de Herrera, Madrid, 2003.
28. Palacios Gonzalo, José Carlos: "The Gothic Ribbed Vault in Rodrigo Gil de Hontañón", in Proceedings of the Second International Congress on Construction History, pp. 2415–2431, Construction History Society, Cambridge, 2006.
29. Pecoraro, Ilaria: "Las bóvedas estrelladas del Salento. Una arquitectura a caballo entre la Edad Media y la Edad Moderna", in Una arquitectura gótica mediterránea, vol. II, pp. 41–65, Valencia: Generalitat, 2003.
30. Pereda, Felipe: "La arquitectura elocuente. El edificio de la Universidad de Salamanca bajo el reinado de Carlos V", Sociedad para la conmemoración de los centenarios de Carlos V y Felipe II, Madrid, 2000.
31. Pérouse de Montclos, Jean-Marie: "L'Architecture a la française", Picard, Paris, 1982, reprinted 2001.
32. Pérouse de Montclos, Jean-Marie: "La vis de Saint-Gilles et l'escalier suspendu dans l'architecture française du XVIe siècle", in L'escalier dans l'architecture de la Renaissance. Actes du colloque tenu à Tours du 22 au 26 mai 1979, pp. 83–92, Picard, París, 1985.
33. Pinto Puerto, Francisco: "Transformaciones. De la línea a la superficie", in Actas del Tercer Congreso Nacional de Historia de la Construcción, pp. 815–826, Instituto Juan de Herrera, Madrid, 2000.
34. Pinto Puerto, Francisco: "Las esferas de piedra. Sevilla como lugar de encuentro entre arte y ciencia en el Renacimiento", Diputación de Sevilla, Sevilla, 2002.
35. Pleguezuelo Hernández, Alfonso: "La Lonja de Mercaderes de Sevilla: de los proyectos a la ejecución", Archivo Español de Arte, no. 249, pp. 13–41, 1990.
36. Rabasa Díaz, Enrique: "Técnicas góticas y renacentistas en el trazado y la talla de las bóvedas de crucería españolas del siglo XVI", pp. 423–433, in Actas del Primer Congreso Nacional de Historia de la Construcción, Instituto Juan de Herrera, Madrid, 1996.
37. Rabasa Díaz, Enrique: "Forma y construcción en piedra. De la cantería medieval a la estereotomía del siglo XIX", Akal, Madrid, 2000.
38. Rabasa Díaz, Enrique: "La transmisión y el aprendizaje de la estereotomía de la piedra", Ars Sacra, no. 22, pp. 92–101, 2002.
39. Rabasa Díaz, Enrique: "The single coursed ashlar vault", in Proceedings of the First International Congress on Construction History, pp. 1679–1689, Instituto Juan de Herrera, Madrid, 2003.
40. Rodríguez Gutierréz de Ceballos, Alfonso: "En torno a Felipe II y la arquitectura", in Real Monasterio-Palacio de El Escorial. Estudios inéditos en conmemoración del IV centenario de la terminación de las obras, pp. 107–125, CSIC, Madrid, 1987.
41. Rokiski Lázaro, María Luz: "La cabecera de la iglesia de Priego (Cuenca). Dibujos y tasación", Cuenca, no. 17, pp. 27–34, 1980.
42. Rovira y Rabassa, Antonio: "Estereotomía de la piedra", Barcelona, 1897.
43. Rovira y Rabassa, Antonio: "Estereotomía de la piedra. Segunda parte que comprende las bóvedas compuestas, bóvedas góticas y escaleras", Barcelona, 1899.
44. Ruiz el Joven, Hernán: "Libro de Arquitectura", manuscript c. 1550. Facsimile reprint with transcription, Fundación Sevillana de Electricidad, Sevilla, 1998.
45. Sanabria, Sergio Luis: "The evolution and late transformations of the Gothic mensuration system", unpublished Ph.D. dissertation, Princeton University, 1984.
46. Sanjurjo, Alberto: "The Vis Saint Gilles Quarrée or the Caracol de Emperadores Cuadrado. A Model Frequently Encountered in Spanish-French Architectural Treatises from the Modern Period", in Proceedings of the Second International

Congress on Construction History, pp. 2793–2814, Construction History Society, Cambridge, 2006.
47. Santana, Alberto (coord.): "Ars Lignea. Las iglesias de madera en el País Vasco", Electa, Madrid, 1996.
48. Serra Desfils, Amadeo: "É cosa catalana: la Gran Sala de Castel Nuovo en el contexto mediterráneo", Annali di Architettura, no. 12, pp. 7–16, 2000.
49. Sobrino González, Miguel: "Barcelona. Las razones de una catedral singular", Goya, no. 307–308, pp. 197–214, 2006.
50. Tamboréro, Luc: "The 'Vis Saint-Gilles', Symbol of Compromise between Practice and Science", in Proceedings of the Second International Congress on Construction History, pp. 3025–3039, Construction History Society, Cambridge, 2006.
51. Tomlow, Jos: "Versuch einer (zeichnerischen) Rekonstruktion des Gewölbes in spätgotischen Kreuzgang des Klosters Hirsau", in Hirsau St. Peter und Paul 1091–1991, Teil I, Zur Archäologie und Kunstgeschichte, Forschungen und Berichte der Archäologie des Mittelalters in Baden- Württemberg Band 10/1, Stuttgart, 1991.
52. Vandelvira, Alonso de. Libro de trazas de cortes de piedras, manuscript c. 1580, Madrid, Biblioteca de la Escuela de Arquitectura. Facsimile reprint with transcription by Geneviève Barbé-Coquelin de Lisle, "Tratado de arquitectura", Caja Provincial de Ahorros, Albacete, 1977.
53. Willis, Robert: "On the construction of the vaults of the Middle Ages", Transactions of the Institute of British Architects, 1842, pp. 1–69.
54. Zaragozá Catalán, Arturo: "El arte del corte de piedras en la arquitectura valenciana del cuatrocientos. Francesch Baldomar y el inicio de la esterotomía moderna", in Primer Congreso de Historia del Arte Valenciano, pp. 97–105, Generalitat Valenciana, Valencia, 1992.
55. Zaragozá Catalán, Arturo: "Naves de arcos diafragma y techumbre de madera en la arquitectura civil valenciana", in Actas del Primer Congreso Nacional de Historia de la Construcción, pp. 551–555, Instituto Juan de Herrera, Madrid, 1996.
56. Zaragozá Catalán, Arturo: "La Capilla Real del antiguo Monasterio de Predicadores de Valencia", in La Capella Reial d'Alfons el Magnànim de l'antic monestir de predicadors de València, vol. 1, pp. 14–59, Conselleria de Cultura, Valencia, 1997.
57. Zaragozá Catalán, Arturo: "Arquitectura gótica valenciana", Generalitat Valenciana, Valencia, 2000.
58. Zaragozá Catalán, Arturo: "Arquitecturas del gótico mediterráneo", in Una arquitectura gótica mediterránea, vol. I, pp. 105–191, Generalitat Valenciana, Valencia, 2003.

CHAPTER NINE

ON LATE-GOTHIC VAULT GEOMETRY

Jos Tomlow

Late-Gothic architecture is rich by its variety in shape and constructive backgrounds. The English heritage is quite distant from the examples found in France or those from Germany and surroundings or those from the Iberian Peninsula. Yet, due to international contacts in detail mutual influences may be recognised, like English influences in Portugal, or French in Spain.

The specific kind of vault geometry in the Late Gothic period, to be presented here, has certain characteristics, which makes it ideal for a rational analysis and description. Throughout central Europe we may find examples with varying patterns, called "reticulated vaulting" (German: *Netzgewölbe, Sterngewölbe*). Early examples of this development of the ordinary Gothic cross ribbed vault, can be found around 1385, in the St. Vitus Cathedral in Prague, which was at that time renovated by *Peter Parler*. The summit of development of the *Netzgewölbe* are the *Schlingrippengewölbe*, which show a rib system of spatial loops and, projected on the floor plan, a circular vault pattern. *Benedikt Ried* built, around 1500, in the Prague castle some of the most remarkable loop rib vaults.[1]

As subject for case studies on the *Netzgewölbe* figure the cloisters of two monasteries, *Kloster Hirsau* and *Kloster Alpirsbach* in the German Black Forest.[2] Both monasteries were initially built in the Romanesque

[1] Muk [8].

[2] Both surveys were done by the author on behalf of the German preservation authority *Landesamt für Denkmalpflege Baden-Württemberg, Aussenstelle Karlsruhe*. The work, aiming mainly at the (theoretical) reconstruction of lost vaults, was done in a research team in preparation of main publications on the two Monasteries (Schreiner [13] and Landesdenkmalamt BW [6]). Many insights, plans and data were due to team members, mainly O. Teschauer, A. Seeliger-Zeiss, U. Knapp (measured drawings of Alpirsbach cloister) and R. Hajdu (photographic documentation of the rib cross elements in Hirsau). Of special importance for understanding the geometric construction of the *Prinzipalbogen* were personal contacts with Werner Müller (+ 2005). Complete results in Tomlow [14] and Tomlow [16].

style. *Kloster Hirsau*, may be regarded as of major importance, both because of its big dimensions, as well as since its name stands for an own building style (*Hirsauer Bauschule*), also applied on the Alpirsbach parts of Romanesque times. The *Netzgewölbe* of the monastery cloister in Hirsau date from 1482–1495[3] and that of Alpirsbach from 1480-approx. 1495.[4] Both cloisters are typical examples of a renewing phase of the Romanesque monasteries, with the aim to substitute the formerly open and windy cloisters by a glassed and vaulted cloister. A certain degree of economic and political stability favoured this process, in which also church building activity in southern Germany increased. (For details, see the pictorial evidences at the end of the chapter.)

From a design theoretical point of view, one could define the Gothic style as one of "intelligent adaptation"[5]. Whereas in Romanesque times, vaults were bound to fit in a rigid grid system, with, as the single element, a square vault plan, the Gothic building master could choose the proportion between the width and length of a vault part. This choice option may lead to intelligent use, when the proportions of vault parts were chosen in such a way, that static equilibrium was established between neighbouring vaults of different span.

Why was the Romanesque vault bound to a square plan? The reason was that the semi circle Roman arch had a fixed proportion of height compared to span in a ratio 1:2. If the radius of one of the arches would be enlarged, instantly problems would occur to define the vault's shape. The ogival Gothic arch—of Arabic source—had the possibility to vary both span and height. The same was possible for the crossing rib.[6] So we have a maximum of freedom for a shape, ready to be used for any kind of purpose. More important was that any of the possible shapes was strictly defined by rules of geometry.

In Late-Gothic times (15th C.) only a reduced number of sciences could be judged to be "exact". Most writings were trivial, speculative,

[3] Schreiner [13].

[4] Landesdenkmalamt BW [6], p. 396.

[5] One has to remember here that the word "Gothic" as well as "stile tedesco" (German style) stand for a barbaric attitude, with clearly negative meaning, as compared to the Italian classic antique tradition, with its strong symmetrical features. However, exactly the freedom in design favoured the structural inventions of the Gothic style.

[6] Already in Roman and Romanesque times the ellipse was part of the shapes of a groin vault. It is a vault produced by the intersection at right angles of two barrel vaults. The diagonal curves are ellipses. One may observe that the ellipse as a geometric shape was not necessarily known by the builders, since it results from the Roman arch shape of the barrel vault during building, resulting in a logical connection.

literary, philosophical or poetic, in which the writer tried to encircle a problem with—in the best case—logical analysis. On the other hand knowledge was based on practical experience, but at least the wise holders of this knowledge lacked mostly a coherent, theoretical hermetic language. Only one exact knowledge existed in all its inherent consequence and this was ready to be applied to practice. This knowledge was "Geometry".

Principal arch (*Prinzipalbogen*)

"How can a rib shape be defined in geometric terms?" A rib in cross section has a symmetric profile, which along the building element stays constant. Since the profile remains of constant shape along the rib, such a single line generating the full rib, exists from a geometrical point of view. "Where to put the single line, the "principal arch" or *Prinzipalbogen*—as it was called by art historians—in connection to the material rib?" The *principal arch* was put on the symmetry axis of a rib section and in the middle of the rib bottom.[7] This choice is evident, because the master builder generates with such a line two strongly related building parts: on the one hand the rib shape and on the other hand the top line of the scaffolding, giving shape to the rib.[8]

The second, more static relevant question is: "What curve is optimal for a vault, a spatial one or a simpler one, which stays within a vertical plane?" If we take as a working hypothesis, that Viollet-le-Duc[9] with his rational approach was quite near the truth, or at least near the opinion of the master builders, when he concluded that the ribs carry the vault membranes, like a skeleton, we have to put the ribs in such position to each other that they find static equilibrium. From vector geometry we know that there is no reason to put the ribs, as projected in plan,

[7] Most rib profiles from Late-Gothic origin, have a lower end which is flat and rather narrow. The ratio between the width of the horizontal bottom compared to the rib width is for example 3 to 12 cm. Sometimes a carved line can be observed, marking in the bottom surface of the rib the symmetry axis (identical to the *principal arch*). The flat surface of the rib is also relevant to the scaffolding type, which was rather narrow in the shape of a curved plank of few cm thickness. The ribs were eventually secured in their position on the scaffolding by more planks, on both sides shaping a ridge.

[8] The scaffolding is necessary for carrying the single rib elements before finishing the vault membranes in between.

[9] Influential for a rational interpretation of the Gothic architecture are Viollet-le-Duc [18] and Ungewitter/Mohrmann [17]. This approach was corrected by Abraham [1].

other than in a straight way.[10] Since the gravity force is always pointing vertically downward, every rib compared to others finds an equal condition, independent from the direction they spring from. Thus we conclude that the Gothic rib, to be described here, projected in plan, is a single straight line, which together with the other rib parts, define a net-like rib pattern of straight lines.

For equilibrium reasons the lines always show knots by a connection with at least two other ribs or—alternatively—the rib starts from a springer at the wall. So the rib curve is now reduced to a line of a single curve which fits in a vertical plane.

More difficult should be the question how the master builders defined the rib curve itself. From static science we know since some time, but much later than the Gothic period, that ideal thrust lines in vaults are depending on the weight distribution. For an arch of similar big blocks this would be the line of a reversed catenary, which is similar but not equal to an ogival arch.[11] But in Gothic times such a complex shape as a catenary could hardly be measured with a satisfactory precision or transformed from one scale into another one.[12] The same may be concluded for other rather complex shapes like ellipse, parabola or hyperbola.[13] Another line, as produced by a thin wooden cane when bent a little, has been reported to be applied in the design of climbing arches, which

[10] Please note that Late Gothic vaults also include patterns by curved lines projected in plan. Examples are both in Portugal, England and in Mid-Europe. Such are for example *Schlingrippengewölbe*, as mentioned before. These should be seen as exceptional cases and they are probably only possible because the vault spans are kept reduced. In the case of the Vladislav's Hall at Prague castle, built by Benedikt Ried with a 16 m span, iron anchors secure the odd vault shape. From a statical point of view the Portuguese example of the Church of the Monastery of the Hieronymites Belém, near Lisbon, might give interesting relationships between the curved rib pattern, the mushroom shaped transition from pillar to the vault, and its statical design concept.

[11] The "catenary" is a curve shape, experimentally defined by a free hanging chain. In reversed way this line, as well as the related "hanging model", is relevant for optimizing statics of stone structures. An introduction to this design tool, developed from approximately 1700 onward, cf. Tomlow [15].

[12] The author rejects the idea that in Gothic times, static theory was developed in such way, that the hanging model could already be used for finding static equilibrium of cathedrals. This opinion is supported by the observation that hanging model shape characteristics differ from Gothic churches. Occasional speculations by colleagues in this direction lack any clear historic proof.

[13] Please let me abstain here from artistic or genius inspired lines, which lack any definition. Anybody familiar with similar historical building methods that are more developed than clay and rubble vaults, will agree that the vault's shape should be definable within very narrow borders and its correct reproduction from the concept or design.

carry stairs (Spain, 19th C.). Such a line may be sufficient for producing a natural harmonic shape. However, it is clear, from profound evidence in numerous case studies, that all ribs of *Netzgewölbe* as discussed here have a curve which is defined by a circle—or better—a circle segment. In other words, whereas many shapes may show a change of curvature, the rib parts discussed here are defined by a constant radius.[14]

As already analysed before, the circle shape was generally applied from early Gothic times in an intelligent way, thus that any proportions between arches are possible. Expressed in a basic, almost archaic way, a quarter of a circle is predestined to shape the intrados of a vault, since the thrust line always has to find a deeper position between the vault's top and the wall that carries the vault. The inclination of a circle segment is continuously growing from the vertical, defined by the wall till the horizontal, defined by the vault's crest. Any position on the circle's path can be reproduced by geometric construction with a pair of compasses, together with a crossing vertical line.

The next question is: "How can a rib pattern be of such order that all rib parts meet in space with enough precision?" For this problem the master builders invented an own method, which ensured that the rib course always will have an inclined downward tangent, starting from the vault's crest. Individual rib parts in a net, were put in straight order in plan, starting from the crest to the springer. Whereas in the materialised *Netzgewölbe* such parts have pronounced angles in plan, the master builder simply designed the rib parts as if they were in a straight row behind each other.[15]

One of the difficulties to understand the method of the *Prinzipalbogen* is our common view on architectural plans and sections.[16] A section

[14] This result of the initial question about the rib shape may be quite logical from an historical point of view, since also the Gothic ogival arch, is a product of the circle, like the Roman arch, from which it was developed.

[15] For a less abstract description please think of a Japanese folding screen. The screen can stand for its own by folding the parts to a certain degree. If laid on the floor the screen will be unfolded, and reaches its total length. If one paints a circle segment in inclined position continuously on three or two screen parts, disregarding the borders, this shape will be distorted when again put upright in a folded position of the screen parts. Nevertheless, following the circle line on the folded screen from top to bottom always will give a tendency downward. This logic system was understood by the master builders and applied to the ribs of *Netzgewölbe*.

[16] The method of representing any object by projecting it in at least two planes has been defined by the Frenchman Gaspard Monge (Descriptive Geometry) around 1800. Many aspects of this method have been applied before, generally with much lesser precision.

drawing is defined by a projection of all visible parts on one vertical plane. This means that, any *Netzgewölbe* will show in a section drawing shortened views on rib parts which have a non-parallel course related to the section plane. In consequence these rib parts are represented as an elliptic segment in the drawing. This is correct in the drawing, but should not be mistaken as a representation of the rib's form along its own plane.

"How many types of the *Principal arch* may exist for a Late Gothic vault?" In a drawing scheme we may recognise some (theoretically) possible, quite differing shapes, that can be generated by changing certain parameters. The proposed plan is a very simple example from the Late-Gothic period.

- The principal arch is a quarter of a circle. In the springer the tangent line is vertical and in the crest the tangent is horizontal. This is quite uncommon.
- Only the lower part of a quarter circle segment is used, to define the two rib parts.
- Only the higher part of a circle quarter segment is used, to define the two rib parts. Please note that in this way the crest's position may be defined as fixed, with as a result that the springer position may vary, in relation to the radius magnitude. This shape deduction is named the "falling arch shape".
- Both arch ends have an inclined tangent.
- The two rib parts have two different radii, but the circles are positioned in such way, that no sharp bend may be seen (compound curve).
- The two rib parts have an equal radius, but differing mid point positions. The result is a pronounced sharp bend in the meeting of the rib parts. The arch shape is still concave (statically correct).
- The two rib parts have an equal radius, but differing mid point positions. The result is a pronounced sharp bend in the meeting of the rib parts. The arch shape this time is convex (statically incorrect).

What characteristics define a Late Gothic vault? One may name five features.[17]

[17] Please compare to a normal Gothic cross ribbed vault: 1.) The rib pattern is a simple diagonal cross, bordered by transverse ribs and the wall. 2.) The rib curve on the springer shows a vertical tangent. 3.) The vault membrane intrados follows parallel

- The rib pattern is shaped as a continuous web.
- The rib curve on the springer shows mostly an inclined tangent.
- The connection line of the vault membrane to the bearing rib is essentially independent of the course of the *principal arch* (rib shape). If necessary the rib may be shaped at the springer like a vertical small wall before it reaches the vault membrane. This implies that the springer needs no filling at its back.
- The vault itself is mostly half-a-brick thick (approx. 8–15 cm) in the case of small vaults.
- The starting point of the ribs in different radial positions in plan tend to be at different heights at the springer.

An interesting result of this type of characteristics is that the vault membrane itself can be concentrated in a horizontal zone of reduced height, near the crest. This feature—which makes the Late Gothic vault rather lightweight—is a rare interesting aspect from a static point of view. In the same way one can observe that the air volume under such a vault is bigger compared to a vaulted space of similar dimensions from earlier times.

"Why has the vault pattern such a great variety?" The question carries the answer. On the one hand it is possible to vary infinitely[18] the vault patterns, because of the innovative design system, which allows many changes without danger to static soundness. On the other hand the late medieval period shows tendencies of decline, a kind of baroque taste, playful but in the same time irrational.[19] So, the answer to the question is: "They built in variation because they were able to do so."

From a design historic approach such a rather trivial source of Late Gothic vaults does not mean that the methodical question, "how" they realised it, would not be of interest.

In the *Alpirsbach* case this question was dealt with in a schematic way, by analysing how the vault patterns could be understood as such and how they met with other patterns, either in the corners or in the aisles, following after a corner vault. As one can see, individual bays

---

the rib profile of constant height. 4.) The vault is of brick or stone. 5.) All ribs spring on the level, mostly defined by a common capital.

[18] I estimate about five hundred rib patterns in literature and executed vaults.

[19] The emerging of protest in the Christian world against the pope and his clergy, ending 1517 with the ban of Martin Luther marks the division point between reformation and the Roman catholic church, on the one hand and the end of Late-Gothic practice on the other hand. This may be culturally connected.

of the four cloister wings are not square, they are transverse oblong (3 : 4). The question why this happened may be answered in a plausible way. The reduced dimensions of *Alpirsbach* cloister would imply a division of the aisle in about 5 or 6 square bays between the corner vaults. To our observation the transverse oblong shape has been chosen in order to avoid a dull division with such a reduced number of bays that can be counted instantly by an observer. In contrast, the actual division with 7 and 8 bays, are suggestive for a rather long depth. A similar situation one can appreciate in Royaumont Monastery in France.

"How exact are Late-Gothic vaults in their dimensions". Rib crossings have a high degree of precision. From my research experience I derived that a worked block should be 2 cm exact, in order to fit well in a rib pattern of 3 or 4 m width. On the other hand the Late Gothic vaulting is very flexible in design and it can adapt to various special circumstances. For instance in *Hirsau* cloister we can observe three aisles with specific dimensions in plan, which we discuss starting with a regular plan.

Hirsau cloister, East wing. Since this wing was the first to be newly built in Late Gothic style the designer did not find restrictions for a constant subdivision. As a result all thirteen bays are exactly square with a depth and width of 360 cm.

Hirsau cloister, South wing. A division of squares of 360 cm did not fit in the total length of the South wing. Instead one found two times 360 cm for the square corner vaults (South-East and South-West) and nine bays with 340 cm depth and one bay—in front of the fountain house—with an enlarged depth of about 560 cm. In order to keep the normal bay width of 360 cm, the master builder seemed to be forced to make the bays rectangular instead of the preferred square shape. With a masterly Late-Gothic design trick, the square shape was kept intact, by adding to the South springers a console, on which the springing point was put before the wall on a 10 cm distance. In this way the South bay patterns measure 340 cm to 340 cm and thus have a square plan condition. The South bay rib pattern shows three capstones for every bay. Since the plan in front of the fountain house has a depth of 5/3 in comparison to a normal South bay, the pattern of capstones may continue throughout the South wing.

Hirsau cloister, North wing. In the North wing the plan is "disturbed" by the transverse ship of the (Romanesque) church, which cuts into the cloister. The possible variety of bay dimensions in such a situation is quite reduced. In this case the master builder chose a long depth, which caused a rectangular bay in the East part of the North wing. The vaults following these have specific dimensions and six vaults span the distance toward the West wing. In this case the depth dimension of these vaults decreased regularly, probably aiming at an almost square vault in the connection to the West wing.[20]

A special vault, unique of configuration, could be reconstructed for the South wing of Hirsau cloister (fig. 9.25). As one may observe from the cloister plan and the detailed drawing, the springer conditions differ on the court (inner) side as compared to the wall on the outer side. Whereas the inner side shows two half springers (marked A) for a bay, from which start transverse ribs, the outer side has two springers on one third point (marked E).[21] As a hypothesis I assume that the master builder tried to avoid point load distribution on the long outer wall of the refectorium, void of buttresses, by doubling the springers.[22]

[20] Similar changes in bay depths may be observed at the East wing in Alpirsbach cloister. In the first bays on the North fixed dimensions between the vault springers occur, because they had to be integrated with existing windows and a door from Romanesque times. The available length until the South-East corner bay forced the designer to choose different depth dimensions of resulting four bays, in an increasing rhythm. Also in the Alpirsbach South wing bay depths differ, because of the fountain house, which with its complex piping could not easily be restructured.

[21] The letters C and F in fig. 9.25 mark an additional springer above the window crests, which functions as a lateral support of the rib configuration.

[22] Further details in Tomlow [14].

Fig. 9.1: Alpirsbach, cloister. East wing, adjacent to the North-East corner vault, with an asymmetric rib pattern.

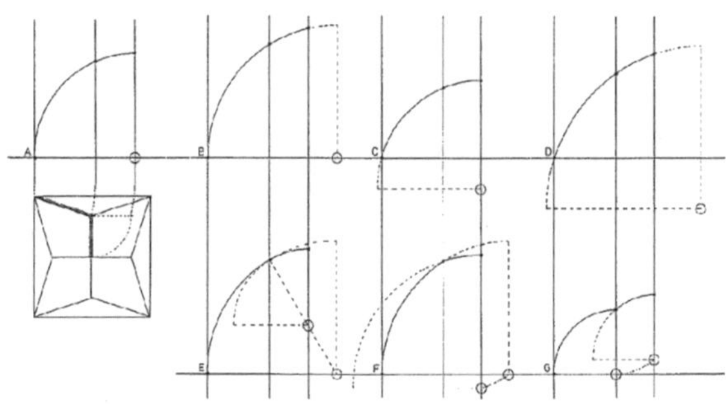

Fig. 9.2: Principal Arch (*Prinzipalbogen*). In order to construct geometrically the rib curve by the "principal arch method", a horizontal plan of a bay, with some rib pattern is connected to a curve, defined by circle segments, in a vertical plane. The rib pattern is such that two types of rib parts are connected in a Y shape. The two parts of one rising rib, starting in the springer (corner) to the crest (centre), are represented on drawing A in such way that the rib length as projected in plan are put in real length behind each other. Image A represents the rib curve (principal arch) in side view with real dimensions. Drawing B till E offer theoretically possible variations of the rib's form generated by different geometric parameters.

g. 9.3a,b,c: Alpirsbach, cloister. North-East corner vault section with an irregular rib pattern. .e door of Romanesque origin leads to the church, which has an elevated level. Thus the thic doorframe cuts into the space which was designated for the vault. In a witty gesture of e master builder some rib connections have been loosed, and these rib parts were relocated o a higher position, in favour of a complete profile of the door frame. Please observe that doorframe and wall rib share the same stone blocks. Drawing Jan Fallgatter.

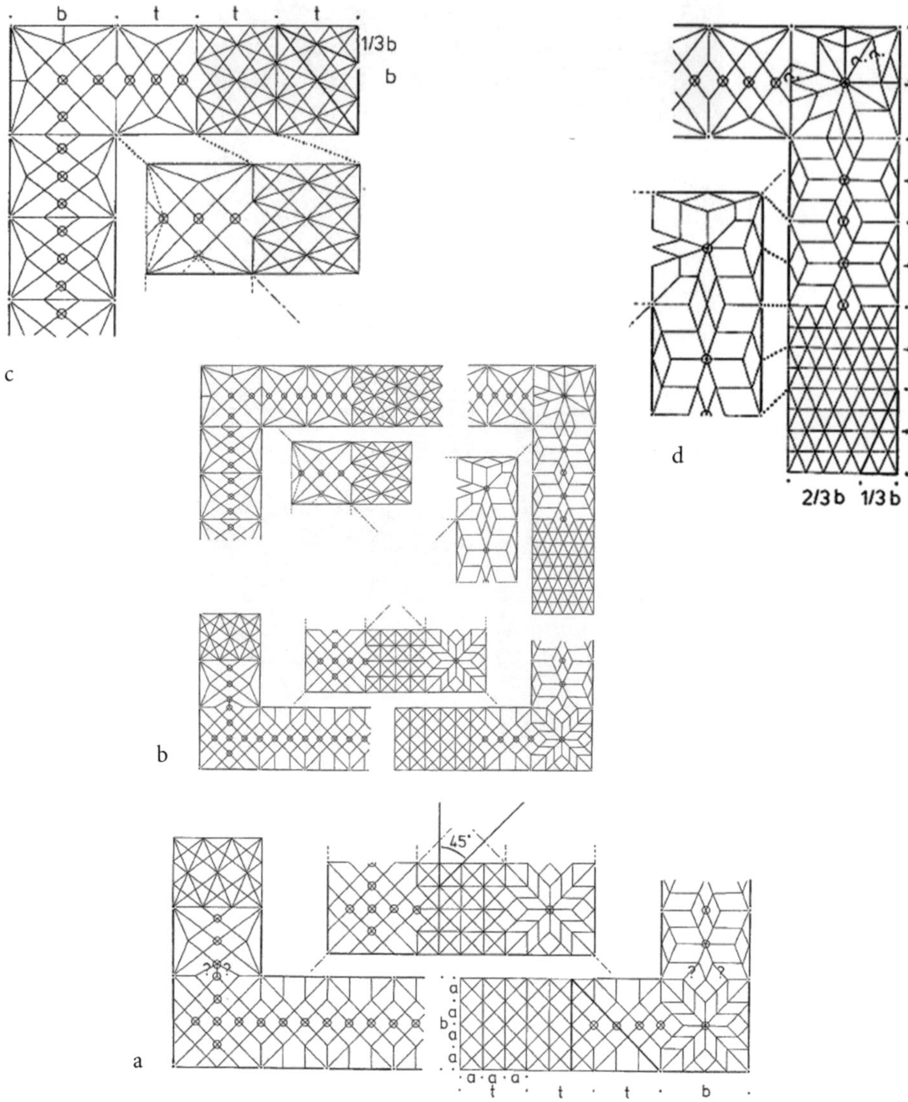

Fig. 9.4: Alpirsbach, cloister. Vault system schemes with deduction of various rib patterns from standard types. Of special interest are:

a) The generation of corner vault-sections—generally square—out of two different cloister wing vault patterns.
b) Transformation of square shape into a transverse oblong vault section (ratio 3:4) in the wing bays.
c) In the South wing a quadratic grid defines the rib positions either under a 45° angle or along the grid lines. In a square bay, the 45° rib, starting in a springer would end in the springer in diagonal position. Allthough the bays, with their 3:4 ratio, are not square, the grid assures a regular and symmetrical rib pattern.
d) Question marks (?) point out a problematic course of a rib section.

Fig. 9.5: Alpirsbach, cloister. South-East corner vault section, with regular rib pattern based on a square grid with 45° diagonals.

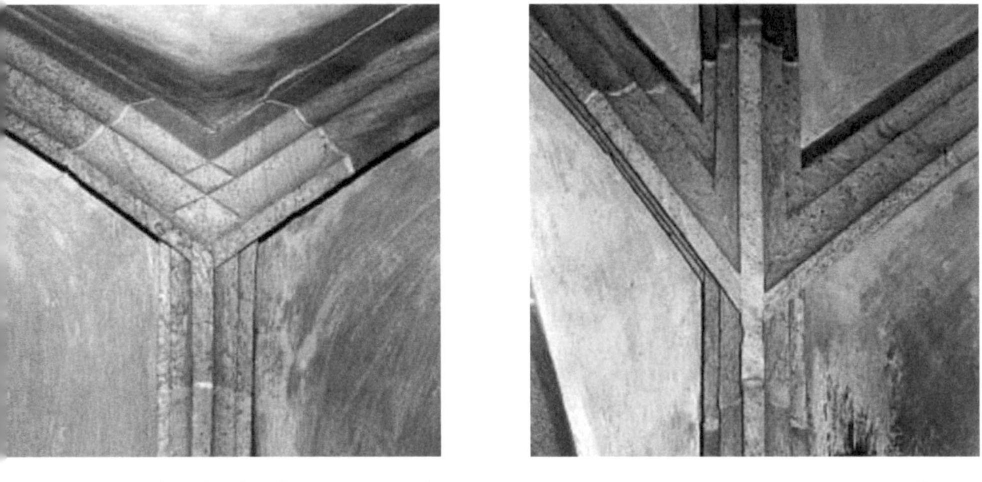

Fig. 9.6a,b: Alpirsbach, cloister. Imperfect connections—in geometric view—, stating low teaching of the responsible building master or stone mason.

Fig. 9.7a,b: Alpirsbach, cloister, South-East corner and South wing. Vault springers are parts of the vault, which are built fully integrated into the wall fabric. For this, they show horizontal joints. The first individual rib element can be recognised by radial joints, normal to the rib curve's tangent. In this case the springers are remarkable because of heavy corrections, due to design changes. On figure b we can see that the wall rib shows two hollow mouldings, whereas the diagonal rib has only one hollow moulding, due to design changes. The stone mason tried to disguise the differences, by re-working the connections.

Fig. 9.8: Alpirsbach, cloister. South wing with 45° diagonal ribs generating a regular looking pattern within bays with depth and width in 3:4 ratio.

Fig. 9.9: Alpirsbach, cloister. North wing / North-East corner bay. The springer marks the connection between the early design phase and a later one. Because yet the later built vault pattern on the left side was not designed finally, the diagonal ribs of the springer copy the shape of the right vault. In the end the later built section on the left had a complete different design. Result: the connection of the rib to the springer shows a pronounced angle and a further rib connection simply was improvised, ending in the bottom part of the vault. The hypothesis is that the vault springer was executed in an early phase, whereas the left vault followed much later.

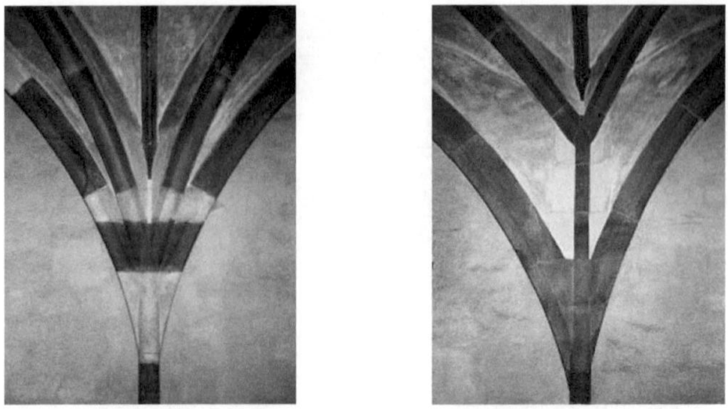

Fig. 9.10a,b: Alpirsbach, cloister. Vault springer, typical for the North wing, left and similar to the West wing (relict in North-West corner). Both vault springers have an identical rib pattern, as projected in plan. This may illustrate the big variation in late-gothic vaults' design at the disposal of master builders.

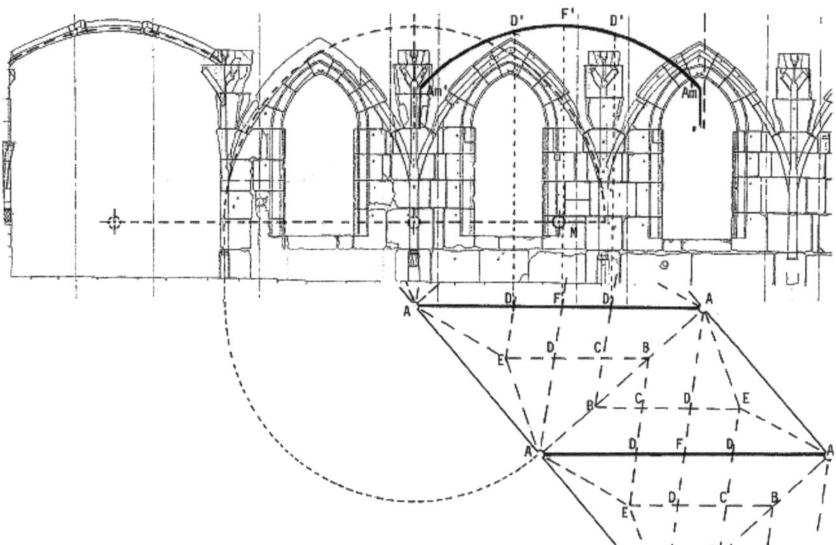

Fig. 9.11a: Alpirsbach, cloister. West wing reconstruction by J. Tomlow, basing on a measured drawing by U. Knapp. Geometric construction of the cross rib shape.

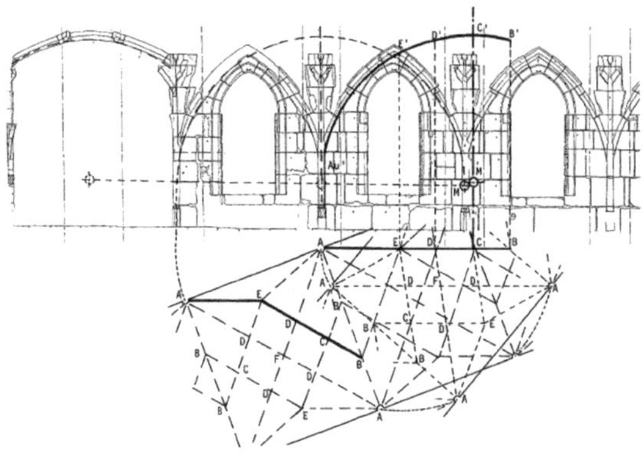

Fig. 9.11b: Shape of rib course Au'-E'-D'-C'-B'.

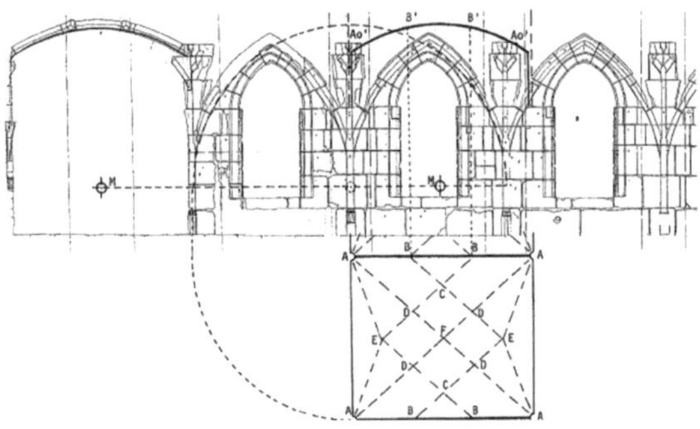

Fig. 9.11c: Transverse rib shape.

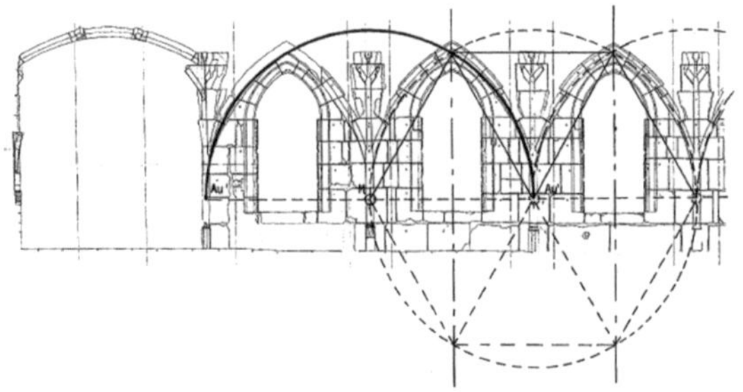

Fig. 9.11d: Wall rib shape.

Fig. 9.12: Alpirsbach, cloister. Interpretation of the West wing vault.

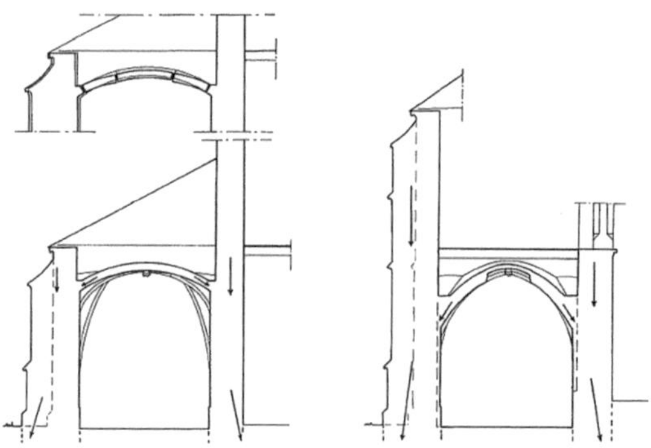

Fig. 9.13: Alpirsbach, cloister. Static interpretation. West wing left with high positioned transverse rib, compared to normal vault of North wing to the right.

Fig. 9.14: Hirsau, cloister, as it appeared during the reconstruction survey, 1986–1988.

Fig. 9.15: Hirsau, inner court wall with Late-Gothic buttress system.

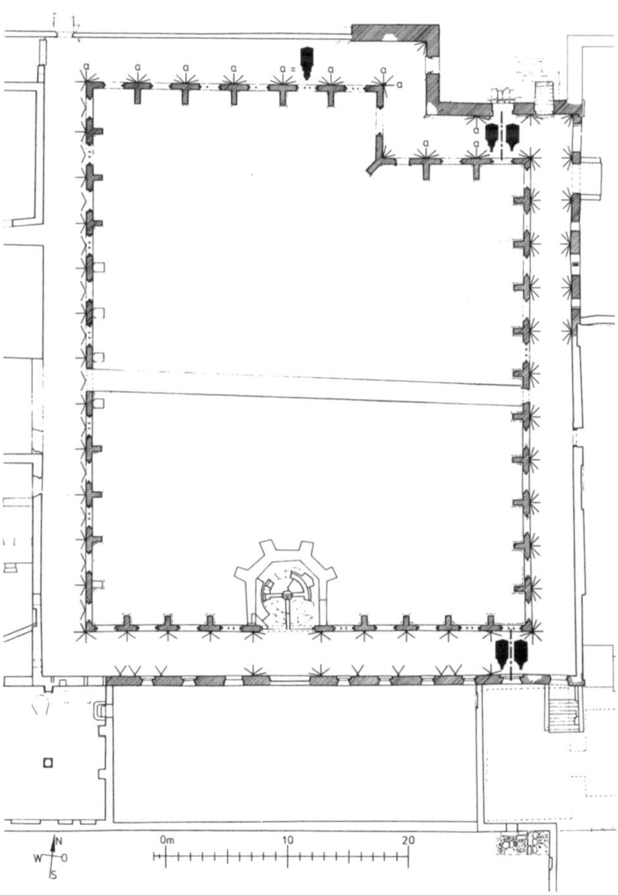

Fig. 9.16: Hirsau, cloister. Documentation drawing, marking springers in Hirsau Monastery ruin as well as the three rib profile types: single moulded (East wing), double moulded and pear-shaped (North wing a = transverse rib).

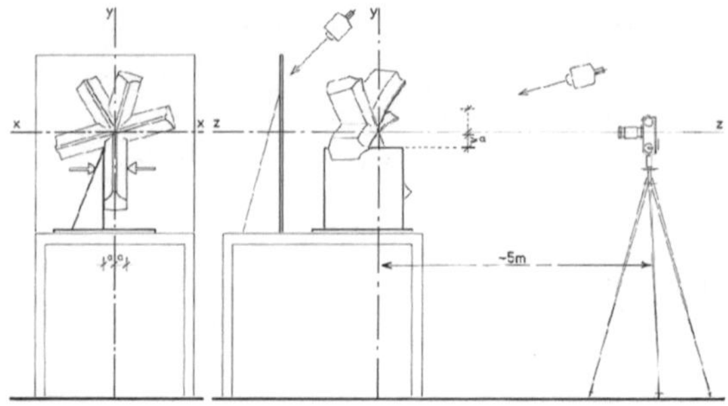

Fig. 9.17: Method of photographic documentation of individual rib crossings. By installing the stone block as indicated in the drawing, its geometrical structure—as projected in plan—is immediately observable.

Fig. 9.18a,b: Hirsau, cloister. Springers in West wing, left, and in the East wing, right.

ig. 9.19a,b: Hirsau, cloister. Springer on North-East corner, left, and in the South wing, in which the crossing point has been transferred to a position in front of the wall plane.

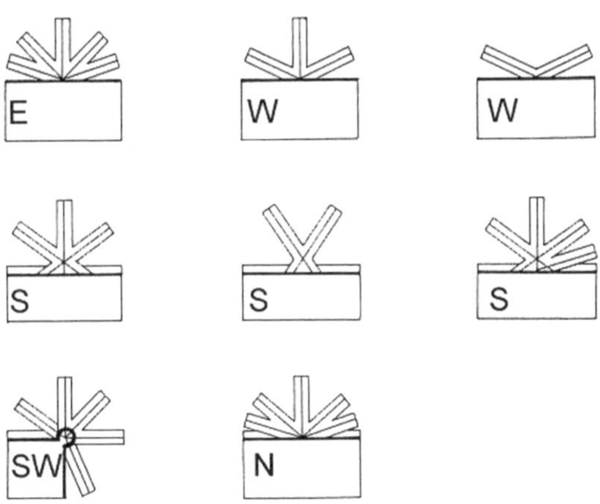

Fig. 9.20: Hirsau, cloister. Documentation drawing, representing schematically the shape of individual springers. Whereas normally the meeting point of the ribs is in the wall plane, in the South wing (S) the meeting point shows a distance of approx. 10 cm in front of the wall.

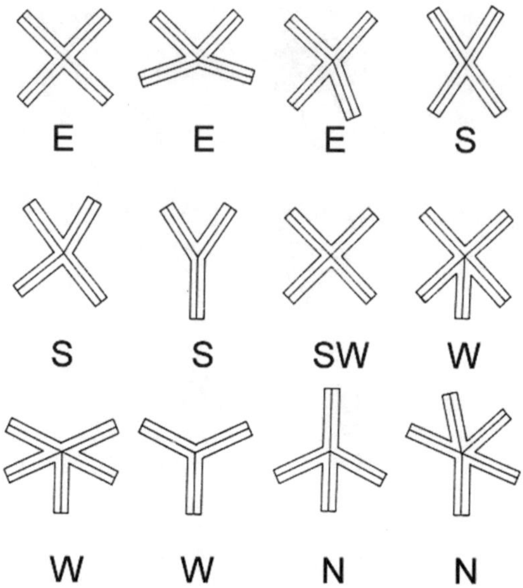

Fig. 9.21: Hirsau, cloister. Documentation drawing, representing schematically the shape of individual rib crossings as found in the remained parts of the ruined monastery (spolia). Their shape is typical for certain rib patterns. The final reconstruction of the vault types was based on data in a statistic relevant quantum, derived from a profusal documentation of rib parts and springers and by research in special literature.

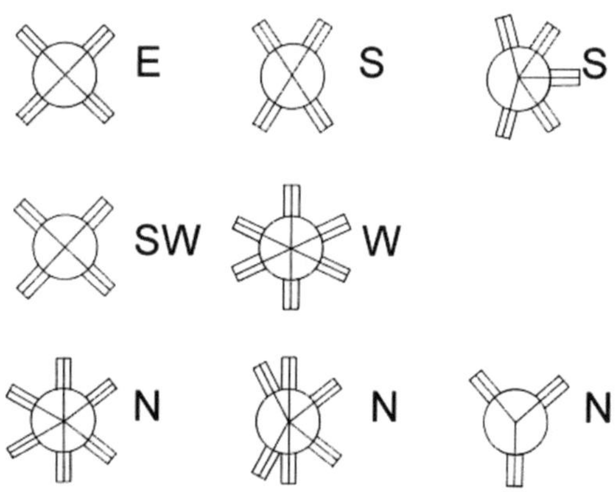

Fig. 9.22: Hirsau, cloister. Geometrical schemes of capstones being assigned to a specific wing orientation.

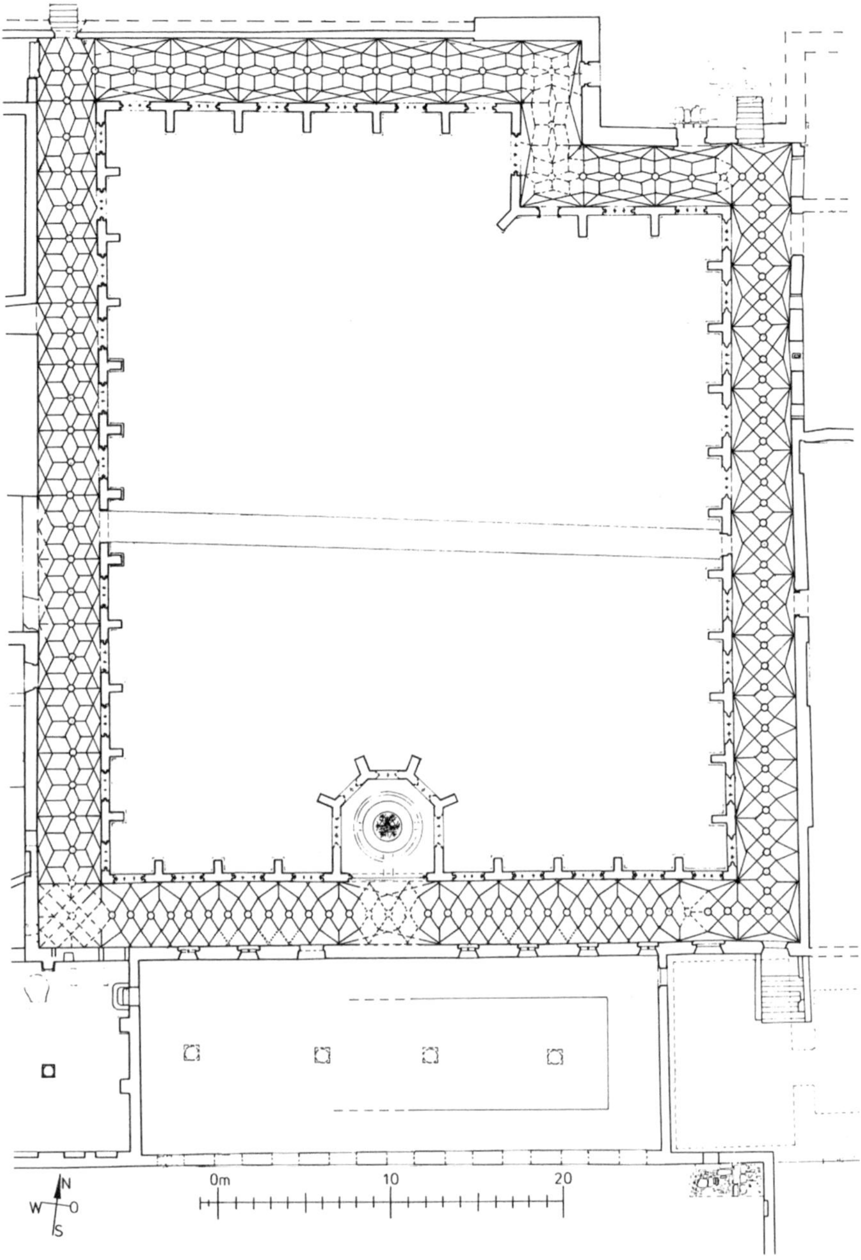

Fig. 9.23: Hirsau, cloister reconstruction. Hypothetic pattern of ribbed vault. Dotted areas are uncertain. Note different depth of North wing, inspired by the odd dimensions of the adjacent Romanesque church.

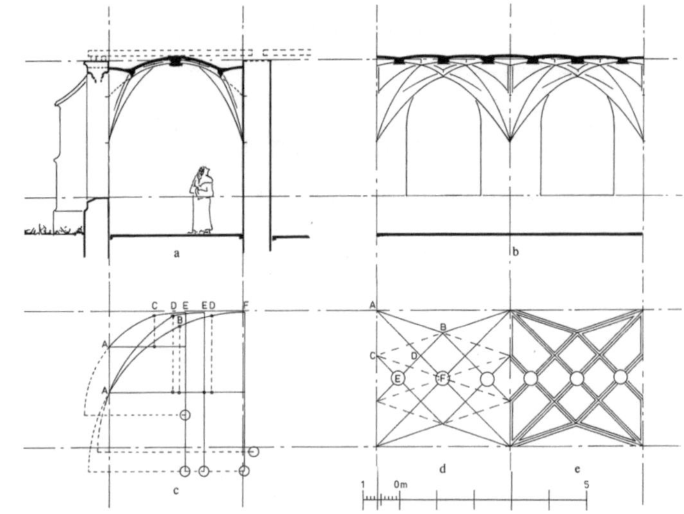

Fig. 9.24: Hirsau, cloister. Reconstruction of East wing vault, based on a on a three dimensional module (360 cm). Plan d shows the basic grid, consisting of space diagonals with the position of the rib crossings, including springer A and capstones E and F. Image c with circle segments shows the principal arches of the ribs, as derived from measured geometric data from the springers and from existent rib crossings (spolia) with fitting characteristics. The vault pattern is very common.

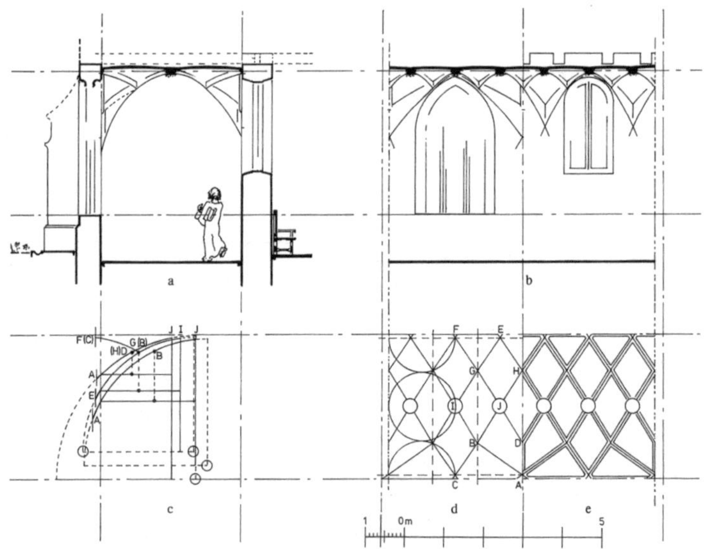

Fig. 9.25: Hirsau, cloister. Reconstruction of South wing vault. Plan d with the basic geometric grid pattern, shows how the relationship between point A, B and C are thus, that they are covered by a half-circle. Thus corner ABC is 90°. Springers C and F are positioned at the window crests and they receive ribs CB and FG, which give lateral support to the ribs during building the vault membranes. This vault has a unique rib pattern.

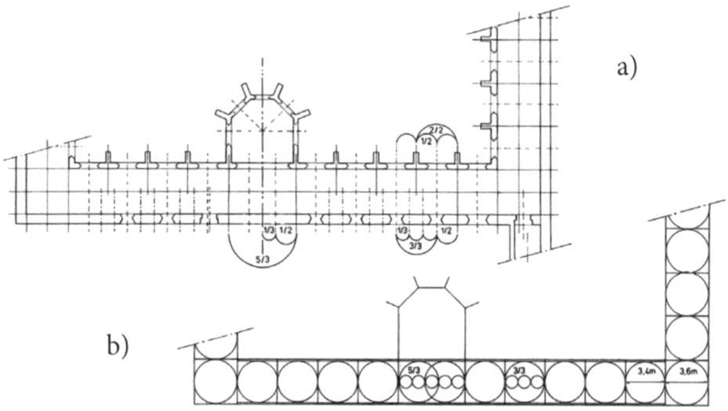

Fig. 9.26: a) Hirsau, cloister. South wing. The scheme shows the asymmetric rythm of the springers. Those springers at the outer wall are twice as many as those at the inner wall (court). The reason could be that the outer wall is quite long, without transverse butressing. The exceeding amount of springers divides the thrust forces on this wall. b) Hirsau, cloister. The geometric proportions of South wing bays (340 cm depth, 360 cm width) differ from the square bays in the East wing (360 cm). In order to keep square conditions for the vault pattern, the springer points have been put 10 cm in front of the wall, with the help of consoles.

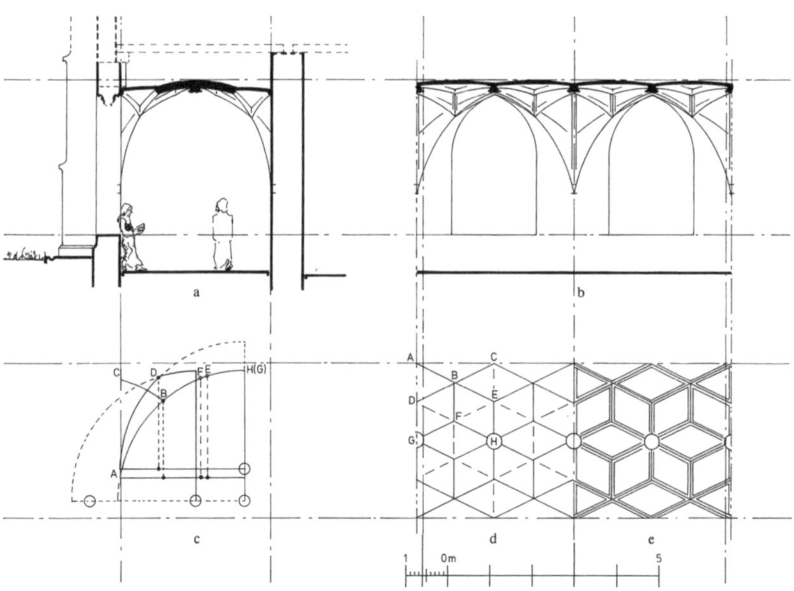

Fig. 9.27: Hirsau, cloister. Reconstruction of South wing vault of a starlike appearance (*Sterngewölbe*).

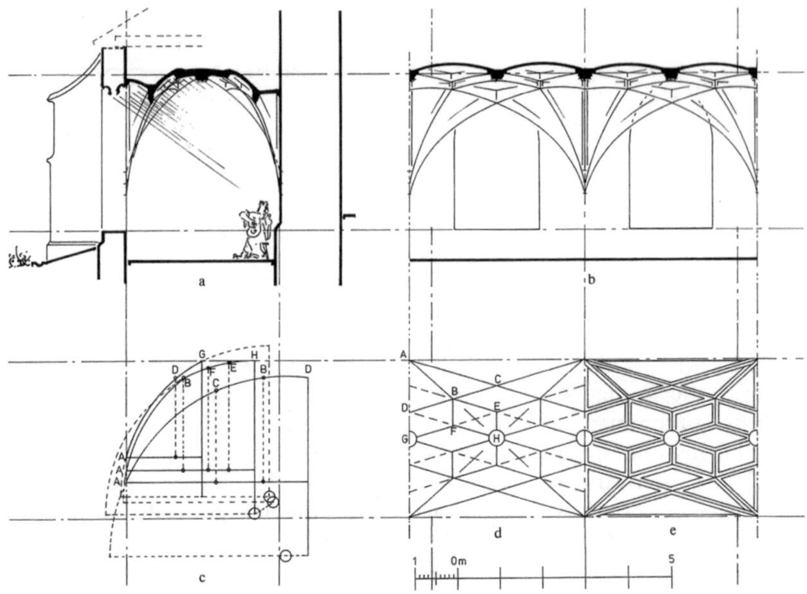

Fig. 9.28: Hirsau, cloister. Reconstruction of North wing vault.

References

1. Abraham. P. Viollet-le-Duc et le rationalisme médiéval, 1934.
2. Binding, G. M. Untermann, Kleine Kunstgeschichte der mittelalterlichen Ordensbaukunst in Deutschland, Darmstadt 1985.
3. Brändle, R., Die Bogenaustragung der spätgotischen Figurierten Rippengewölbe des Kreuzganges Kloster Alpirsbach. Das Prinzipalbogenmodell, architectura—Zeitschrift für Geschichte der Baukunst, Band 35, 2/2005, S. 138ff.
4. Kolb, G., Benediktinische Reform und Klostergebäude, in: Blätter für Würtembergische Kirchengeschichte, Jahrgang 86, 1986, 231–298.
5. Kolb, G., Kloster Alpirsbach, München, Zürich 1990.
6. Landesdenkmalamt Baden-Württemberg (ed.), Alpirsbach—Zur Geschichte von Kloster und Stadt, Forschungen und Berichte der Bau- und Kunstdenkmalpflege in Baden-Württemberg Band 10, Stuttgart 2001, 427–448.
7. Meckel, C. A., Die Konstruktion der figurierten Gewölbe in der deutschen Spätgotik, in: Architectura, Jahrbuch für Geschichte der Baukunst I, 1933, 107–114.
8. Muk, J., Die Gewölbe des Benedikt Ried, in: Geschichte des Konstruierens IV, Wölbkonstruktionen der Gotik 1, Konzepte SFB 230 Heft 33, 1990, S. 193–205.
9. Müller, W., Von der Norm zur Form—die Konstruktion spätgotischer Zierkonsolen, in: R. Graefe (Hrsg.), Zur Geschichte des Konstruierens, Stuttgart 1989, 81–91.
10. Müller, W., Grundlagen gotischer Bautechnik, München 1990.
11. Müller, W., N. Quien, Virtuelle Steinmetzkunst der österreichischen und böhmisch-sächsischen Spätgotik, Petersberg 2005.
12. Seeliger-Zeiss, A., Studien zur Architektur der Spätgotik in Hirsau, in: Hirsau St. Peter und Paul 1091–1991. Teil I, Zur Archäologie und Kunstgeschichte. Forschungen und Berichte der Architektur des Mittelalters in Baden-Württemberg Bd. 10/1, Stuttgart 1991, 265–363.
13. Schreiner, K. (ed.), Hirsau St. Peter und Paul 1091–1991. Teil I, Zur Archäologie und Kunstgeschichte. Forschungen und Berichte der Archäologie des Mittelalters in Baden-Württemberg Bd. 10/1, Stuttgart 1991.
14. Tomlow, J., Versuch einer (zeichnerischen) Rekonstruktion des Gewölbes im spätgotischen Kreuzgang des Klosters Hirsau, in: Hirsau St. Peter und Paul 1091–1991. Teil I, Zur Archäologie und Kunstgeschichte. Forschungen und Berichte der Archäologie des Mittelalters in Baden-Württemberg Bd. 10/1, Stuttgart 1991, 365–393.
15. Tomlow, J., Der geometrische Faktor beim Konstruieren, in: Prozeß und Form "Natürlicher Konstruktionen", Der Sonderforschungsbereich 230, hrsg. v. K. Teichmann, J. Wilke, Berlin 1996, 75, 77–83.
16. Tomlow, J., Neue Erkenntnisse zur Baugeschichte des Kreuzgangs—Entwurf und Ausführung eines mittelgroßen spätgotischen Gewölbes, in: Alpirsbach—Zur Geschichte von Kloster und Stadt, hrsg. v. Landesdenkmalamt Baden-Württemberg, Forschungen und Berichte der Bau- und Kunstdenkmalpflege in Baden-Württemberg Band 10, Stuttgart 2001, 427–448.
17. Ungewitter, G., Mohrmann, K.: Lehrbuch der gotischen Konstruktionen, Leipzig 1901, 4. Auflage.
18. Viollet-le-Duc, E., Dictionnaire Raisonné de l'Architecture Française du XIe au XVIe siècle, Paris 1858–1868.

MODERN ERA

CHAPTER TEN

THE SHIPBUILDING TEXT OF MICHAEL OF RHODES

David McGee

Written in 1434 and 1435, the Michael of Rhodes manuscript contains the first known Treatise of Shipbuilding [25]. The original manuscript was lost for centuries and only recently resurfaced, but the contents of Michael's section on shipbuilding have been known since 1840, when the copy of Michael's text known as the *Fabrica di galere* was published by the great French scholar Augustin Jal [13] [17].

Many things have been said about the *Fabrica di galere* and what it represents, but all previous commentary took place in the absence of Michael's original manuscript and the context it provides. The purpose of this brief article is to provide some of that context by placing the manuscript in the context of Michael's life, by describing the contents of the manuscript as whole, and finally by discussing Michael's shipbuilding text and the drawings its contains.

Focusing on a single manuscript, my methodology will be quite different from that recently applied to Venetian shipbuilding texts, where the trend has been to bring all the extant texts together in order to derive from them a generic technique that is thought to be applicable to the design of galleys between 1400–1680 [1] [4] [5]. Given how difficult it is to understand the early texts, the results of these studies have been impressive. Nevertheless, the "generic" approach typically involves the extraction of isolated facts from texts that are centuries apart, which are then combined in a way that, while it may contribute to an understanding of the design of galleys, tends to ignore the nature of the individual sources themselves. Many of the most puzzling features of the early manuscripts remain unexplored and unexplained.

The Michael of Rhodes manuscript allows us to take a different approach. Now that we have the original, and the rich context it provides, we can ask new questions and get new answers. These answers, I hope to show, are both surprising and challenging. We will discover, for example, that the first known Treatise of Shipbuilding was not written by a shipbuilder, and not written for shipbuilders. We will discover that the first known Treatise of Shipbuilding has its origins, not in the shipyard, but in medieval schools of commercial mathematics. We will find that there is very little evidence for the use of proportions, or of any other formal geometry in the text. We will discover that Michael's drawings are not really "design" drawings—in the sense that the they were to be used to determine the shape of ships. Rather, they are best understood as "graphic lists," whose purpose was to make confusing written lists of measurements understandable to lay persons.

The Life of Michael of Rhodes

We know a great deal about the life of Michael of Rhodes. The reason we know so much is that Michael included his service record in his manuscript [25, ff. 90[2]b–93b], stating the position he held, his destination, and the officers with whom he served, for each of the annual voyages he made from 1401 to 1443 [Figure 10.1]. From these details we have been able to turn to other sources and learn a great deal more [34].

Michael's first entry gives his name as Michali da Ruodo and states that he joined the Venetian navy as a lowly oarsman on June 5, 1401. Over the course of the 34 voyages he made between 1401 and 1434, however, Michael rose through the ranks to hold the highest position a non-noble like himself could hold in the Venetian navy. He remained an oarsman until 1407, when he was hired on as *nochiero*, meaning a sort of apprentice mariner whose task was to learn, by doing, everything he could about ship operations. He learned well enough that he made *paron* in 1414, becoming one of only three senior officers on board, with a particularly responsible for fitting out. In 1421, he achieved the position of *comito*, second only to the noble-born Venetian captain (called the *patron* on commercial ships and *sopracomito* on warships). Since these captains were not always the masters of practical ship operations, the *comito* was often the *de facto* commander of a galley—the officer who issued orders directly to the crew concerning sailing, rowing, and navigation, and who apparently also directed the crew in battle. During his time as *comito*, Michael was twice appointed *amiraio* in the military fleet. This officer commanded the Admiral's flagship, and was responsible for the movements of the fleet as a whole, as well for the welfare and discipline of the crews [15] [19, pp. 344–48] [32].

Michael's rise through the ranks show that he was both ambitious and determined. The high offices to which he was appointed show that he was both talented and trusted. But where did he sail?

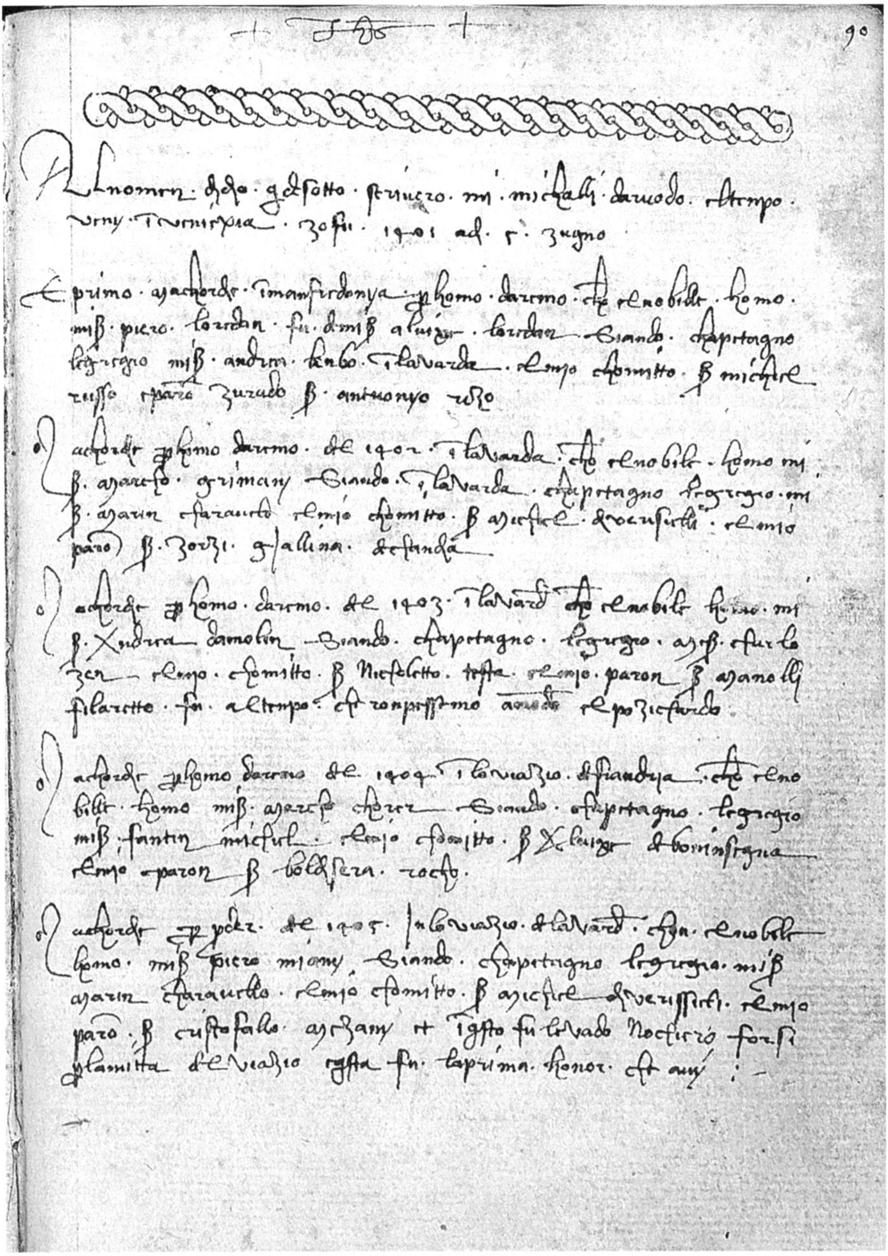

Fig. 10.1: The Michael of Rhodes, folio 90[2]a, the first page of Michael's service record. Courtesy of the Burndy Library.

Fifteen of the annual voyages Michael made up to 1434 were in merchant galleys built in the Venetian Arsenal and auctioned off to the highest bidder for annual commercial convoys to established destinations. Eight times Michael sailed on the long, grueling voyage to Flanders, twice to Alexandria. Three times he sailed to Constantinople and on to either Tana or Trebizond in the Black Sea. In 1434 he sailed for the first and only time to Aigues Mortes on the southern coast of France [12] [19] [35].

The fact is, however, that up to 1934 Michael spent more time fighting than trading, sailing 19 times with the Venetian guard fleet—and he fought a great deal. In 1403, for example, he took part in a long campaign that ended with a bloody victory over the Genoese off Modone [27]. In 1407 he took part in the capture of Lepanto [21]. In 1416 he took part in the first major sea battle between Venice and the Ottoman Turks near Gallipoli, after which 1100 "Turkish" prisoners were massacred, most of them actually skilled, seafaring Greek Christians [28]. In 1429, Michael returned to Gallipoli as *amiraio*, commanding the fleet in a spectacular, but failed, assault on the harbor [26]. In 1431 he fought against the Genoese in a two hour battle that saw hundreds killed or drowned, during which he himself was so badly injured, as he wrote in his service record, that he was forced to return home over land "wounded and broken" [7] [25, f. 92b].

The number of voyages Michael survived indicate how tough he was. The amount of fighting in which he was involved reveal a man who literally knew the sharp end of a stick, a man who had seen many people die, who had killed people himself who had almost been killed in return. This is important, because it seems highly unlikely that such a man would take two years to write a manuscript for the sake of his own amusement. On the contrary, it is far more likely that Michael had a definite purpose in mind, and we can make an educated guess as to the purpose based on progress of his career after he was wounded in 1431—an event which marked a turning point in his life. He served once more in the military fleet in 1432, while the war with Genoa was still on, but that was his last military voyage and he then focused his attention on obtaining the two highest posts available to him in the commercial galley fleet.

The higher of these two posts was that of *amiraio* of the commercial convoys. The second highest was that of *homo di conseio*. Unfortunately, we don't know exactly what the duties of this second officer were in Michael's time. We do know the position existed only in the merchant galleys, where there was one *conseio* for each ship in each of the five main commercial convoys, and Lane has suggested that the *conseio* had something to do with navigation [18, p. 344]. However, we know the *conseio* was *not* in the direct line of command from *patron* to *comito* to *paron* to the crew (in part because Michael never names the *conseio* in his list of commanding officers, unless he himself held the post). Instead, as the word "conseio" would suggest, this officer appears to have been an official "advisor," either to the captain or to the merchants who formed a sort of council on board. [18, p. 342 and 348] [32].

We also know the Venetian government thought the posts of *amiraio* and *conseio* in the commercial fleet important enough that officers were elected to these positions by the members of a government committee that included the doge, some of his senior councilors, and an ever-changing group of other noble merchant citizens [34]. We know further that the posts of *amiraio* and *conseio* were so important to senior non-noble officers like Michael that they continually put their names forward for election.

Finally, we know that winning these elections was not easy. Each year, Michael had to compete against a group of about 30 other master mariners, all with much the same skills and experience, and most of whom were Venetian citizens. He was not always victorious. When Michael first entered the elections in the 1420s, for example, he lost. When he started to compete again in the early thirties, he lost. It was he lost that he was forced to sail as *comito* on the very dangerous voyage to Alexandria in 1433, and on the unusual (for him) voyage to Aigues Mortes in 1434—the year he began his manuscript.

The context therefore strongly suggests that Michael wrote his manuscript to influence the annual elections for shipboard office and thereby continue to advance his career. What did he write?

The Manuscript

Michael original manuscript contains 204 numbered folios, or 408 pages, in a codex that is not much bigger in size than a thick, modern, hardcover novel.

The first 180 pages concern mathematics [25, ff. 1b–90[2]a]. This is the largest section of the manuscript and begins with a practical problem on the profits to be made from a cargo of pepper purchased in Alexandria [Figure 10.2]. Much of the text is taken up with practical commercial

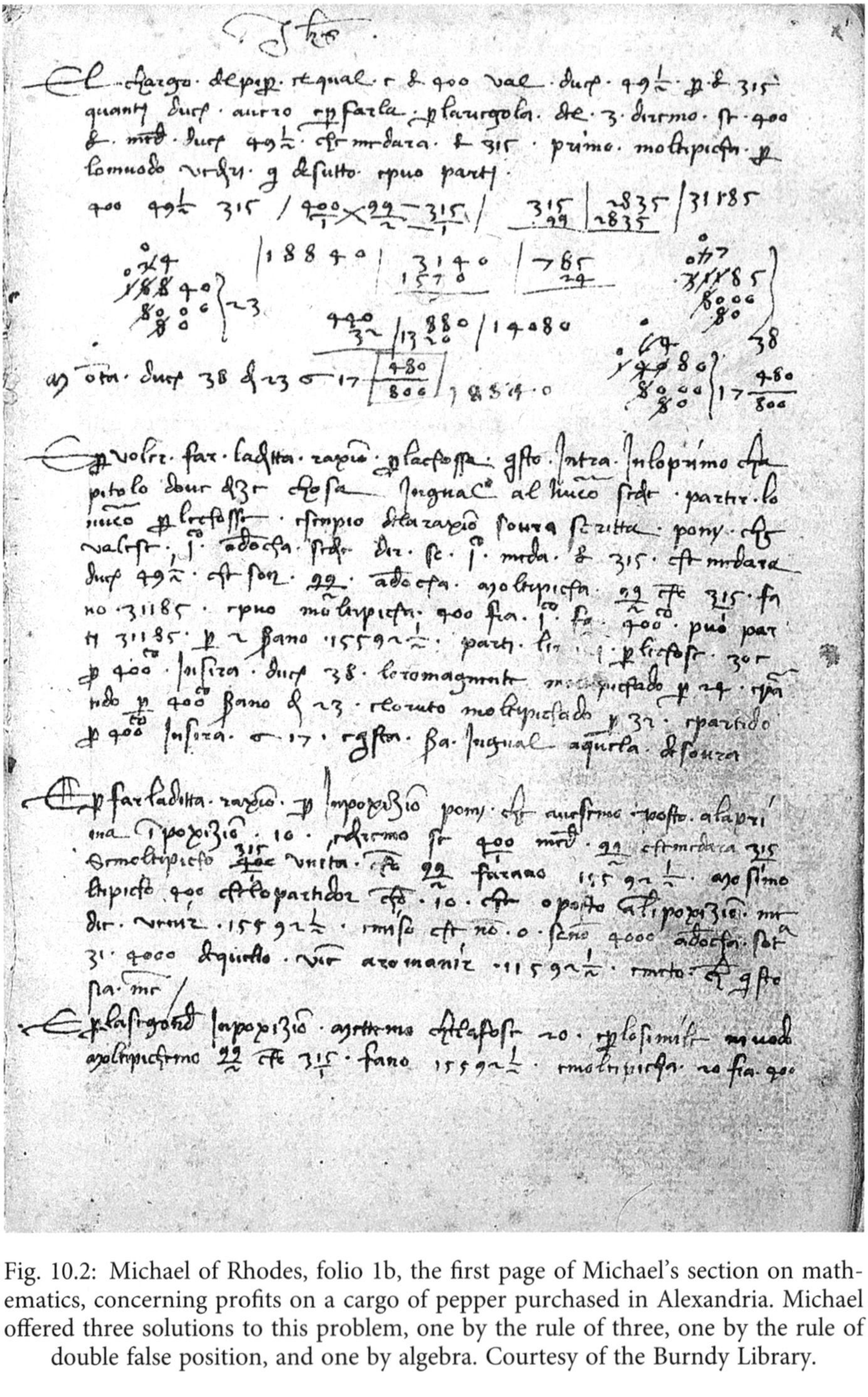

Fig. 10.2: Michael of Rhodes, folio 1b, the first page of Michael's section on mathematics, concerning profits on a cargo of pepper purchased in Alexandria. Michael offered three solutions to this problem, one by the rule of three, one by the rule of double false position, and one by algebra. Courtesy of the Burndy Library.

problems of this kind, concerning profits, loans, partnerships and so on. However, the text also contains an equal amount of what might be termed theoretical material explaining the basis of algebra, radicals, cube roots, operations with fractions, the rule of three, and more. Surprising as it may seem, Michael not only understood this math and needed it for his work, he was highly interested in the subject. This is shown by the fact that he normally provided three different solutions to the same problem (when one solution would do), and worked through the computations himself.

The first and most important point to be made about Michael's section on mathematics is that it reflects what is known as the "abacus" tradition, so-called because its origins can be traced back to the *Liber abaci* published by Leonardo Pisano in 1202 [33]. Pisano had lived in various Arab lands and his book was essentially a compilation of Arabic advances in mathematics, including the use of decimal places, Arabic numerals (borrowed from the Hindus), algebra and so on. Pisano particularly stressed the practicality of Arabic methods, and Italian merchants quickly grasped their usefulness. Abacus schools sprang up all over Italy to teach the new math, along with reading and writing [14].

A second point is that Michael's mathematical skills make it clear that he either attended one of these schools or engaged an abacus master to teach him mathematics during his early years at sea (another measure of his determintation to better himself). This is important because it was at the abacus schools that merchants and nobles received their cultural formation, just as they do in high schools and universities today. That is to say, by going to abacus school, Michael was set out to learn the language and absorb the culture of those above him in society.

Approximately three hundred manuscripts associated with the abacus schools survive [39]. These are generally two types. One contains more or less didactic texts associated with teaching in the schools. The other type are *zibaldoni*, or notebooks, containing a miscellany of practical commerical problems, as well as information on exchange rates, astrology, calendars, date of Easter, and so on [11] [36]. Different subjects were copied from one manuscript to another according to the interest of their owners after they had left school. Michael's mathematical section reflects the didactic type of text. But his manuscript as a whole is written in the *zibaldone* tradition [14]. It contains many pages of material on time reckoning, the date of easter, astrology and navigation, all in addition to his pages on mathematics [25, ff. 95a–135a, 18a–27a, 192b–192a].

In short, Michael wrote his manuscript in 1434 and 1435—including the text on shipbuilding—in the tradition of the abacus schools, in one of the main literary genres of the class above him in society and the very class on whom he was now dependent for election to high office.

The Shipbuilding Text

Turning to Michael's section on shipbuilding, perhaps the first point to make (for the beenfit of historians of shipbuilding) is that Michael's text is not identical to the *Fabrica*. It lacks many of the additional notes about ships that are found at the end of the *Fabrica*, and particularly any reference to Theodore Baxon [6] [19, pp. 56–58]. It contains no diagrams or references to the *partison* method of determing hull frames. It does, however, contain the same descriptions of five ships, beginning with three galleys and ending with two sailing ships [25, ff. 135b–182b].

Each of these ships is treated in the same way, following the pattern of the first ship described, which is the galley of Flanders [25, ff. 135b–164a]. This text begins with the didactic comment that:

Here are the instructions for a galley of the Flanders mould, and of doing all things and procedures until it is fit for going by sail or by oar, that is to make it complete, to fit it out and rig it, as will be explained to you below item by item [23, f. 35b].

What follows is indeed a tedious, item by item list of measurements for various dimensions, expressed in often obscure technical terms, beginning with the principal dimensions, followed by measurements relating to the wales, other frames, and deck structures, ending with a picture of the galley on the stocks. Next comes a long list of dimensions relating to the ribbands, accompanied by several diagrams, ending with a picture of the hull afloat. This picture is followed by pages containing more lists of equipment and their dimensions, starting with the masts and spars and continuing with information about the boats, rigging, sails, anchors, the capstan and the rudders. Each of these lists is illustrated [Figure 10.3]. The whole section on the galley of Flanders ends with a magnificent picture of the ship under sail at at sea [Figure 10.4].

The other ships in Michael's manuscript are dealt with in an abbreviated manner, but in the same order. His account of the galley of Romania, for example, begins with a didactic statement, followed by a list of principal dimensions that ends with a picture of the ship on the stocks. Next is a list of dimensions for wales and ribbands, ending

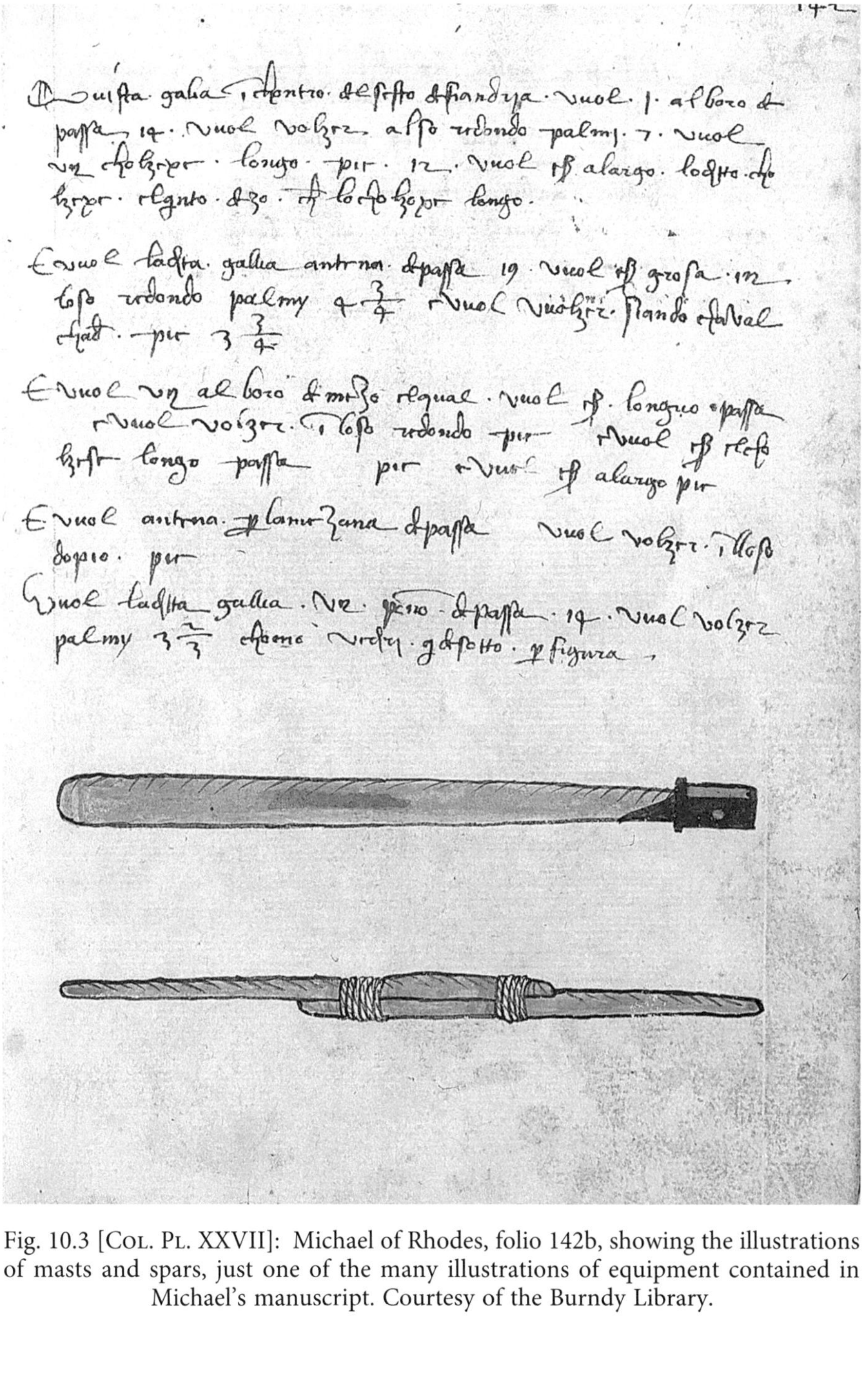

Fig. 10.3 [Col. Pl. XXVII]: Michael of Rhodes, folio 142b, showing the illustrations of masts and spars, just one of the many illustrations of equipment contained in Michael's manuscript. Courtesy of the Burndy Library.

Fig. 10.4 [Col. Pl. XXVIII]: Michael of Rhodes, 145b, showing the galley of Flanders under sail. Note the detail of the rigging, about which Michael was an expert. The ropes and lines for the rigging are described at length in the text. Courtesy of the Burndy Library.

with a picture of the hull afloat. This is followed by the dimensions of boats, rudders, sails, and ends with another magnifcent picture of the ship at sea [25, ff. 148a–156a] [Figure 10.5]. Michael describes the "light", or military galley in the same order, giving lists of dimensions up to the ship afloat, then details and pictures of the equipment, the sails and the galley under sail [25, ff. 157b–164a]. The two sailing ships Michael describes are a lateen-rigged ship and a square-rigged ship. They are discussed in the same order as the galleys, although only one page of each is devoted to dimensions of the hull. All the other pages are devoted to lists of sails and equipment. The lists for the sailing ships were supposed to be illustrated, and labeled spaces were left in the text for the pictures [25, ff. 164b–168a]. However, none of these drawing were made, except for the final picture of the *nave quadra* at sea under sail [25, ff. 168b–182b].

It is believed that Michael copied much his shipbuilding material from existing documents. The question therefore arises as to how he can be considered an author and how his text can be thought of as a treatise. The answer is provided by a comparison with the roughly contemporary text of Zorzi da Modon and the somewhat later manuscript entitled *Ragioni Antiche*. Both were written in the *zibaldone* tradition. Both contain descriptions of ships similar to those described by Michael. But the descriptions in both texts are decidely haphazard, written in no particular order, in a variety of formats, including texts that are sometimes long, sometimes short, that sometimes include drawings and sometimes don't [10] [22] [23]. By comparison, Michael's text is a paragon of orderliness. He describes all five ships in one place, in one basic format, following the same order, paying a great deal of attention to the layout, and particularly to the organization of the text around the illustrations.

In short, Michael expended a considerable effort organizing his materials. It is in this way that he can be considered to be an author, and it due to the organization that Michael's text deserves to be called a treatise.

Design Issues

Turning now to design issues, it should first be made clear that Michael's text is not about design *per se*. He does not discuss the way in which shapes or dimensions are determined prior to contruction. He does not

Fig. 10.5 [Col. Pl. XXIX]: Michael of Rhodes, foliò 156a, the galley of Romania at sea. Courtesy of the Burndy Library.

Fig. 10.6 [Col. Pl. XXX]: Michael of Rhodes, folio 182b, the *nave quadra*, or square rigged ship at sea. This is the only illustration of Michael's two sailing ships that was completed. Courtesy of the Burndy Library.

discuss the way in which shapes or dimensions are determined during construction in the shipyard. Rather, he gives lists of dimensions that record the shapes of ships that have already been built. His manuscript *stores* the design of ships. Indeed, it reveals three different approaches to design storage.

The first approach can be seen in Michael's descriptions of the principal dimensions of the *nave latina* and the *nave quadra*. Here is what he says about the *nave latina*:

> We wish to make a lateen ship whose keel is 12 paces long. We want the floor to be as many feet as the keel is long in paces, and one-quarter less. This will be 9 feet.
>
> And this ship with a keel of 12 paces and a floor of 9 feet should have for its breadth at three feet above the keel as many feet as the keel is long in paces and one-third more. The keel is 12, a third more would be 4, making 16 in all.
>
> And this ship with a keel of 12 paces and a floor of 9 feet and a breadth at three feet of 6 feet needs a maximum breadth equal to that at three feet, and half again. That is 24.
>
> And this ship with a keel of 12 paces and 9 feet in the floor, and a breadth at three feet 16 feet, and a maximum breadth of 24 feet, should have a deck height of as much as the floor and a half-foot more, that is 9 1/2 feet.
>
> And this ship with a keel of 12 paces and 9 feet in the floor, and a breadth at three feet of 16 feet, and a breadth of 24 feet, and a rise of 9 1/2 feet, should have deck length equal to the keel and one-quarter more. It will be 15 paces [25, f. 164b].

This list provides a series of rules which are easy to remember, in which the principal dimensions are all derived from the length of keel. This, it may be argued, reflects an older, traditional method of design storage as verbal rules in the minds of shipbuilders.

When assessing these rules, however, it is important to guard against a certain tendency to romanticize the medieval craftsman, which often takes the form of statements to the effect that they worked according to age-old proportional rules, using geometrical techniques that involved lines, centers, arcs of circles, and so on. For Michael does *not* express his rules for sailing ships in terms of the nice, clean ratios that the romantic tendency would lead one to expect. What he actually provides is a series of *ad hoc* statements about dimensions that involve changing units of measure and have additional distances arbitrarily tacked on. We are told, for example, that the floor of the *nave latina* should be one foot for every pace, less one quarter of the number of paces in feet, while the height is described as the being same as the length of floor—and a half foot more. There is no talk of one distance being one third this, or one sixth that. In this connection, it should be mentioned that fractions were not interpreted as equivalent to proportions at this time (as is made clear by the way Michael treats fractions throughout his text).

Numerical analysis of the *nave latina* does reveal irrational numbers, which might be taken to imply a proportional, geometrical approach. But when the analysis is extended to include all Michael's ships, no obvious proportional relationships emerge. In other words, the evidence provided by Michael's manuscript does not support the idea that such ships were designed using geometrical procedures on the basis of proportions [8]. The additional notion that shipbuilders used formal geometric diagrams, as put forward by Chiggiato, seems fanciful.

The second method of recording designs in Michael's manuscript is the list. There are so many of them and, in the case of the galleys, they are so long and tedious, it is safe to say that no one could possibly remember them all. On the contrary, it would seem that the only place so many details could be stored is on paper. This use of paper lists points to the bureaucratic context of the Venetian Arsenal where both military and commercial galleys were constructed at public expense, but the use of such lists in such a context is not new. John Pryor has published a document containing the orders of Charles, Count of Anjou, for the construction of galleys in the year 1202 [31]. This document begins with a long list of measurements and continues with a list of masts, sails and rigging in a manner that is a clearly a forerunner of the kind of document copied by Michael. As with the Count Anjou, so with Michael. The lists of dimensions provide no hint of an underlying geometrical or proportional method of design.

The third method of storing designs found in Michael's manuscript is the used of drawing. These are examined in the next section.

Michael's Drawings

The kind of drawings used by Michael can be seen in Figures 10.7 and 10.8, which show the midship bend, as well as stem and stern diagrams for his galley of Flanders.

Before examining these drawings, however, it should be noted that they reflect a pictorial language that had already been established in

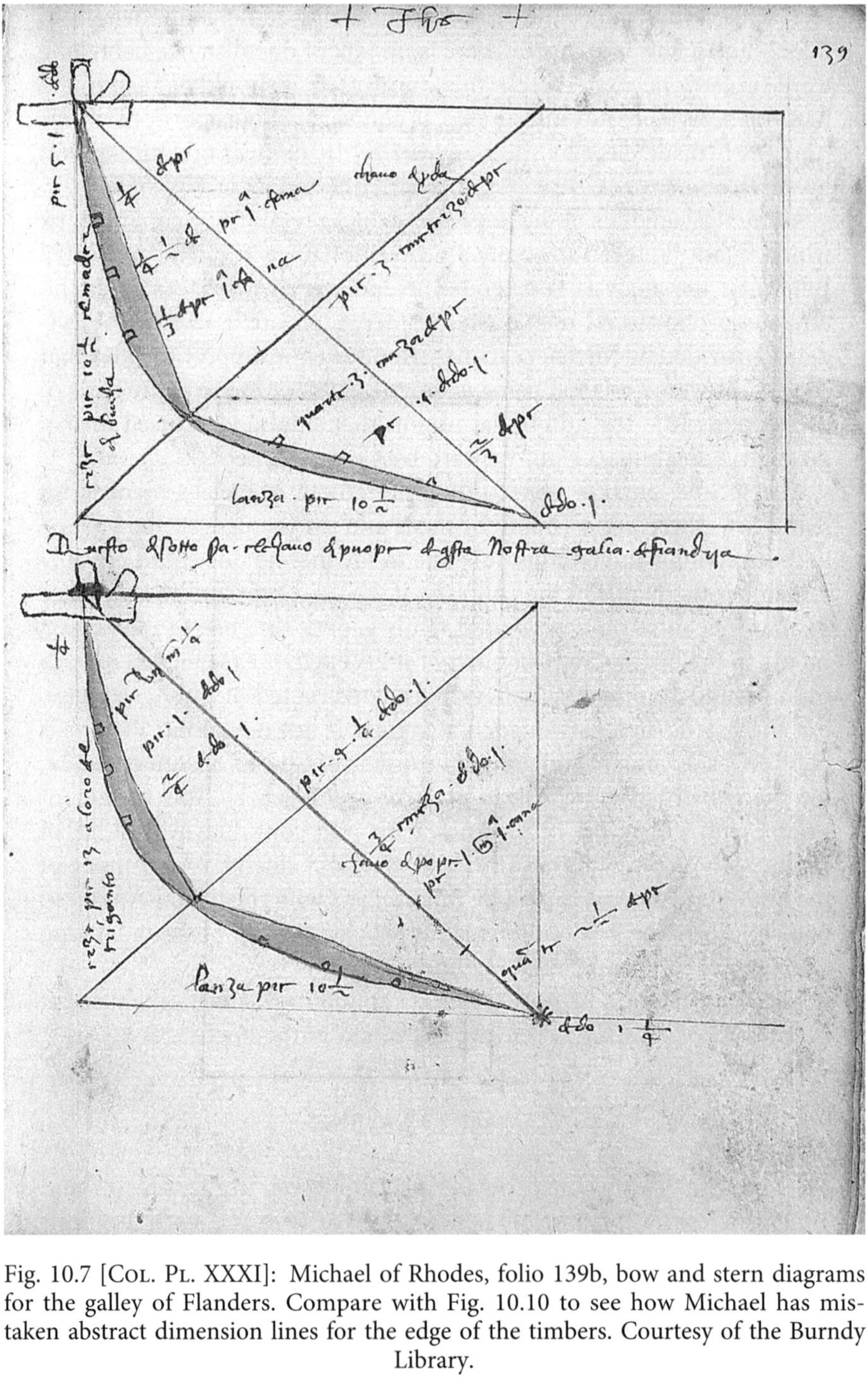

Fig. 10.7 [Col. Pl. XXXI]: Michael of Rhodes, folio 139b, bow and stern diagrams for the galley of Flanders. Compare with Fig. 10.10 to see how Michael has mistaken abstract dimension lines for the edge of the timbers. Courtesy of the Burndy Library.

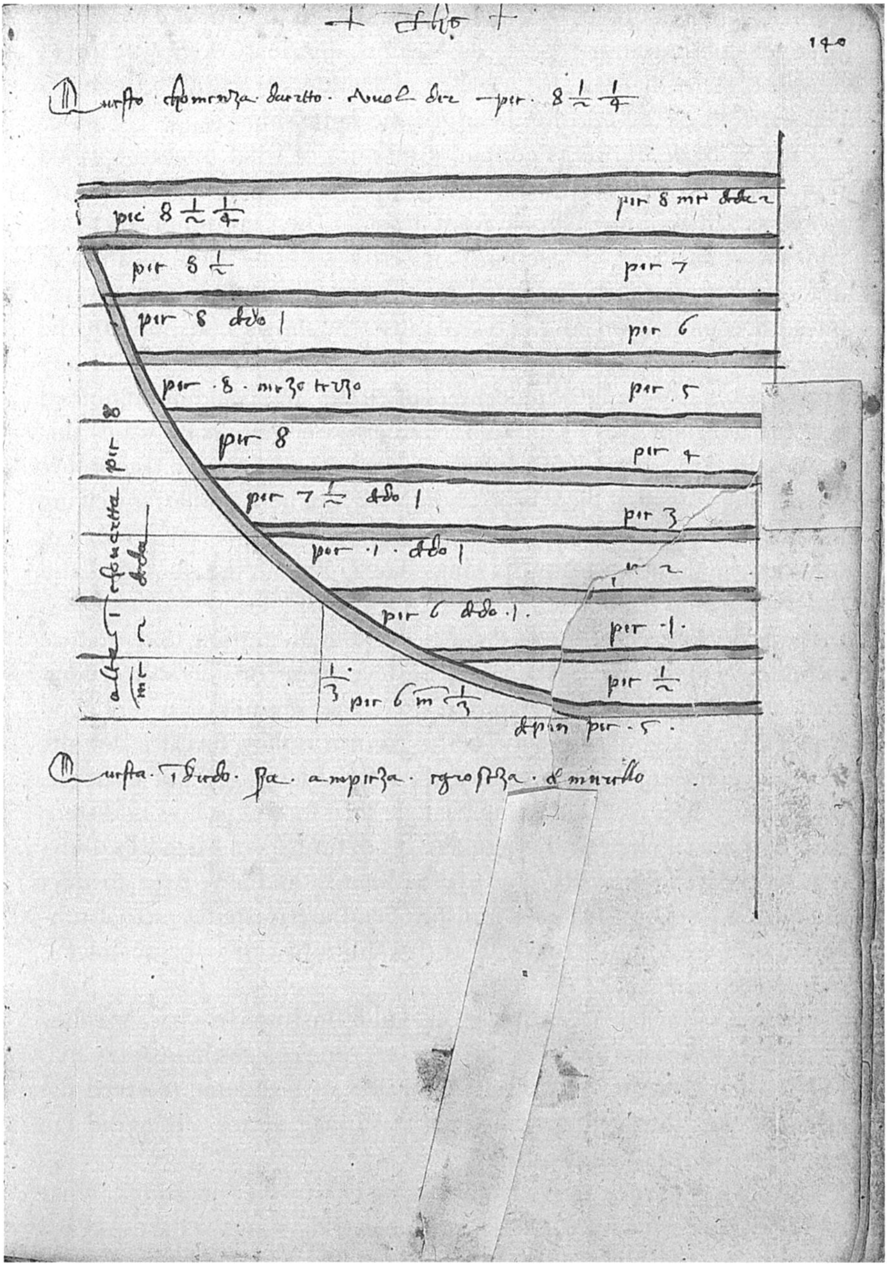

Fig. 10.8 [Col. Pl. XXXII]: Michael of Rhodes, folio 140b, the midship bend of the galley of Flanders. Compare with Fig. 10.11 to see how Michael has mistaken abstract dimension lines for actual timbers. Courtesy of the Burndy Library.

Venice, better examples of which can be seen in the work of Michael's younger contemporary, Zorzi da Modon. Figure 10.9 shows Zorzi's bow and stern diagram for a galley of Flanders. Figure 10.10 shows two versions of the midship bend for the same ship [22].

Each of these drawings contain a mixture of what might be called representation and abstraction. The timbers, for example, are represented as actual timbers shown from the side. These are simply pictures, and not in any way "projections" or "cross-sections". The location of different measures are indicated by abstract dimension lines. Actual physical examination of Zorzi's manuscript shows that none of the curves in these drawings were made with a compass, nor with any formal geometry for locating centers of circles that could be combined into the irregular arcs of the hull. Furthermore, there is no scale. This would appear to prove (as the lack of geometrical drawing techniques already suggests) that there is no underlying idea here about transferring the lines on the paper to another medium, which is to say no idea of transferring the lines from drawings like Zorzi's to the dockyard and thus to the ship itself.

It was drawings like Zorzi's that Michael was attempting to reproduce in his own manuscript. Michael's, however, are considerably cruder. They are not drawn in the standard style. They are not even very good copies of the standard style. To the contrary, they reveal a definite misunderstanding of the established pictorial language. For example, the abstract lines of the aligning triangle seen in Zorzi's bow and stern diagrams are taken to be the actual edges of timbers in Michael's drawings (compare Figures 10.7 and 10.9). What should have been abstract dimension lines in Michael's midship bend are treated as actual timbers (compare Figures 10.8 and 10.10). In both cases, the abstract is represented as real.

Michael's "mistakes," however, are quite instructive—because they weren't mistakes for him. On the contrary, one can easily imagine that Michael felt that his drawings of the galley of Flanders preserved the essential function of these drawings as storage devices. They did the job. But if so, what job was that?

The answer may be arrived at by asking two further questions. What *could* be transferred from Zorzi's much better drawings to the dockyard and thus to the ship? What is it that could be transferred even from Michael's terrible drawings to the yard? The answer is the numerical data they contain, which turns out to be the same information data given in the standard storage device, namely the list of dimensions. We

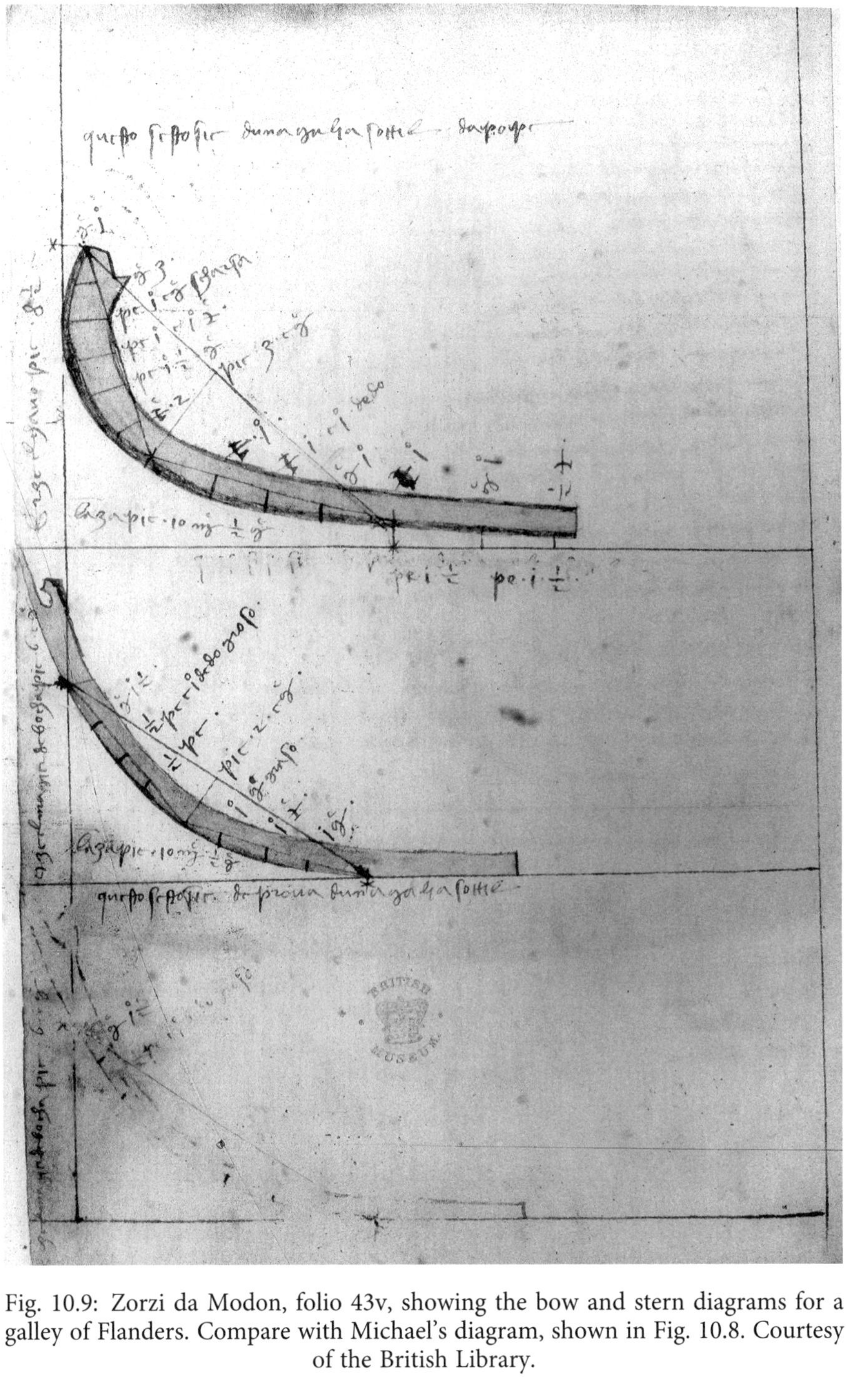

Fig. 10.9: Zorzi da Modon, folio 43v, showing the bow and stern diagrams for a galley of Flanders. Compare with Michael's diagram, shown in Fig. 10.8. Courtesy of the British Library.

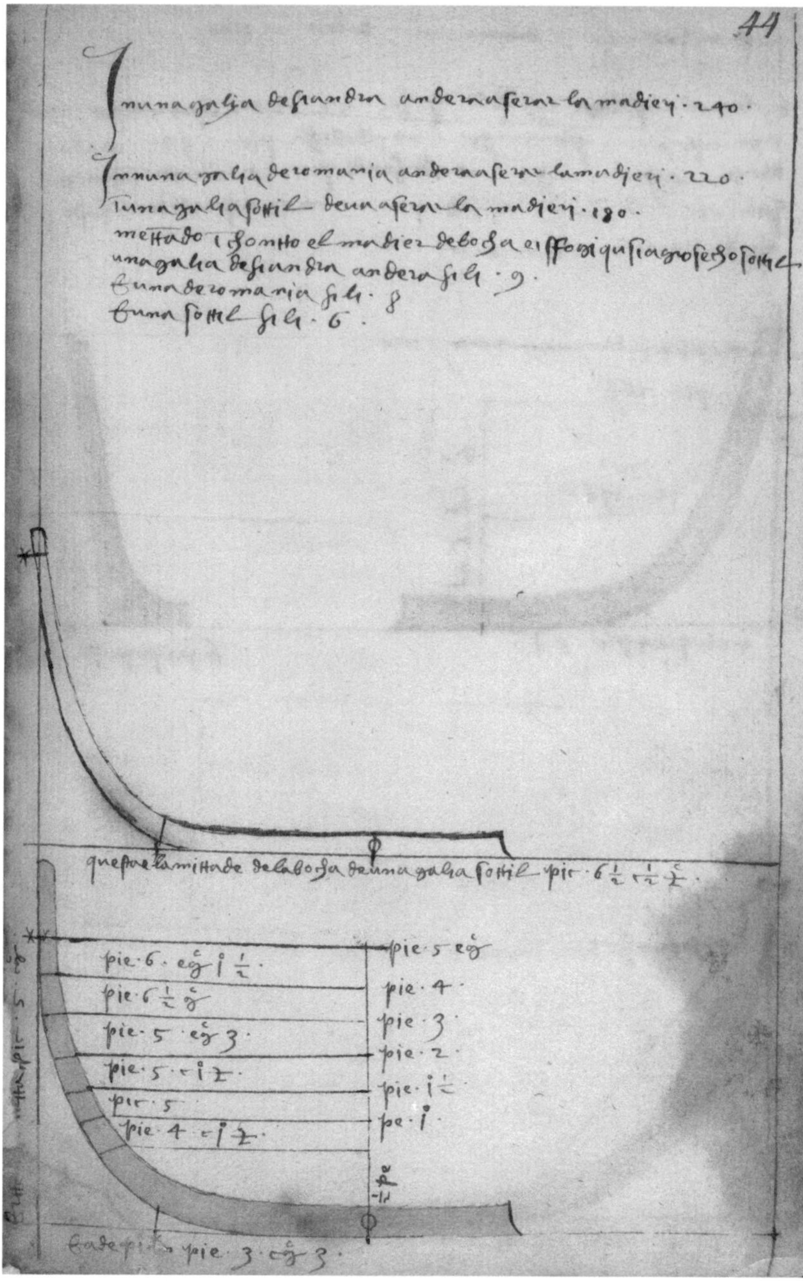

Fig. 10.10: Zorzi da Modon, folio 44r, showing two diagrams for the midship bend for galley of Flanders. Compare with Michael's diagram, shown in Fig. 10.9. Courtesy of the British Library.

may then ask what additional information is provided in the drawing, compared to the list? The answer is, the diagrams tell you what the dimensions given in the list refer to.

It is in this sense that early Venetian ship diagrams are best described a "graphic lists". The detailed dimension given by Michael are bewildering to modern scholars. Even the simplified list of "principal dimensions" would be bewildering to the most elite of modern citizens. There is no reason to believe these lists were much less bewildering to Michael's elite Venetian contemporaries. But Michael's graphic lists make the meaning of the dimensions clear.

Conclusion

We now know that the first Treatise of Shipbuilding was not written by a shipbuilder, but rather by a master mariner named Michael of Rhodes. The purpose of the manuscript, I have argued, was not to aid shipbuilders, but to influence the elections to office on which Michael's livelihood depended.

From this perspective it seems clear that Michael's intended audience was ultimately the merchants and nobles who served on the election committees. But it is not likely that Michael actually showed the manuscript to the electors at election time in an effort to prove his competence. His skills were already well-known from his years of service to the state. Instead, it seems that Michael intended to use his book aboard ship, where he spent much of his time with traveling merchants. We know that these merchants were often poorly informed about the sea. We also know, from pilgrim accounts, that they were eager to learn from the mariners with whom they sailed [16]. It seems that Michael's strategy was to show his book to as many merchants and nobles as he could and thus spread his reputation through the class above him, not by teaching *per se*, but by demonstrating that he shared the culture of the ruling classes, even to the extent of owning a sumptuous, illustrated manuscript that he had written himself.

An audience of merchants and nobles helps explain the text as a whole, written in the genre of the abacus schools, which was a genre of the class he was trying to impress. It explains the long section on mathematics, which was the lingua franca of the commercial class. It explains the inclusion of the material on time reckoning and navigation, subjects that were not only essential for keeping to the schedules on which galley convoys sailed, but which were also very common in abacus schools texts.

This audience of merchants and nobles also helps to explain many features of the shipbuilding text itself, which was of course of no use to shipbuilders, but did cover the five kinds of ships merchant nobles could expect to own, command, travel or ship their goods in. This audience also helps to explain Michael's emphasis on lists of rigging and equipment, which were very expensive and which merchants had to pay for when fitting out their ships for a voyage. These lists are just the kind of thing that merchants and nobles would expect to see in official documents and contracts. I've tried to indicate that the contents of these lists were often quite baffling. This helps to explain the purpose of the drawings of equipment, which were of no use to mariners, but very helpful to those who were new to the sea. The same argument applies to the ship diagrams. They illustrate what is meant by the lists.

If impressing members of the upper classes to secure elections was Michael's goal, his plan worked. With one exception, he sailed either as *amiraio* or *homo di conseio* on each of the voyages he took from 1435 to 1443, with the exception of two voyages he made as *comito* to Constantinople in the service of the Pope.

Beyond Michael and his life, there are other issues to be addressed. One has to do with the origins of the Venetian tradition of shipbuilding manuscripts. As Michael's text shows, this tradition has its origins not in the shipyard, but in the medieval abacus schools. Zorzi's manuscript as well as the *Ragioni Antiche* were also written in the abacus tradition. These facts need further investigation, particularly from the point of view of the mathematical culture of the time and the missing geometry from the earliest shipbuilding text we know.

A final issue is the relationship between this early Venetian approach to ship construction and "naval architecture." According to the Oxford English Dictionary, the phrase "naval architecture" refers to the both the design of ships and the superintendence of their construction. This definition makes "naval architecture" more or less synonymous with shipbuilding. Historically, however, the phrase has a more precise meaning. It came into use around 1600, as pointed out by by Richard Barker, to describe a new approach to the design and construction of warships associated with the use of measured, three-view, architectural-style plans [3] [20] [26].

Making these plans involved a considerable amount of mathematics, but ultimately relied on geometry, proportions, and the use of the arcs of

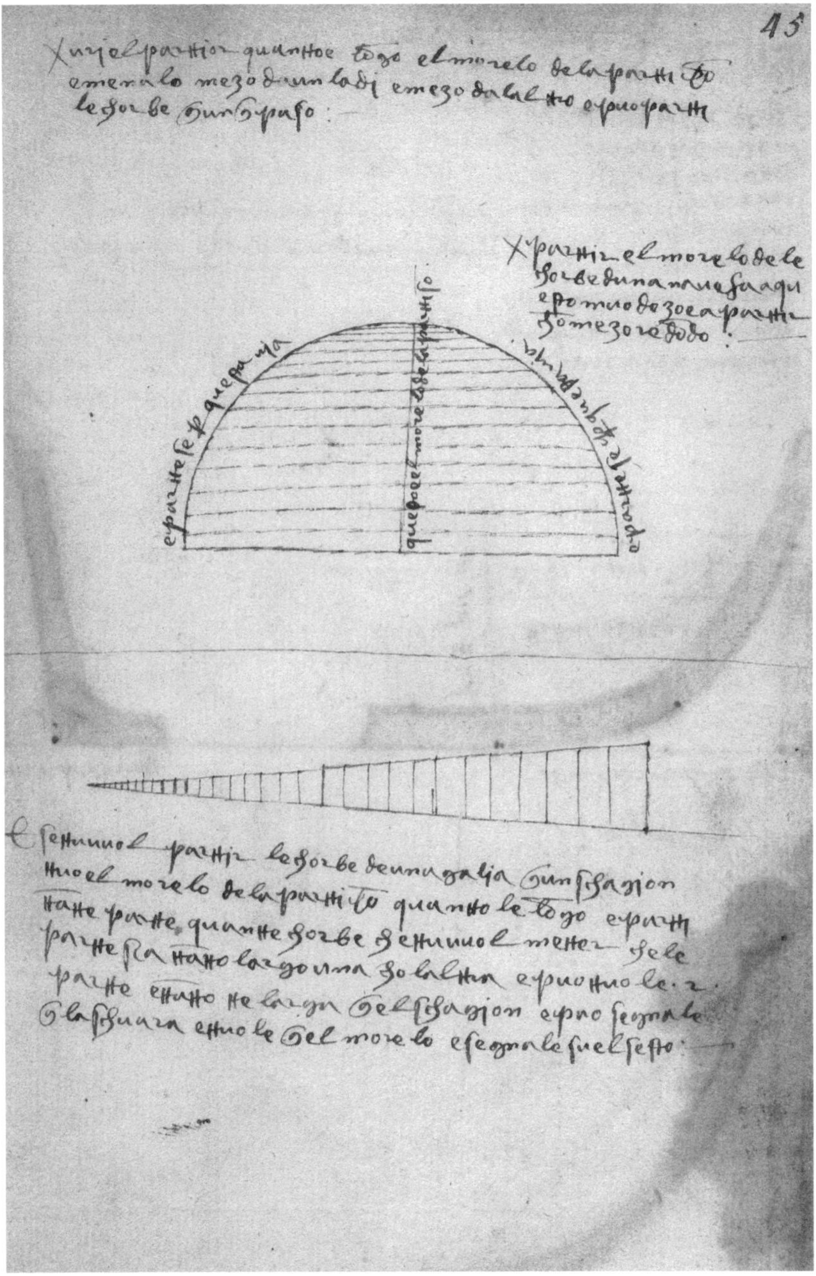

Fig. 10.11: Zorzi da Modon, folio 45r, showing the *mezzaluna* and another geometrical device for determining the shape of a ship's frames. Courtesy of the British Library.

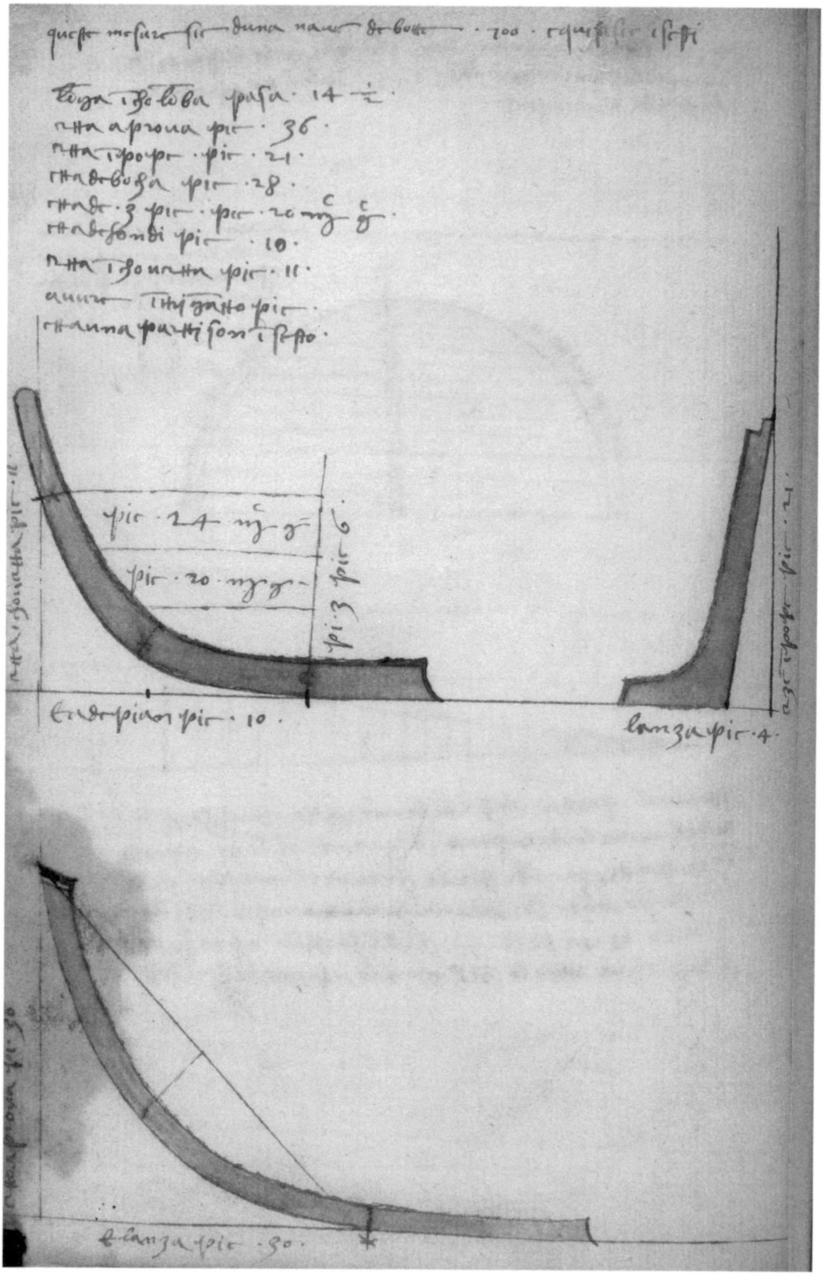

Fig. 10.12: Zorzi da Modon, folio 44v, giving the standard diagrams of a sailing vessel. Note that the drawing to the upper left is not the bow but the midship bend. The bow is shown below. Courtesy of the British Library.

perfect circles to form the irregular arcs of a ship's hull and its individual frames. Behind these procedures, however, is an idea that is very easy for us moderns to overlook. This is the idea that there is a geometrical link between the visual and the real. With such an idea in place it is possible to think that a line on a piece of paper can be transferred to the dockyard and become a line in the ship itself—because it is the *same* line. Thus it becomes possible to think about designing ships on paper, using graphic representations to determine real dimensions.

These ideas do not seem to have taken hold in the early Venetian tradition, as we can see by looking at Zorzi's *mezzaluna*, a diagram used to help determine the changing shape of a galley's many frames [Figure 10.11]. It is not a *representation* of the frames (indeed, in relation to the shape of frames it is upside down). But the point of the diagram is not to visualize curves that could then be applied to timbers. The point is to take linear dimensions from the diagram that could then be marked on measuring sticks. Another page from Zorzi shows the lack of interest in visualization more starkly. It shows the standard diagrams of a sailing ship [Figure 10.12]. On the top right is the stern post. To the left of that is a curved timber which we almost automatically interpret as the bow. It is actually the midship bend. The bow is the curved timber below. Nothing could better show the lack of interest in providing a visual representation than this.

Given this lack of interest in geometry, proportion, and visualization, I think we will have to look elsewhere for the origin of a naval architecture based on the use of three-view plans, and not in the early Venetian manuscripts.

References

1. Alertz, Ulrich, "The Naval Architecture and Oar Systems of Medieval and Later Galleys," in *The Age of the Galley: Mediterranean Oared Vessels Since Pre-Classical Times*, ed. Robert Gardiner (London: Conway Maritime Press, 1995), 142–247.
2. Anderson, Roger Charles, "Jal's 'Memoire No. 5' and the Manuscript 'Fabrica di Galere'," *Mariner's Mirror* 31 (1945): 160–167.
3. Barker, Richard, "Many May Peruse Us: Ribbands, Moulds and Models in the Dockyards," *Revista da Universidade de Coimbra*, Vol. XXXIV 1988, pp. 539–559.
4. Bellabarba, Sergio, "The Ancient Methods of Designing Hulls," *Mariner's Mirror* 79 (1993): 274–292.
5. Bondioli, Mauro, "La galea" in Graziano Arici, *La galea ritrovata: Origine delle cose di Venezia* , Venice: Marsilio, 2003, pp. 62–65.
6. Bondioli, Mauro, and Gilberto Penzo, "Teodoro Baxon e Nicola Palopano proti delle galee sottili. L'influsso greco nelle costruzioni navali veneziane della prima metà del XV secolo," *Archeologia delle Acque* 1 (1999): 67–80.

7. Cappellini, Guido, "Lo scontro di Rapallo, 27 agosto, 1431," *Nuovo Archivio Veneto*, n.s. 6 (1903): 69–131.
8. Chiggiato, Alvise, "Contenuti delle architetture navali antiche," *Ateneo Veneto* n.s. 29 (1991): 141–211.
9. Dellaville le Roux, Joseph, *La France en Orient au XIVe siècle: expéditions du maréchal Boucicaut*, Paris: E. Thorin, 1886.
10. Dosio, Giorgetta Bonfiglio (ed.), *Ragioni antique spettanti all'arte del mare et fabriche de vasselli: Manoscritto nautico del sec. XV*, Venice: Comitato per la Publicazione delle Fonti Relative alla Storia di Venezia, 1987.
11. Dotson, John E. (ed.), *Merchant Culture in Fourteenth Century Venice: The Zibaldone da Canal*. Binghamton, N.Y.: Medieval and Renaissance Texts and Studies, 1994.
12. Doumerc, Bernard "Le galere da mercato," in Alberto Tenenti and Ugo Tucci (eds.), *Storia di Venezia, Vol. 12: Temi: Il Mare*, Rome: Istituto della Enciclopedia Italiana, 1996, pp. 357–395.
13. Florence. Biblioteca Nazionale Centrale Cod. Magliabechiano, cl. XIX, cod.7 (*Fabrica di Galera*).
14. Franci, Raffaella, "Mathematics in the Manuscript of Michael of Rhodes," in Pamela Long, David McGee and Alan Stahl (eds.), *The Book of Michael of Rhodes: A Fifteenth-Century Maritime Manuscript, Vol. 3*, Cambridge, MA; MIT Press: forthcoming, 2009.
15. Hocquet, Jean Claude, "I Genti di Mare," Alberto Tenenti and Ugo Tucci (eds.), *Storia di Venezia, Vol. 12: Temi: Il Mare*, Rome: Istituto della Enciclopedia Italiana, 1996, pp. 481–526.
16. Hyde, J. K., "Navigation of the Eastern Mediterranean in the Fourteenth and Fifteenth Centuries According to Pilgrim's Books," in H. McK. Blake, T. W. Potter, and D. B. Whitehouse (eds.), *Papers in Italian Archaeology 1: the Lancaster Seminar: recent research in prehistoric, classical and medieval archaeology* (Oxford, British Archaeological Reports, 1978), pp. 521–537.
17. Jal, Augustin, "*Memoire No. 5,*" in *Archéologie Navale*, Paris: Arthus Bertrand, 1840.
18. Lane, Frederic C., *Venice: A Maritime Republic*, Baltimore: Johns Hopkins University Press, 1973. pp. 344–348.
19. Lane, Frederic, *Venetian Ships and Shipbuilders of the Renaissance*, Baltimore: Johns Hopkins University Press, 1934.
20. Lavanha, João Baptista, *Livro Primeiro da Architectura Naval*, translated by Richard Barker, Lisbon: Academia de Marinha, 1996.
21. Lazzarini, Vittorio, "L'acquisto di Lepanto, (1407)," *Nuovo archivio veneto* XV (1898): 267–287.
22. United Kingdom, London, British Library, Ms. Cotton Titus A. XXVI, *Libro di Zorzi Trombetta da Modon*.
23. United Kingdom, London, National Maritime Museum, Ms. NVT 19, *Ragioni antique spettanti all'arte del mare et fabriche de vasselli*.
24. United Kingdom London, British Library, Additional Manuscript 38655, *L'Architettura Navale di Steffano de Zuanne de Michel Vice Proto de Marangoni*.
25. Long, Pamela. O., David McGee and Alan Stahl (eds.), *The Book of Michael of Rhodes: A Fifteenth-Century Maritime Manuscript, Vol. 3*, Cambridge, MA; MIT Press: forthcoming, 2009.
26. Madrid, Library of the Royal Academy of History, Salazar Codex No. 63, Lavanha, João Baptista, *Livro Primeiro da Architectura Naval*.
27. Manfroni, Camillo "Lo scontro di Modone: Episodio della lotta veneto-genovese 1403," *Rivista marittima* 30 (1897): 75–90 and 319–341.
28. Manfroni, Camillo, "La marina Veneziana all difesa di Salonicco, 1423–1430," *Nuovo Archivio Veneto*, n.s. XX (1910): 5–68.

29. Manfroni, Camillo, "La Battaglia di Gallipoli e la politica veneto-turca (1381–1420)," *Ateneo Veneto* 25 (1902): 3–34, 129–169.
30. Michael of Rhodes Manuscript.
31. Pryor, John, "The galleys of Charles I of Anjou king of Sicily: ca. 1269–84," *Studies in Medieval and Renaissance History* 14 (1993): 34–103.
32. Sacerdoti, See Alberto, "*Note sulle galere del mercato veneziane nel XV sec.*," *Bollettino dell'Istituto di Storia della Società e dello Stato Veneziano* 4 (1962): 80–105.
33. Sigler, Laurence E., *Fibonacci's Liber Abaci* (New York: Springer-Verlag, 2002).
34. Stahl, Alan, "Michael of Rhodes: Mariner in Service to Venice," in Pamela Long, David McGee and Alan Stahl (eds.), *The Book of Michael of Rhodes: A Fifteenth-Century Maritime Manuscript, Vol. 3*, Cambridge, MA; MIT Press: forthcoming, 2009.
35. Stöckly, Doris, *Le système de l'Incanto des galées du marché à Venise (fin XIIIe–milieu XVe siècle* (Leiden: E. J. Brill), 1995.
36. Stussi, Alfredo, ed. *Zibaldone da Canal: Manoscritto mercantile del sec. XIV*. Venice: Comitato per la Pubblicazione delle Fonti relative alla Storia di Venezia, 1967.
37. Surdich, Francesco, *Genova e Venezia fra Tre e Quattrocentro*, Genova (Fratelli Bozzi), 1970.
38. Tucci, Ugo, "Le conseil des douze sur les navires vénitiens," in *Le navire et l'économie maritime du Moyen Age au XVIIIe siècle, principalement en Méditerranée,* ed. Michael Mollat (Paris: S.E.V.P.E.N., 1958), 119–125.
39. Van Egmond, Warren, *Practical Mathematics in the Italian Renaissance: A Catalog of Italian Abbacus Manuscripts and Printed Books to 1600*, Florence: Istituto e Museo di Storia della Scienza, 1980.

CHAPTER ELEVEN

NAVAL ARCHITECTURE DIGITALIZED INTRODUCING ARITHMETIC AND GEOMETRY INTO LATE MEDIAEVAL SHIPWRIGHTRY

Ulrich Alertz

Introduction

Nowadays naval architecture is divided into several fields: the task of designing is separated from the trades of constructing and building. Medieval shipwrightry combined these tasks in one single trade. The shipwright was concerned with building the ship's body. But he was not responsible for sawing the timbers, for caulking and painting, for rigging, for the galley's oars or for any other equipment. His most distinguished task was to design the hull.

A new kind of naval architecture using regulations, calculations and drawings originated in the late mediaeval Mediterranean. It spread throughout the western and northern shores of Europe (see for instance Barker [11], p. 442. Rieth [26], pp. 9–16. Barker [12]). This revolution in ship design can be called the first pan-European naval architecture. It is definitely the first period in ship design accessible to us by written evidence. Venetian shipwrights, sailors and merchants were the first to write down the new methods, later on called system of *partisoni*. In the 16th and 17th century *partisoni* designs can be found in Iberian, French, English and German manuscripts and books (e.g. Manuel Fernandes [17], 1616; "Fragments of Ancient English Shipwrightry"[1] c. 1570–1630; Fernando Oliveira [22], c. 1580; Joseph Furttenbach [18], 1629). So we are rather well informed about this period when anonymous naval architects began to improve their work systematically until they could exercise their craft in a fairly abstract way without using their working tools just by writing or drawing on a piece of paper or a calculating board.

[1] Cambridge, MCL, Ms. Pepys 2820.

Fig. 11.1: Nave and tarida (galley for horse transport) ordered by Saint Louis, king of France in 1246 and his brother Charles of Anjou, king of Sicily, in 1274. (Models designed by the author and built by the workshop of the International Castle Research Society ICRS/GIB)

Design Concept

In contrast to most objects of civil architecture and engineering, ships of those days consisted of bent pieces of timber. When regarding the hull of a medieval *nave* or a *galera* there can hardly be found any straight line or any right angle (Fig. 11.1).

The master shipwright facilitated his task of designing these forms by reducing three-dimensional shapes to planes. A small set of curved lines, which could easily be drawn on a large parchment or on the floor of a loft, completely defined the hull. Exactly four curved lines were

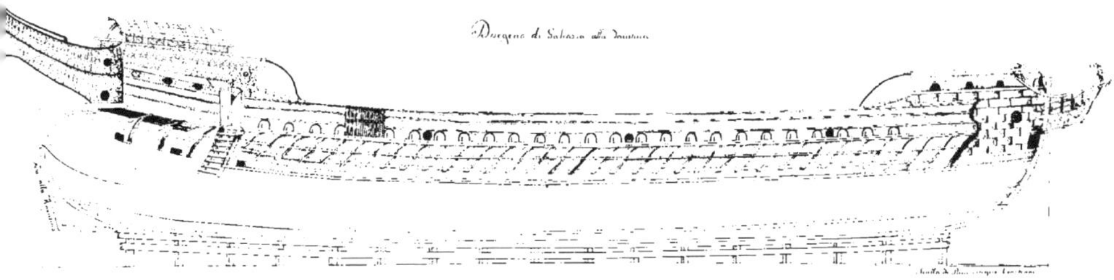

Fig. 11.2: Galeazza alla Faustina (Steffano de Zuanne [9], f.27).

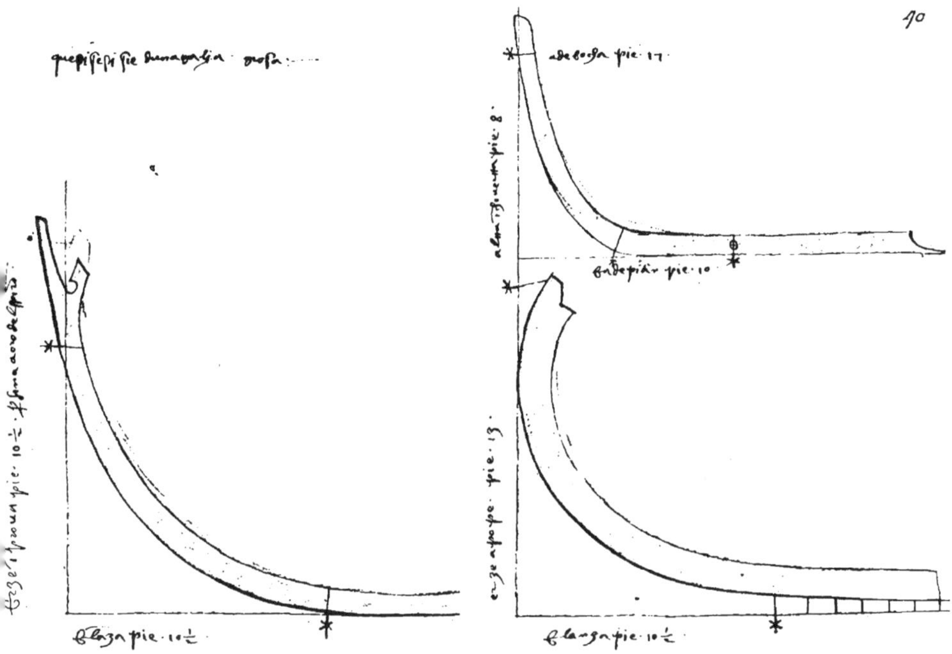

Fig. 11.3: Stem, sternpost and main frame of a Flanders Galley (Trombetta [2], f.39v–40r).

necessary to design any ship's hull: Keel, stem and sternpost outlined the midship plane of the lateral plan, whereas the widest frame defined the vertical cross section. All these curve properties were noted down numerically as sets of linear dimensions—of offsets or radii.

Keel

The keel follows the curve of the wharf, which is defined by its two endpoints and a perpendicular offset. This *sentina del cantier* is the vertical measure for the central lowering of the wharf (see Fig. 11.2).

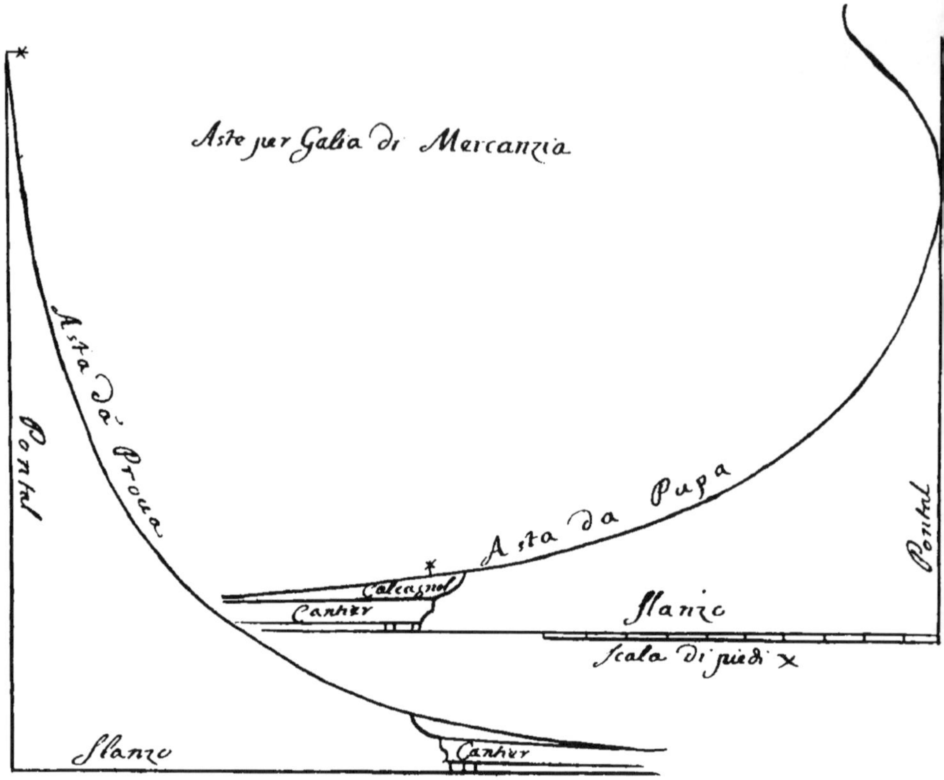

Fig. 11.4: Stem and sternpost of a Merchant Galley (Steffano de Zuanne [9], f.19v).

Stem and Sternpost

The lines of the moulds for stem and sternpost are drawn freehand or with a compass, so that they would fit into the right angle of *pontal* and *lanza*.

Pontal are called the perpendicular lines from the foremost point of the stempost and the aftermost point of the sternpost to the bottom. Sometimes they are drawn from the height of the deck. *Lanza* is the horizontal line from the end of the keel to the perpendicular.

A straight line joins the *calcagnol* (Figs. 11.4, 11.5) at the keel's ending to the uppermost point of the perpendicular line. This oblique line forms the hypotenuse and is used as a base to fix some points on the curves of stem and sternpost.

In naval manuscripts of the 15th century calliper measurements in equal distances perpendicular to the oblique line divided the integral of the curves of stem and sternpost.

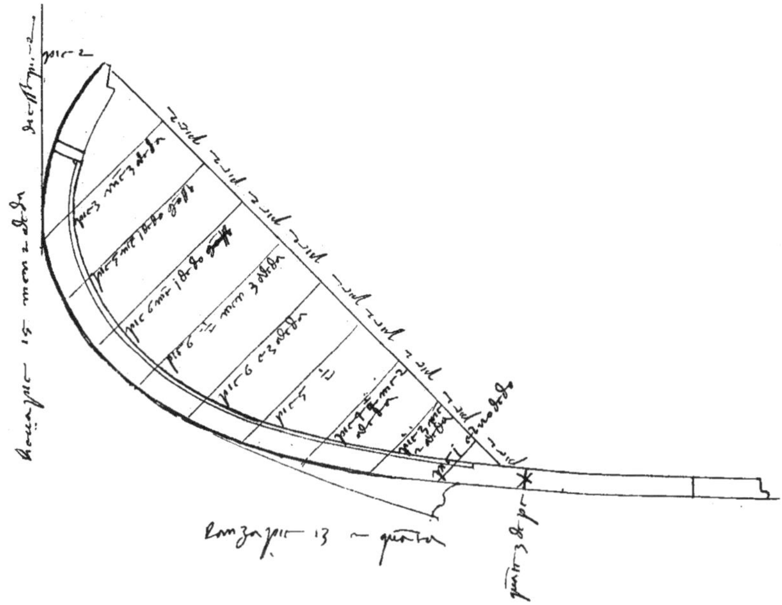

Fig. 11.5a: Asta da pope—sternpost (Ragioni antique [3], f.54r).

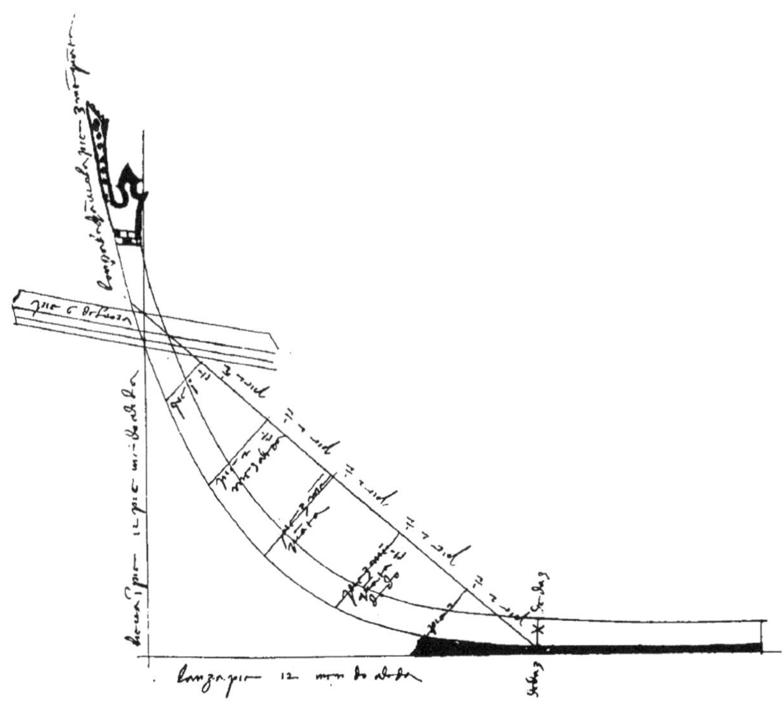

Fig. 11.5b: Asta da prova—stempost (Ragioni antique [3], f.54r).

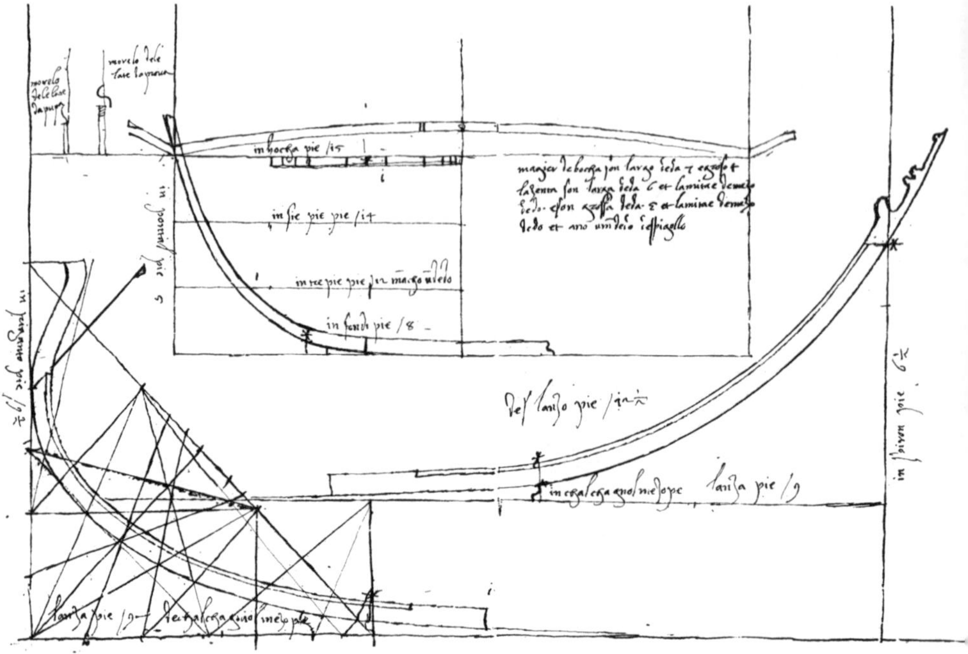

Fig. 11.6: Aste, corba maistra, lata, scala, morelli—stem and sternpost, main frame, deck bea
scale, adapted scales (Pre Theodoro [5], BNM, f.8v–9r).

Concave endings of 16th century galleys' sternposts brought about problems (Fig. 11.6). Additional base lines and offsets had to be introduced. Shipwrights, like Pre Theodoro (= Pretodaro), at times used radii to define these curves. The lateral outline is thus complete.

Main Frame

The vertical cross section is defined by the *sesto*, the midship frame's mould. It is determined in the same way with offsets at different heights. A faired line links the points thus found. Michalli da Ruodo measured the mould at set intervals of 1 foot (Fig. 11.7). But since the second half of the 15th century four offsets were sufficient for the same purpose regardless of the absolute height of the frame.

Using this less complicated procedure, which is based more on the shape of the frame than on its height, a single value determines the width of the bottom (*in fondi*) and the spacing of the frame top ends (*in bocha*). Between them two calliper measurements are made at 1/3 and 2/3 of the frame's height in *tre pie* and in *sie pie* (Fig. 11.8). Pre

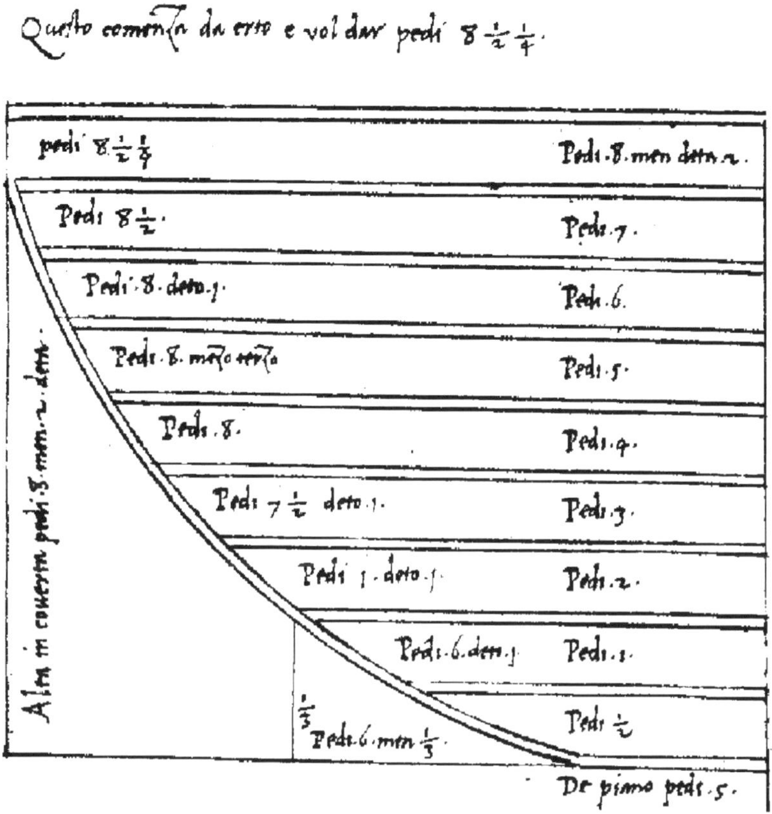

Fig. 11.7: Corba de mezzo—midship frame (Michalli da Ruodo [1], BNCF, f.6r).

Theodoro names a rule of thumb for measuring *tre pie* and *sie pie* based on the width of the floor ([5] BNM, f.36v; ASV, f.38v).

Combining Lateral Plan and Vertical Cross Section

First of all the *cantier* or wharf is erected (Fig. 11.9A). Its curvature is given by the *sentina del cantier* (Fig. 11.9B), the perpendicular at lowest point of the wharf. *Lanza* and *pontal* added to the ends of the *cantier* form two triangles with additional offsets (Fig. 11.9C). They define the outline of the midship plane with keel, stem and sternpost (Fig. 11.9D). There is a gap at both ends between wharf and keel, which is called *calcagnol* (heel, Figs. 11.4, 11.9E). A second vertical plane, orthogonal to the first is defined by the contours of the main frame (Fig. 11.9G).

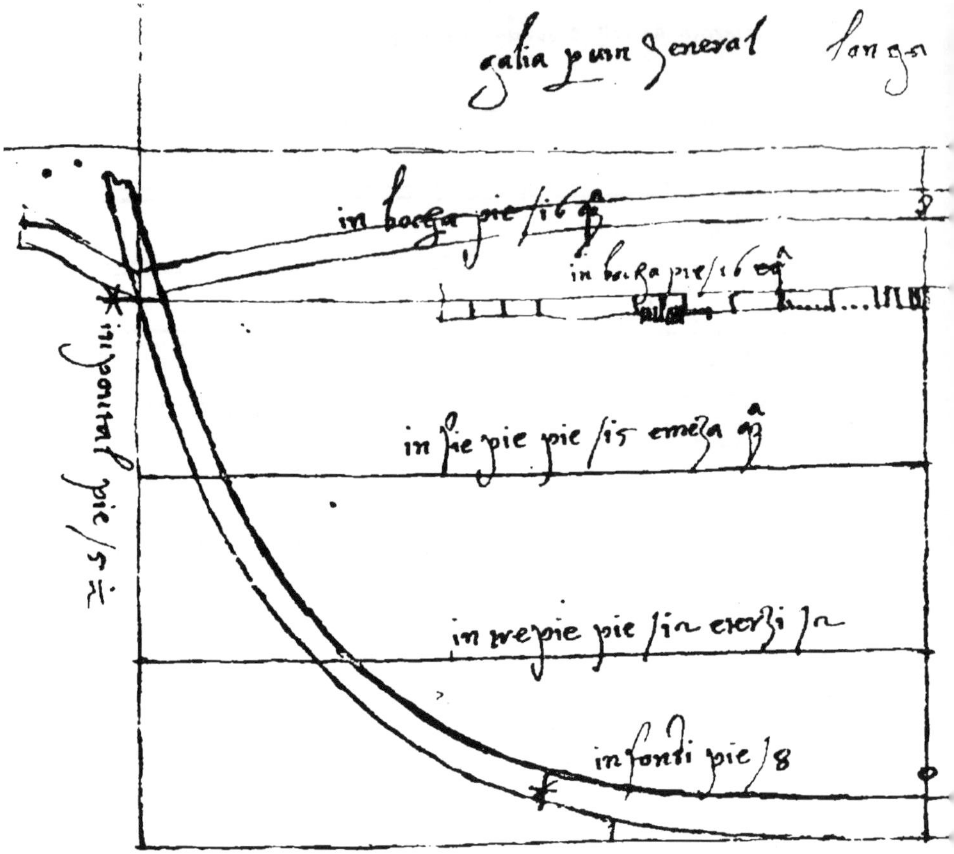

Fig. 11.8: Corba de mezzo with its lata—midship frame with deck beam (Pre Theodoro [5] BNM, f.4v).

Up to this point the shaping of a ship is quite straightforward. But what now follows makes naval architecture to a task completely different from civil architecture.

"Partisoni" or Shaping the Frames

> *Delle cose necesarie et importanti per fabricare la partizion del fondi, stella, scorer del forcame e ramo sono quatro cose le piú necesarie in una fabrica da remo che si possano praticare, e nelle Navi cusi aggiunge anco l'archetto e si come sono le piú necesarie cosi anche sono le piú dificile si da formare, che da por in opera, e da intendere anco dalla magior parte di quelli della professione non che da altri, che non sono fabricanti,…* (Steffano de Zuanne [9], f.18r)

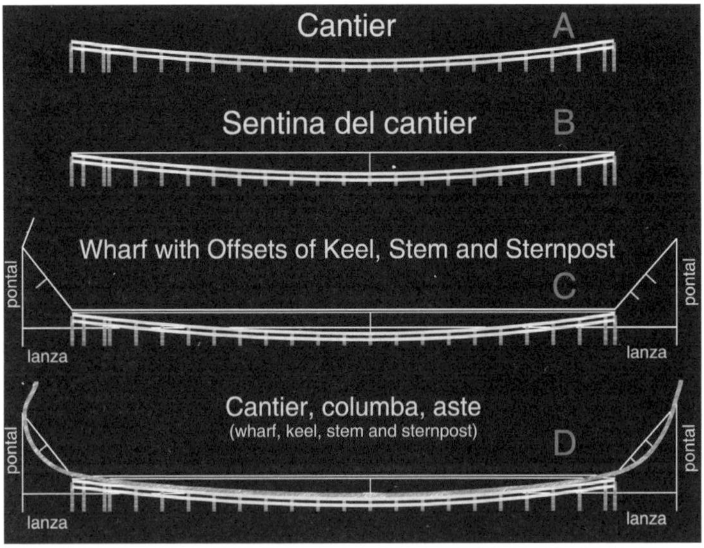

Fig. 11.9a [Col. Pl. XXXIII]: Shaping the lateral plan A–D.

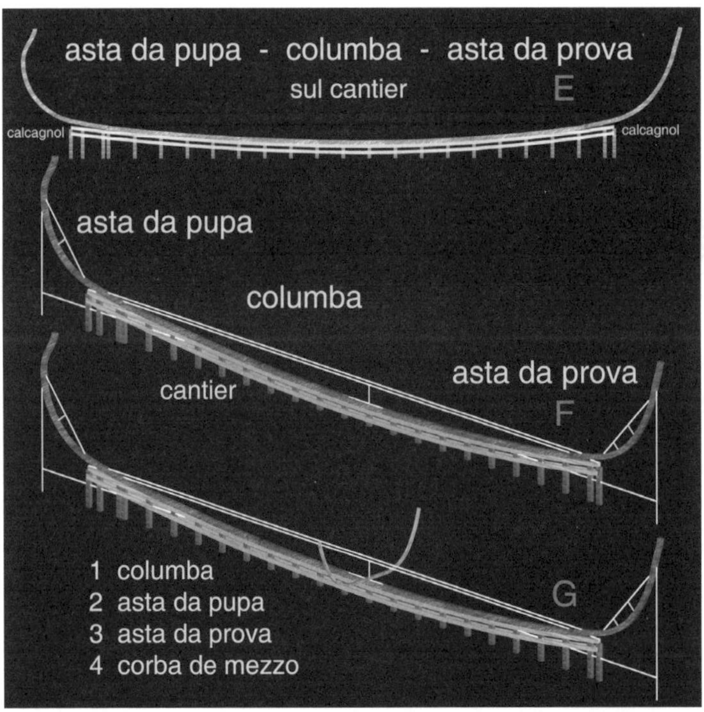

Fig. 11.9b [Col. Pl. XXXIII]: Shaping the lateral plan E–G.

Once the main frame was defined, a series of three or four variations—known as *partisoni* in Venice[2]—could be used to derive the shape of all the other frames.

The first step, *partison de fondo* (Fig. 11.10), gave its name to the whole procedure. It shortened the *fondo* and allowed the bottom of the hull to taper inward at the ends.

In a second step the mould was put on a plinth, called *stella* (Fig. 11.11), so that the deck rose slightly and the bottom waterlines ran at a more pronounced angle fore and aft.

The third step, the *legno in ramo* (Fig. 11.12), increased the frame side angle by turning the *sesto* around the *poselexe del fondo*, which is marked here with a small star at the bottom of the *sesto* or frame (Figs. 11.10–11.12). Compared with the U-shaped midship frames the front and rear frames were more V-shaped.

Some Venetian naval authors mention a fourth step, called *scorrer del sesto*, which was applied to the *chomedo del fondo* (floor-heads) (Pre Theodoro [5], BNM, f.15v , ASV, f.17v) to move the above mentioned rotating axis marked by the *poselexe del fondo* along the line of the frame.

"Pavion"—*the Plan*

The *sesto* is moved according to uniquely different values for each *partison*-step and for each frame. Every frame that is calculated with the *sesto*, is called *corba in sesto—"frame in the mould"*. A Great Galley of about 1550 had 85 of these frames. They are limited fore and aft by *cavi de sesto—"heads of the mould"*. Only for these two moulds numerical values were stated and written down by the naval architect. The shipwright calculated the *partisoni* of the other 82 frames by means of a geometrical procedure.

Fig. 11.13 shows Steffano de Zuanne's parchment with two triangular diagrams divided into smaller segments. A set of such drawings is called *pavion*. This technical term characterizes the naval architecture of those days. The Venetian word *pavion* is related to *pavese—"shield"* and *pavimento—"floor"*. The *pavion* is actually the plan for shaping

[2] According to B Drachio, in ASV, Arsenale, busta 25, Visione, f.6; published and translated by Lehmann [20]. These Mediterranean design techniques are elaborately and profoundly discussed by Rieth [24].

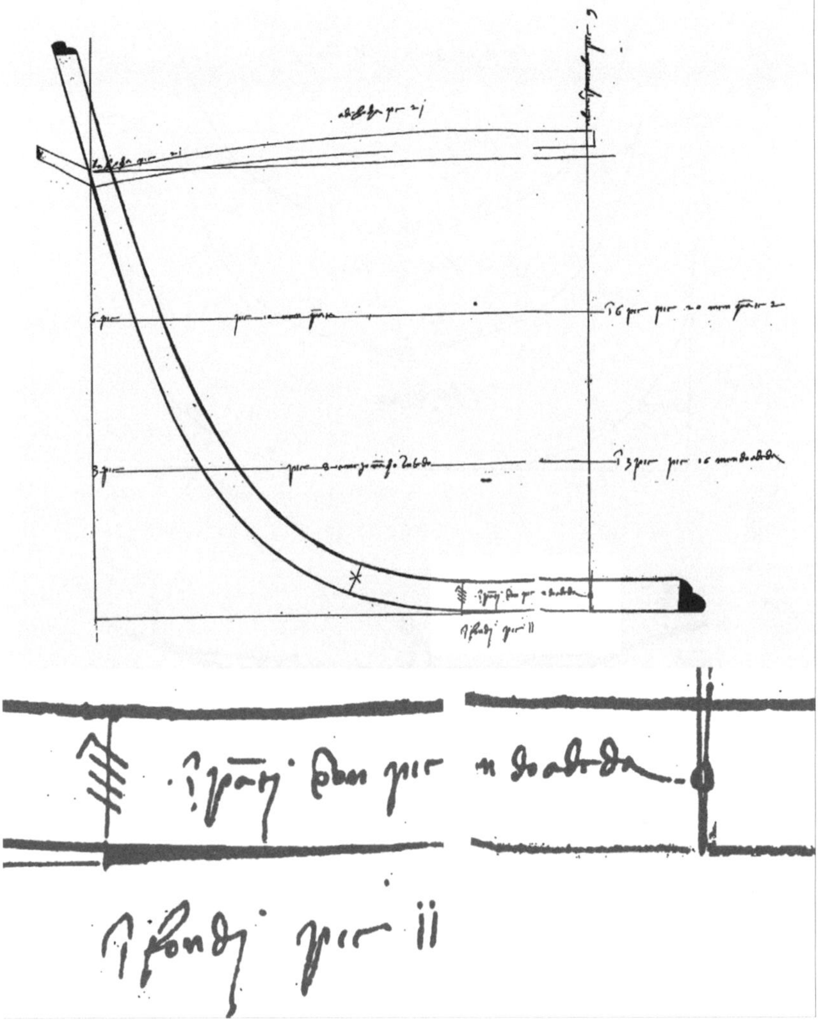

Fig. 11.10 [COL. PL. XXXIV]: Partison del fondo (Ragioni antique [3], f.66v).

the ship, which is drawn in full size on the floor of the mould-loft. Like a modern slide rule or reckoning table it is used for shaping not only the frames but a vast variety of elements, such as deck beams, parts of the outrigger, sails and others. However the *pavion* had to be calculated and traced out individually for every single ship. Different figures were drawn: *"Skilled mathematicians and others have developed many methods for drawing the lines"*—Steffano de Zuanne writes—*"but*

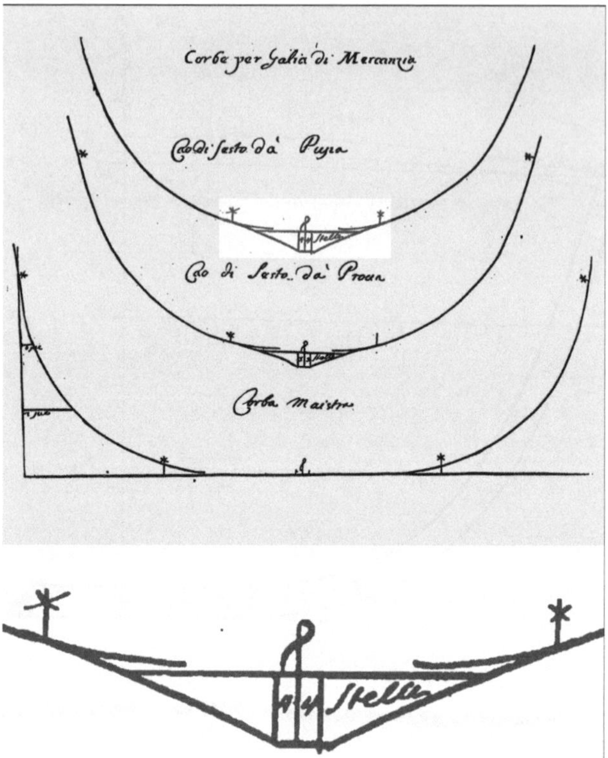

Fig. 11.11 [Col. Pl. XXXV]: Stella (Steffano de Zuanne [9], f.20r).

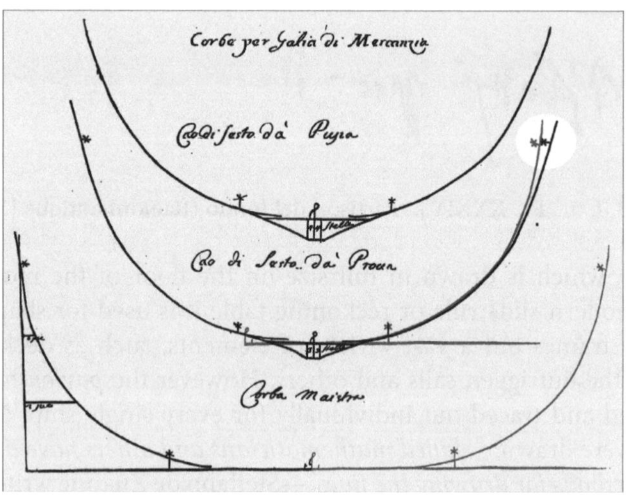

Fig. 11.12 [Col. Pl. XXXVI]: Legno in ramo (Steffano de Zuanne [9], f.20r).

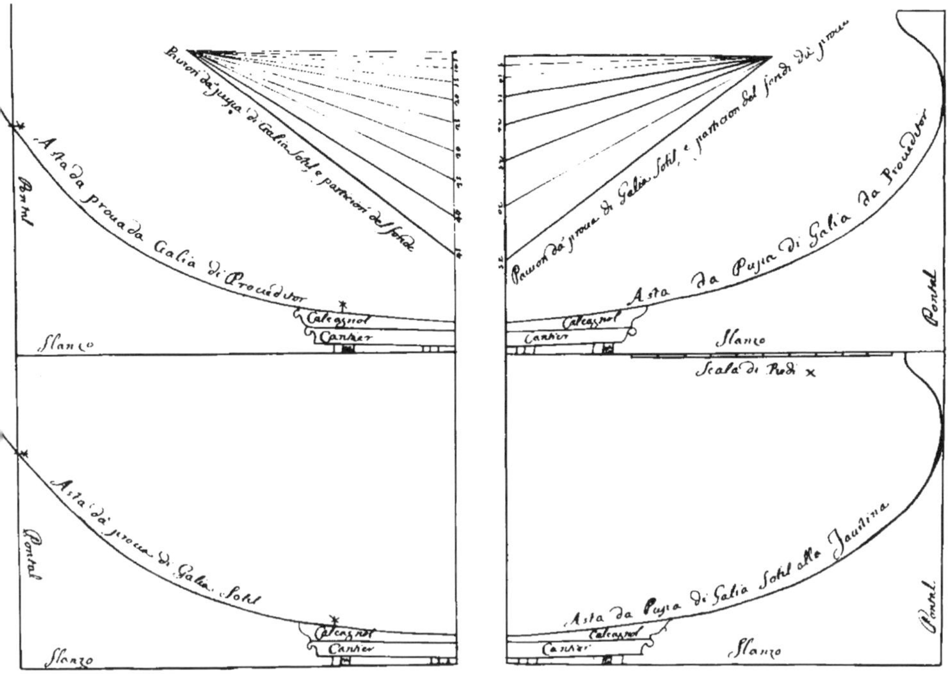

Fig. 11.13: Pavion of a galley together with stem and sternposts (Steffano de Zuanne [9], f.8v–9r).

the simplest and most reliable is the freehand method with a sheet of paper"—he suggests (BM, Add Mss 38655, Steffano de Zuanne [9], f21r). *"Io ho fatto vedere con una semplice strica di carta nella mani senza altra cosa far partizion bellissime, e curiose, si per vaseli tondi che da Remo."* (Steffano de Zuanne [9], f. 21r)

Steffano de Zuanne outlined triangles that were divided according to the formula shown in Figs. 11.14 and 11.15. Other shipwrights preferred oval lines or the so-called *mezzaluna—"half-moon"* (in red) for the same purpose. Results differed in detail, as can easily be seen in Fig. 11.14. But these approaches are all based on the same procedures. All *pavioni* always divided a given linear measure into unequal parts of evenly growing intervals. Triangles, circles, half-circles or any other kind of *pavioni* were the means to design scales, called *morelli* (Fig. 11.6 top left and Fig. 11.15) used for adjusting the mould.

The shipwright saved most of his reckoning work with these small scales. He usually calculated every fourth or fifth frame accurately and marked its measure with a line on the *morello*. These frames, called *onze*, that is *"aligning frame"*, were erected first.

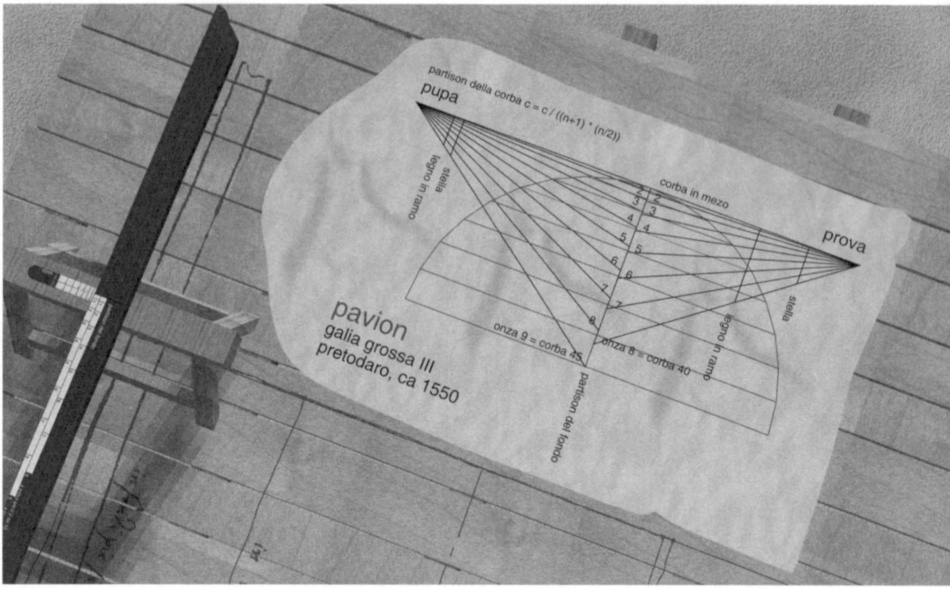

Fig. 11.14 [Col. Pl. XXXVII]: Pavion.

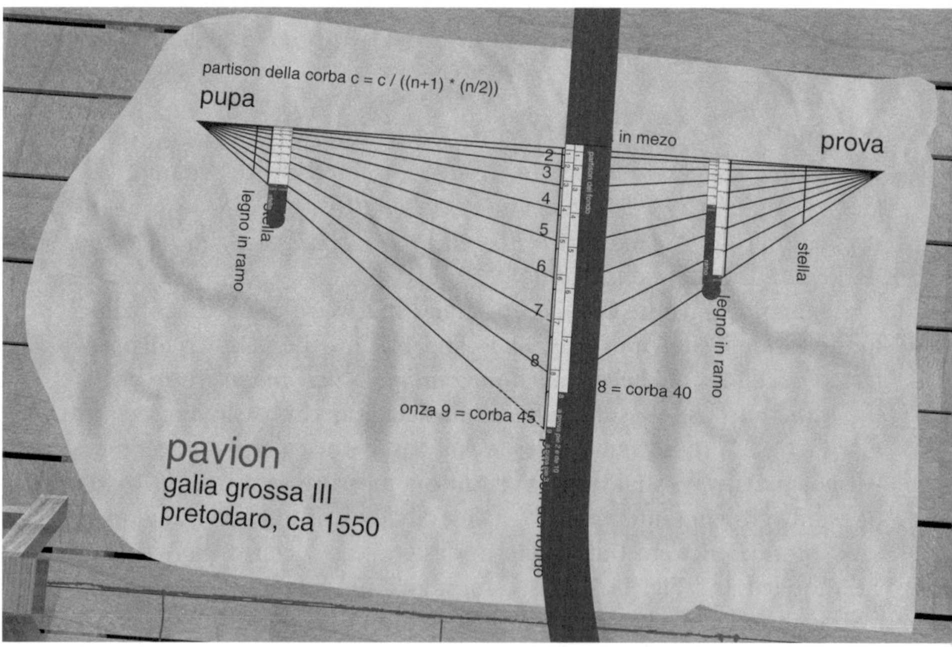

Fig. 11.15 [Col. Pl. XXXVIII]: Morelli—adapted scales.

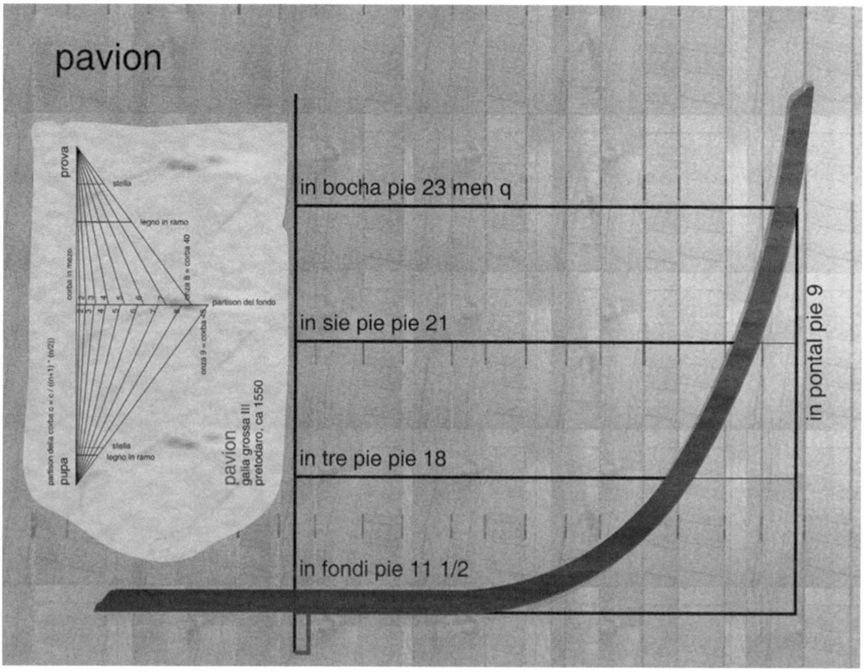

Fig. 11.16a [Col. Pl. XXXIX]: Pavion (left), sesto and its offsets.

Shaping the frames is the shipwright's foremost task, which characterizes his work and distinguishes naval architecture fundamentally from civil architecture. Therefore one should have a closer look at this very special procedure:

- The naval architect determines width and height of the midship bend, divides its height into three equal parts and draws its outline according to these divisions (Fig. 11.16a).
- The *sesto* is shaped to this line.
- On the left the *pavion* with the scales is put on the floor to transfer the *partison del fondo* to the mould (Fig. 11.16a).
- All *partisoni* are applied to the line of the main frame.
  1 *partison del fondo* (Fig. 11.16b first row).
  2 *stella* (second row).
  3 *legno in ramo* (third row).
- Mould and keel have to be moved almost one foot upwards—considering the effect of the *sentina del cantier*. Having done this, the line of frame no 45, that is the *cavo di sesto*, can be drawn (Fig. 11.16b last row).

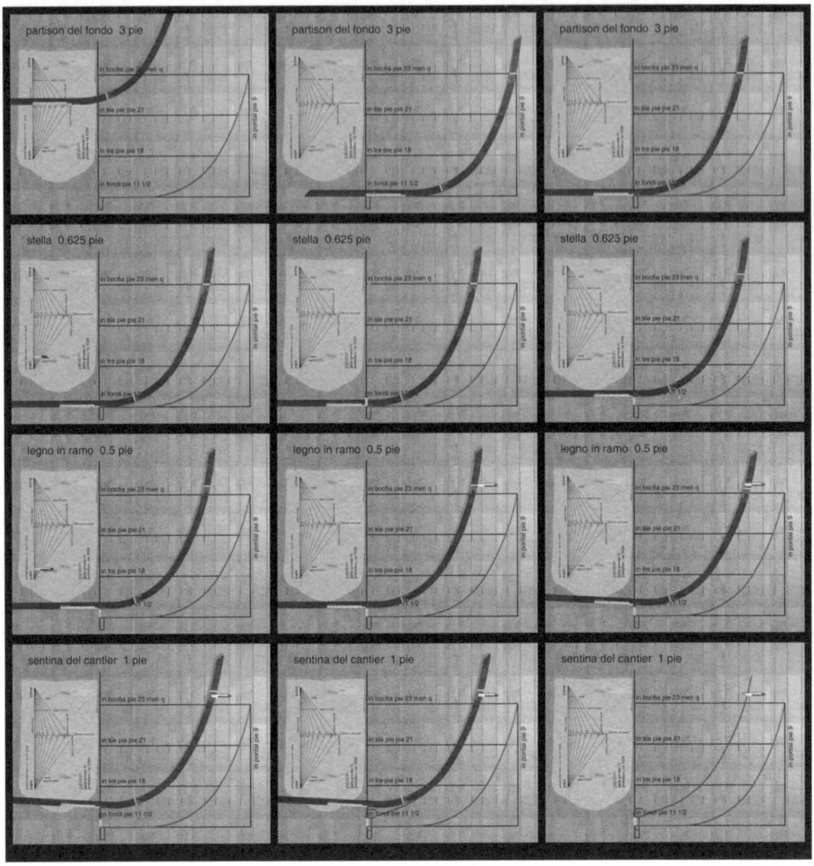

Fig. 11.16b [Col. Pl. XL]: The effects of: partison del fondo, stella, legno in ramo, sentina del cantier (from top).

Such design techniques have not yet become extinct. In a few traditional Venetian boat yards *gondole* are still being built according to *sesti* and *partisoni*.

Fig. 11.17 gives two moulds with their *partisoni* scales in the *squero Tramontin*. Their use is quite different in detail, but the procedure is similar to that of the galley-builders 500 years ago. One can imagine what the floor of Pre Theodoro or Steffano de Zuanne might have looked like (Fig 11.18). There are his tools for his shaping task: *sesto*, *pavion* and *morelli*.

Fig. 11.17 [Col. Pl. XLI]: Sesti (moulds) in the Squero Tramontin (Tramontin's wharf).

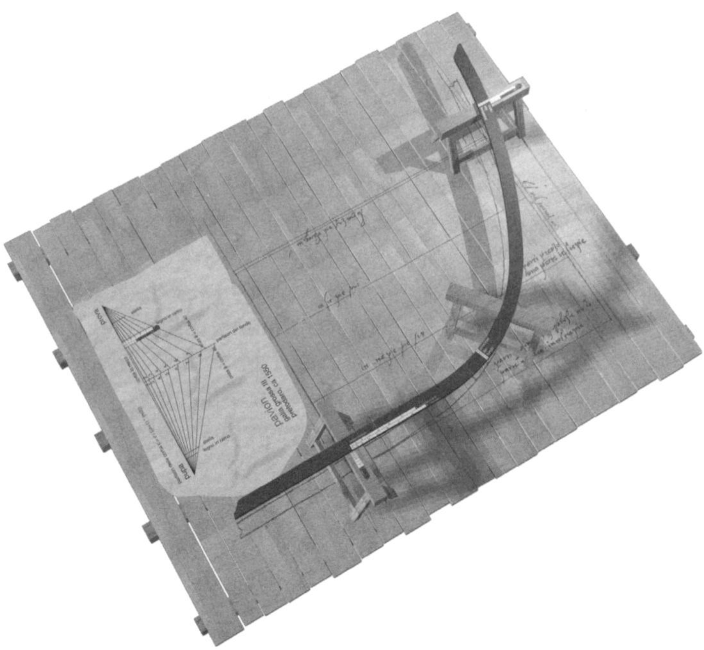

Fig. 11.18 [Col. Pl. XLII]: Tools in the mould loft.

Objections

Naval Architecture Really Digital?

One might perhaps object that there was rather little evidence for digital shaping: no numbers, no calculating, no arithmetic at all, but drawings of triangles, of half circles, of scales, of a mould, of stem and sternpost. Drawings seem to have been much more important for the job of a Venetian *marangono* (shipwright) than numerical mathematics.

Drawings and sketches indeed supported the carpenter's work when he was assembling the vessel. But what we are dealing with is the design of the ship. Among these drawings there are no drafts of the projected hull, no line drawings that would give an idea of the ship's shape, not even drawings of parts of the vessel. They were not necessary because the carpenter could easily outline the shape of every single frame with *sesto*, *morelli* and *partisoni*. These tools where the real objects of ship design. Not the ship's shape but the shape of the *sesto* was determined by the naval architect. He completed his work by setting the measures for three or four *partisoni* which allowed him to draw his *pavion* according to the sine-function or to similar calculations.

The shipwright's rules were certainly devoid of any scientific or mathematical foundation. Their existence was due solely to the simple fact that common shapes could be reproduced easily by their means. Nevertheless the masters might well have been influenced by the then very popular ideas of 14th's century *calculatores*.

Thomas Bradwardine's theories ("Tractatus de proportionibus velocitatum in motibus", 1328) on arithmetically definable proportions of movements and his mathematical approach to nature in general was elaborately discussed for example at the university of Padova—well observed by scholars of mechanics like Giovanni Fontana (see several essays of Giovanni Fontana listed by Birkenmajer [13], and Thorndike [27]), who begins his treatise (Fig. 11.19): *"Dimensions of lines, areas and bodies, sites and distances are measured by technicians and engineers in multiple different ways..."*[3] Contemporary engineers like Conrad Gruter[4]

[3] Fontana's treatise on metrology (Bologna, BUB 2705, f.85r–105, "Metrologum de pisce, cane et volucre") will soon be fully translated and published by Horst Kranz.

[4] Roma, BAV, Vaticanus latinus 5961, Conradus Gruter de Werdena: "Considerando quantum et quamdiu mechanicis insudari studiis,...", Venezia 1424, cap. 7, 20, 24, 26, published by Lohrmann / Kranz / Alertz [21].

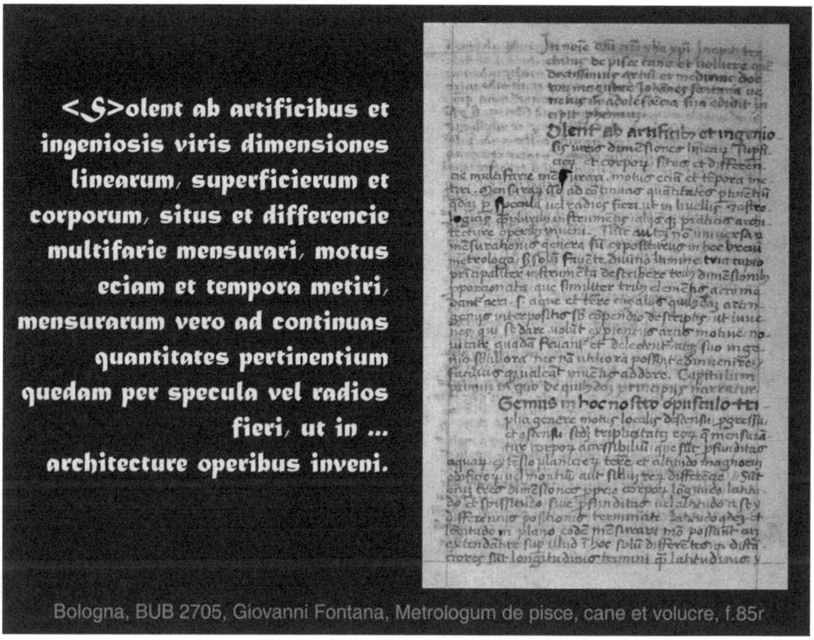

Fig. 11.19 [Col. Pl. XLIII]: Treatise on Metrology of Giovanni Fontana, Introduction.

were also convinced, that mechanical functions directly depended on the right proportion of numerical values. Why should Venetian naval architecture of the 14th and 15th century not have been affected likewise—even if Bradwardine or any of the innumerable comments on his 'De proportionibus' are not explicitly mentioned?

Naval Manuscripts Far from Real Work?

A second objection might occur concerning the role of the naval manuscripts. They have sometimes been regarded as being far away from the real work. They would merely deal with idealised numbers and calculations that represented more the author's abstract will than the needs of ship-design, was said.

These arguments are obviously refuted by the notes of naval architects such as Pre Theodoro or by Steffano de Zuanne's handbook. But it seems to me, that even the non-professionals in naval architecture like Michalli da Ruodo, Zorzi Trombetta and Joseph Furttenbach were much closer to the real matters of shipwrightry than one would probably expect. An experiment might help to investigate this assumption.

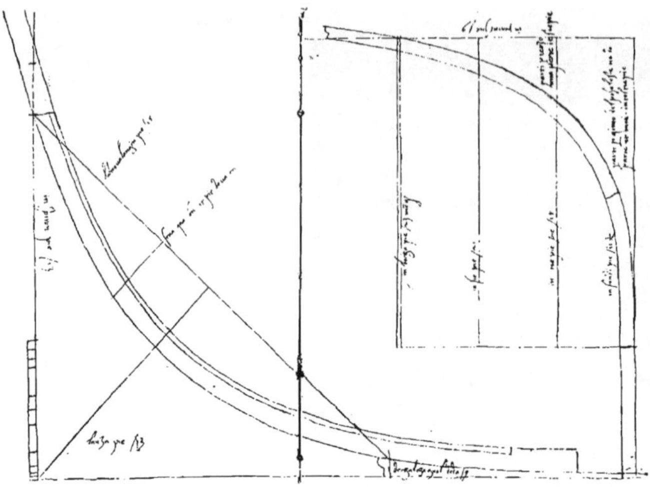

Fig. 11.20a: Offsets of asta da prova and sesto of Pre Theodoro's third galia grossa ([5], BNM, f.IIr–IIIr, 36v).

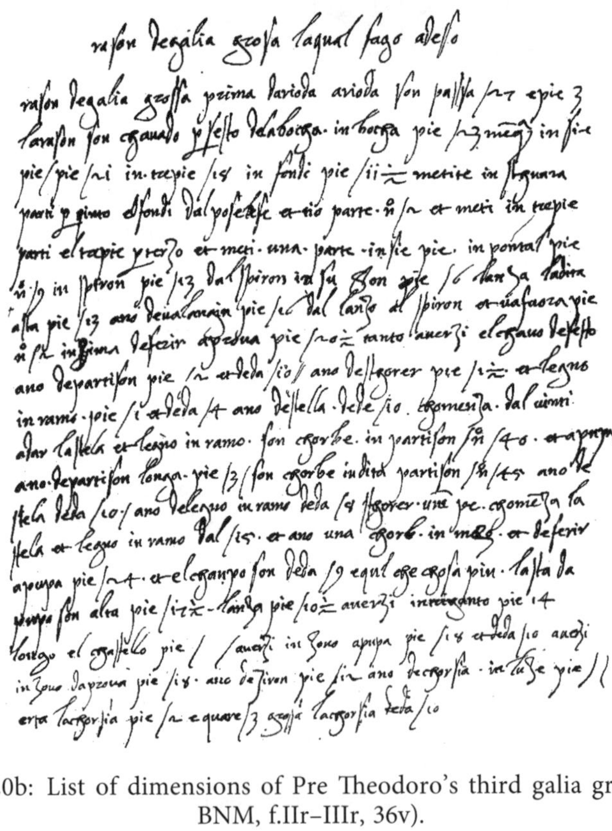

Fig. 11.20b: List of dimensions of Pre Theodoro's third galia grossa ([5], BNM, f.IIr–IIIr, 36v).

```
*********************************************
GALIA GROSSA C di Pre Theodoro
in piedi x 10
*********************************************
(setq Titl "GALIA GROSSA C di Pre Theodoro")
METHOD OF DESIGN
(setq MezLun nil)                  ;'mezza luna' or triangle
(setq KnikSp nil)                  ;defines the kind of frames
HULL
(setq LnRR   1380)                 ;length between perpendiculars
(setq PM       10)                 ;number of parallel midship frames
(setq KWL      80)                 ;waterline
KEEL AND STRAKES
(setq SenCan   10)                 ;'sentina del cantier'
(setq Cantir '( 6.4   7   ))       ;thickness, height of 'cantier'
(setq Columb '( 6.4   9   ))       ;thickness, height of 'columba' (keel)
(setq Parmez '( 7.5   9.5 ))       ;thickness, height of 'paramezal'
PLANK ABOVE THE CENTA (WALE)
(setq MDB    '( 3.75 10))          ;thickness, height of 'magier de boca'
PLANKS
(setq fili   12)                   ;no of planks                 (estimated)
MAINFRAME
(setq Fondo   115)                 ;'in fondo'
(setq trepie  180  )               ;'in trepie'
(setq siepie  210  )               ;'in siepie'
(setq boca    227.5)               ;'in bocha'
(setq Corb   '( 4.375 4.375))      ;dimension of frame
(setq Lata   '( 3.75  4.375))      ;dimension of deckbeam
(setq ponM     90  )               ;height of frame
(setq bolson   12.5)               ;'bolson della coverta'
(setq Colo  bolson)                ;length of 'colo della lata'
DECK
(setq Corda  '( 4     5))          ;breadth, height of 'corda'    (estimated)
(setq SCorda '( 5     7.5))        ;breadth, height of sotto corda
(setq Spina  '( 3.75  7.5))        ;breadth, height of spina
(setq SSpina '( 5     7.5))        ;breadth, height of sotto spina
(setq Co-Co    40)                 ;distance 'corsia' - 'corda'
(setq ScaLat   22)                 ;no of the 'lata' close to the 'schaza'
(setq ScaLen    5)                 ;length of 'schaza' - in late  (estimated)
VOGA
(setq DZovo     4.75)              ;dimension of 'zovo'           (estimated)
(setq HPave    28  )               ;height  of 'pavesata'
(setq Post   '( 5     5.7  ))      ;breadth, height of 'postiza'
(setq Banda  '( 3.33  9    ))      ;breadth, height of 'banda'
(setq CadCol '( 4.375 6.875))      ;breadth, height of 'chadena da cholo'
(setq Banchi   30)                 ;number of 'banchi'
(setq fogoNo   27)                 ;no of banco at the 'fogon'
(setq P-P      35)                 ;distance of 'postizzi'
(setq S-S       7.143)             ;distance of 'schermi'
(setq Z-PP (+ P-P S-S ))           ;distance: 'zovo' - 'primo postizzo'
(setq BacBac    2)                 ;distance between 'baccalari' (in frames)
```

```
;------------- different data fore and aft -------------
;LIST OF ONZE
 (setq Onze   '((-1 ONZE)
               (O (list 0 5 10 15 20 25 30 35 40)
                  (list 0 5 10 15 20 25 30 35 40 45))
               (Z (list 45 50 52.5 53.75)
                  (list 50 55 57.5 60 62.5)) ))
;CAO DI SESTO (ALIGNING FRAME)
 (setq CaoDS  '((-1 CAODISESTO)
               (stel    6.25  6.25)  ;'stella'
               (part   26.25 30   )  ;'particion dal fondi'
               (ramo   12.5   5   )  ;'ramo'
               (Fpa     2     2   )
               (ferir 205   240   ))) ;'ferir'
;STEM AND STERPOST
 (setq Calcag '((-1 CALCAGNOL)        ;'calcagnol'
               (H       5.625 5.625)
               (L      90   110   )))
 (setq Asta   '((-1 ASTA)             ;stem and sternpost
               (H      130   175  )  ;height
               (Lanzo 130   105   )  ;'slanzo'
               (HLot-   0   -60   )  ;intersection: perpendicular-stem below H
               (LDeck   0   -25   )
               (HDeck   0     0   )  ;deck above stem or sternpost
               (L+     14   -24.8 )  ;angle of stem / sternpost
               (H+     50     0   )  ;height of stem or sternpost above H
               (Vala   28    30   )  ;'valangin'
               (ValaO   0.3   3   )  ;upper 'valangin' 'alla Faustina'
               (ValaOO  0     0   )  ;upper part of upper 'valangin'
               (ValaUU 23     0   )));lower 'valangin'
 (setq Trigto '((-1 TRIGANTO)         ;randsom
               (B       0     5)     ;thickness of the 'triganto'
               (H       0     6)     ;height
               (L       0   140)     ;length or breadth
               (BOLM   -1     0.3)   ;reduces 'bolzon' for 'triganto', 'zovo'
               (voga        nil) ))  ;width of poop platform
;CENTA
 (setq Centa  '((-1 CENTA)
               (B       6.25)        ;thickness 'per il morso'
               (H       5   )        ;height
               (MDB+    0    18   )  ;enlarges distance 'mdb'-'centa'
               (Ende    S     S) ))  ;end of 'centa' (sternpost or deck)
;VOGA
 (setq Voga   '((-1 VOGA)
               (B      296  )        ;breadth between 'postizze'
               (H       32.5)        ;height above deck
               (B-       0     0)    ;diminishes breadth
               (H-       1.25  5)    ;diminishes height
               (Colo+   10    10)    ;enlarges length of 'colo'
               (BaPo     4.2   4.2)  ;distance: 'banda'-'postizza'
               (palm   100   145)))  ;'palmeta'                    (estimated)
 (setq Corsia '((-1 CORSIA)           ;gangway
               (S        5.625)      ;thickness of 'corsia' side planking
               (B       27.5)        ;breadth of the gangway
               (H       25)          ;height above deck
               (B-       0     0)    ;diminishes breadth of the 'corsia'
               (H-       3.75  0) )) ;diminishes height
```

Fig. 11.20c [Col. Pl. XLIV]: Dimensions of the galia grossa C, noted in LISP.

Experiment

We have got here in Fig. 11.20a and Fig. 11.20b the main dimensions, *partisoni* and measurements from Pre Theodoro's notes for shaping the *galia grosa laqual fago adeso*—"*…which I am just building*"—as he tells us in the first line (BNM, f.36v). This vessel, his third Great Galley, is called here *galia grossa C.*

The question will be: Can such a list of measures be sufficient to build a specific ship according to the system of *partisoni*?

Transformed into a different layout Pre Theodoro's measures are accessible to the LISP-interpreter of the CAD-programme. These about 100 lines (Fig. 11.20c) have all the data needed to draw the galley. Most of this list is commentary (in red and blue). Dimensions are black. Some measures, missing on the original folio, had to be added from Pre Theodoro's other *galere grosse* or from contemporary sources on similar vessels, such as the '*Misure di vascelli et cetera di…proto nell'Arsenale di Venezia*' of 1547 or had to be estimated, which is mentioned in the list's commentary (blue). But all the dimensions essential for shaping the galley's hull could be based on Pre Theodoro's text.

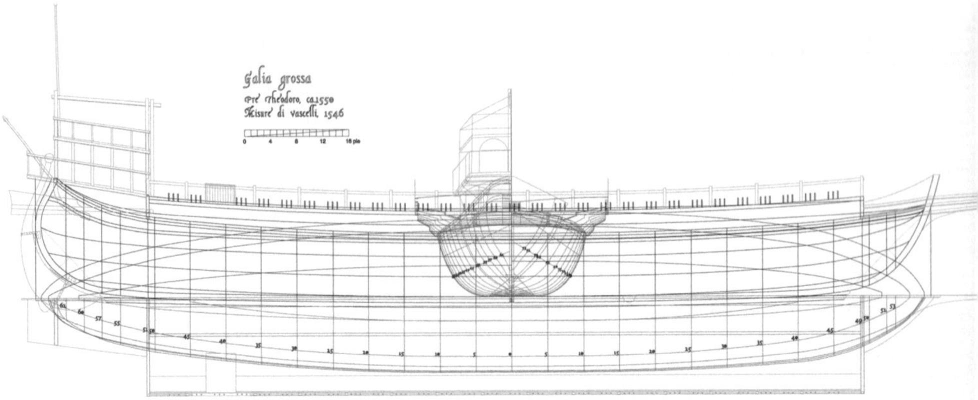

Fig. 11.21: CAD lines-plan of Pre Theodoro's galia grossa C.

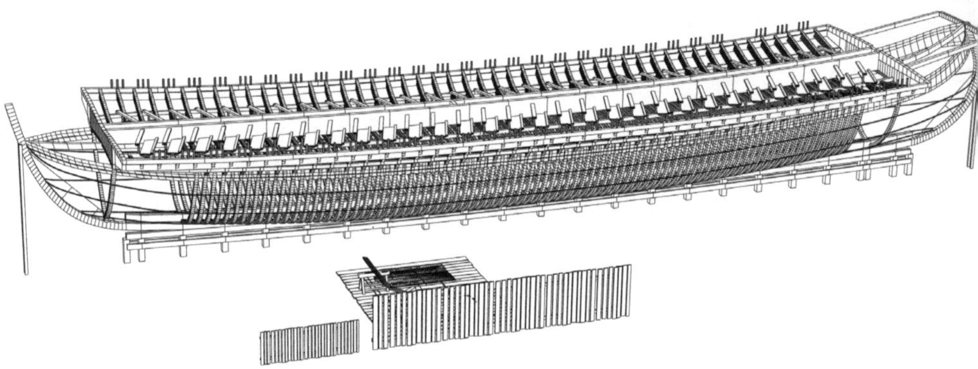

Fig. 11.22: 3D vectors without hidden lines of the galia grossa.

The CAD-programme opens an empty window in order to load some 100 pages of LISP-programme-code which then will read and interpret this data list to build the galley exactly to the dimensions of Pre Theodoro's third *galia grossa*.

Modern ship designers usually get an image of their projected vessels by drawing the lines of the hull (Fig. 11.21). The lateral plan is black. The frames are outlined in red, and the green lines show the hull from above. The blue lines follow the contours of the diagonals as projected into their own oblique plane. Everything not comprehended in the system of *partisoni* is drawn manually in cyan.

Such a drawing is quite appropriate for designing on paper. The computer however is not limited to two dimensions. Its vector space

Fig. 11.23 [Col. Pl. XLV]: Galere grosse (great galleys) and their boats.

can easily be extended to three dimensions, which might give a far more realistic image of the vessel.

Yet the result would look quite confusing, because lots of coloured, curved lines would seem to disintegrate the shape of the galley's hull. We can however imagine the areas between the lines as defining surfaces of the structural elements. Removing all lines which are not visible behind these surfaces will partly resolve this problem, but the drawing still looks rather technical (Fig. 11.22).

Therefore the CAD-programme will apply colour and patterns to the surfaces of this grid model. The pattern is for example a photograph of the veining of wood. Some sources of light will give shadow effects to the scene. All these calculations would take a while—sometimes several days on a small computer.

Finally we should have a look at a Venetian shipyard (Figs. 11.23 and 11.24). First of all, the keel beams are put on the wharf (Fig. 11.24 foreground left); stem and sternpost are erected and then the main frame is set upright together with two *cavi di sesto* (in the background). After having examined his layout the shipwright will put all calculated frames in place (ship in the foreground). The master then attaches the

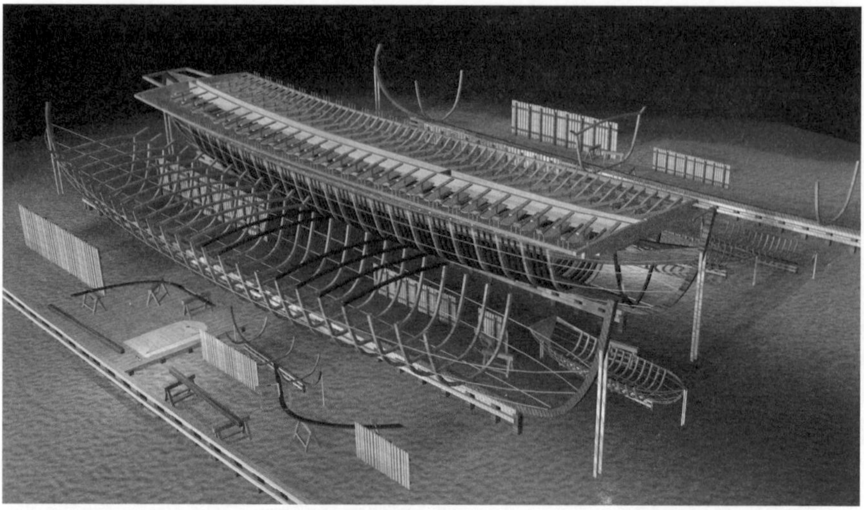

Fig. 11.24 [Col. Pl. XLVI]: Squero—Building a galia grossa.

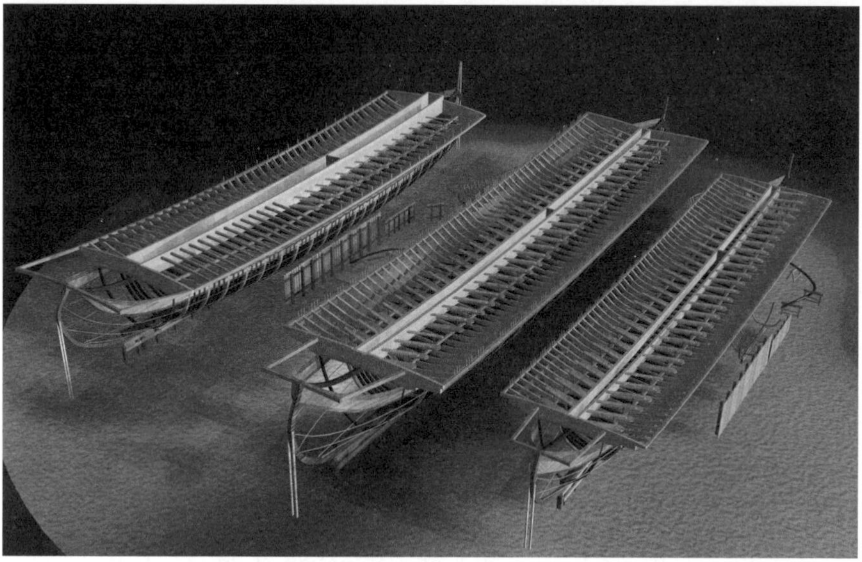

Fig. 11.25 [Col. Pl. XLVII]: Merchant galleys of 1434, 1550 and 1686 (from right to left) on the squero.

maistre—flexible ribbands, which determine the hull curvature—before all the other 82 *corbe in sesto* are set up (galley in the middle—see e.g. Damianidis [16], pp. 218–219).

The *voga* takes up almost the entire deck. Two *zovi* (yokes) are connected by the *corsia*—the gangway. Finally the *bacchalari* and *banchi* (thwarts) will be installed. This was usually done when the galley was already afloat. But the planking is left out here because it is of no importance for the design.

Conclusion

Ship design of this period was completely based on calculation and variation. Those shipwrights had no intention to get visual images of what they were planning. Draughts or layout plans did not exist until the lines drawings of the 17th century came into use. But one should not forget, that the *"lines"*, which gave their name to this kind of drawings, originated in outlining the results of *partison* calculations.

To my tentative understanding the naval architect's work differed in parts quite substantially from civil architecture and engineering. These trades were—as far as I believe—not necessarily capable of exchanging their trade rules.

References

A. Venetian Naval Manuscripts

1. Michalli da Ruodo, 1434–1444. Firenze, Biblioteca Nazionale Centrale, codice Magliabecchiano classe XIX, 7 (123 fols), "Fabrica di galere" = "Libro di Marineria" (Late 15th- or early 16th-century copy); Wien, Österreichische Nationalbibliothek, Coll. Marco Foscarini, Codex 6391 (122 fols) "Arte de far vasselli" (16th century copy). (The recently rediscovered original will be published by David McGee, Pamela O. Long and Alan Stahl; see David McGee's article here).
2. Zorzi Trombetta da Modon, 1441–1449. London, British Library, Cottonian Manuscripts, Titus A.26 (62 fols), (partly published by Anderson [10]).
3. Anonymous, 1470–1529. Greenwich, National Maritime Museum, NVT 19 (67 fols), "Ragioni antique spettanti all'arte del mare et fabriche de vasselli", (with partial copy of Michalli da Ruodo's text), (published by Bonfiglio Dosio / van der Merwe / Chiggiato / Proctor [15]).
4. Anonymous, 1546. Venezia, Archivio di Stato, Archivio Privato Pinelli, busta 2 (21 fols), "Misure di vascelli" (published by Tucci [28]).
5. Pre Theodoro de Nicolò, c. 1550. Venezia, Biblioteca Nazionale Marciana, Manoscritti italiani, classis IV, 26 (=5131) (42 fols) "Instructione sul modo di fabricare galere" (partly published by Lane [19]); Archivio di Stato, Archivio Proprio Contarini 19 (39 fols) (copy of the Marciana original).
6. Allessandro Picheroni della Mirandola, c. 1550. Venezia, Biblioteca Nazionale Marciana, Manoscritti italiani, classis VII, 379 (=7588) (21 fols), "Disegni di biremi, triremi, quadriremi".

7. Nicoló e Hieronimo di Bernardin Secula, c. 1550. Venezia, Archivio di Stato, Memorie per servire ai vacui dei Commemoriali, vol. I, "Navilii sotili, cioè vele latine, galie grosse, galie sotili zoé quinquireme, quadrireme, trireme, bireme, unireme, galeotte, fuste, bergantini et fregate. Galee grosse. Il modo di far galee grosse de Nicolò Secula" (published by Bondioli [14]).
8. Baldissera Drachio Quinto, 1599. Venezia, Archivio di Stato, Archivio Proprio Contarini 25 (15 fols), "Visione del Drachio" (published by Lehmann [20]).
9. Steffano de Zuanne de Michel, 1686. London, British Library, Additional Manuscripts 38655 (72 fols) "L'Architettura Navale di Steffano de Zuanne de Michel Vice Proto de Marangoni".

B. *Other References*

10. Anderson, Roger Charles: "Italian Naval Architecture about 1445", in: The Mariner's Mirror 11, 1925, pp. 135–163.
11. Barker, Richard Alan: "Shipshape for Discoveries, and Return", in: The Mariner's Mirror 78, 1992, pp. 433–447.
12. Barker, Richard Alan: "Whole-Moulding: a Preliminary Study of Early English and Other Sources", in: Nowacki / Valleriani [23], pp. 33–65.
13. Birkenmajer, Aleksander Ludwik: "Zur Lebensgeschichte und wissenschaftlichen Tätigkeit von Giovanni Fontana (1395?–1455?)", in: Isis—An International Review Devoted to the History of Science and its Cultural influences 17, Chicago 1932, pp. 34–53.
14. Bondioli, Mauro: "Larte della costruzione navale veneziana tra il XV e il XVI secolo: riflessioni e nuovi documenti", in: Navalia. Archeologia e storia a cura di Furio Ciciliot. The International Propeller Club, Port of Savona, Savona 1996.
15. Bonfiglio Dosio, Giorgetta / van der Merwe, Pieter / Chiggiato, Alvise / Proctor, David V: "Ragioni antique spettanti all'arte del mare et fabriche de vaselli—Manoscritto nautico del secolo XV", Fonti per la Storia di Venezia, Sez.V, fondi vari, Il Comitato Editore, Venezia 1987.
16. Damianidis, Kostas: "Methods used to control the form of the vessels in the Greek traditional boatyards", in: Rieth, Éric (Ed), Concevoir et construire les navires, pp. 217–244.
17. Fernandes, Manuel: "Livro de traças de Carpintaria por Manuel Fernandes, 1616", Programa Nacional de Edições Comemorativas dos Descobrimentos Portugueses, Academia de Marinha, Lisboa 1989.
18. Furttenbach, Joseph: "Architectura navalis. Das ist: Von dem Schiff Gebäw / Auff dem Meer und Seekusten zugebrauchen....", Saur, Ulm 1629.
19. Lane, Frederic Chapin: "Venetian Naval Architecture about 1550", in: The Mariner's Mirror 20, 1934, pp. 24–49.
20. Lehmann, Louis Th.: "Baldissera Quinto Drachio—Visione del Drachio, translated and annotated by L.Th. Lehmann", Amsterdam 1992.
21. Lohrmann, Dietrich / Kranz, Horst / Alertz, Ulrich: "Konrad Gruter von Werden, De machinis et rebus mechanicis. Ein Maschinenbuch aus Italien für den König von Dänemark 1393–1424", 2 vols. Studi e testi 428–429, Città del Vaticano 2006.
22. Lopes de Mendonça, Henrique (Ed): "O Padre Fernando Oliveira e a sua obra nautica. Livro da fabrica das naos", Memorias da Academia Real das Sciencias de Lisboa, nova serie, tomo VII, parte II, Lisboa 1898.
23. Nowacki, Horst / Valleriani, Matteo (Eds): "Shipbuilding Practice and Ship Design Methods From the Renaissance to the 18th Century. A Workshop Report, Unrevised Second Edition—2005", Max-Planck-Institut für Wissenschaftsgeschichte, Preprint 245, Berlin 2005.
24. Rieth, Éric: "Le maître-gabarit, la tablette et le trébuchet. Essai sur la conception non-graphique des carènes du Moyen Âge au XXe siècle", Mémoires de la section

d'histoire des sciences et des techniques 9, Éditions du Comité des travaux historiques et scientifiques (CTHS), Paris 1996.
25. Rieth, Éric (Ed): "Concevoir et construire les navires. De la trière aux picoteux", Editions Erès, Ramonville Saint-Agne 1998.
26. Rieth, Éric: "First Archaeological Evidence of the Mediterranean Whole Moulding Ship Design Method: The Example of the Culip VI Wreck, Spain (XIIIth–XIVth c.)", in: Nowacki / Valleriani [23], pp. 9–16.
27. Thorndike, Lynn: "A History of Magic and Experimental Science, vol 4: Fourteenth and Fifteenth Centuries", New York 1934, pp. 150–182.
28. Tucci, Ugo: "Architettura navale veneziana. Misure di vascelli della metà del cinquecento", in: Bollettino dell'Atlante Linguistico Mediterraneo 5/6, Firenze 1963–1964, pp. 277–293.

CHAPTER TWELVE

PREGNANT COLUMNS. FROM WORD TO SHAPE

Antonio Becchi

> Questa tale Entasis acompagnandosi come vedi in la littera M con epsa columna quale pare come uno corpo di femina prægnante, vel de homo che patisse la præssione de uno grave carico
>
> Cesare Cesariano

The association between *Architectura civilis* and *Architectura navalis* has long fed the imagination of historians. From the times of Noah's ark—house-building-world, when architecture became amphibious, thus becoming a saviour—metaphors, inferences and conjectures have persistently merged with historiographical interpretation. Their proliferation has constantly coincided with a hasty, often bizarre, analysis of sources available, giving rise to hypotheses which are sometimes rashly mistaken for theses. Entasis has played a marginal, though significant, role in this phenomenon.

Elsewhere[1] I considered the interpretation of the Vitruvian passage on *entasis*, as it appears in the major architectural treatises published between the sixteenth and the seventeenth centuries, by examining the association between *quaestio adiectionis* and *quaestio diminutionis*. In the present paper, however, I focus on the geometric constructions that have been proposed to convey the curvature assigned to a column's profile.

The few lines dedicated by Vitruvius to entasis[2] in the third volume of his *De architectura* have led interpreters to 'fabricate' images from words and find a satisfactory means of translating them into 'built architecture'. Before the publication of the *editio princeps* of Vitruvius's *De*

[1] See [7].

[2] See ([37], book III): "De adiectione quae adiicitur in mediis columnis quae apud graecos Entasis appellatur: in estremo libro erit formata ratio eius quemadmodum mollis et conveniens efficiatur subscripta". See also ([37], book IV) about doric columns: "De adiectione eius quae media adaugetur uti in tertio volumine de ionicis est perscripta ita & in his transferatur".

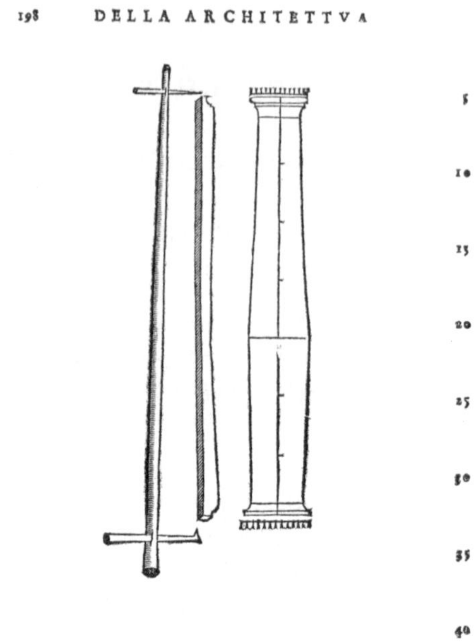

Fig. 12.1: The entasis in one of the images presented by Cosimo Bartoli ([3], p. 198]) in his Italian translation of Leon Battista Alberti's *De re aedificatoria* [1].

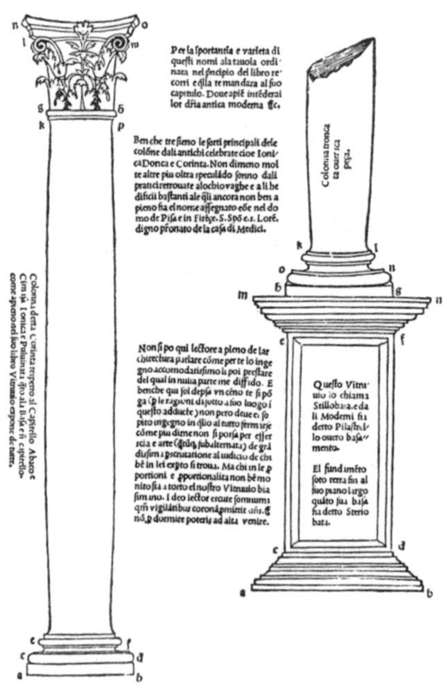

Fig. 12.2: Examples of entasis in Luca Pacioli's *De Divina Proportione* [29].

architectura [37], Leon Battista Alberti dealt with the subject in his *De re aedificatoria* [1]. Alberti outlines the Vitruvian question in somewhat ambiguous terms, mostly because it lacks illustrations ([1], book VI): "Itaque his quas recensuimus perscriptionibus linea quae finitor dicitur composita est. Ad cuius lineae modum tabula gracilis deformabitur, qua fabri lapicidae iustum columnae ambitum et finitiones captent atque terminent." Several years later, Cosimo Bartoli translated the text and published it, together with illustrations [2]. Bartoli's text makes clear reference to the template with which the desired curvature is set and the figure ([2] and [3], p. 198; Fig. 12.1), that goes with it, leaves no room for doubt, but the original Latin passage is nowhere as decisive on this point. Alberti writes about "tabula gracilis": This expression might either refer to the tracing of the curve with a thin and flexible "table," as suggested by other authors, or to the template provided to guide stonecutters on site.

Between the late fifteenth and early sixteenth centuries, other authors dealt with the issue, but failed to make any significant contribution. Luca Pacioli ([29]; Fig. 12.2) and Antonio da Sangallo il Giovane marked the curved profile of entasis clearly, but provided no specific tracing method. Not until Cesare Cesariano's superb, tormented edition of the Vitruvian *Treatise* [16] did the words and the images acquire any precision. Cesariano repeatedly returned to the subject and finally, gave a crystal clear explanation, the term entasis being translated into a particularly eloquent word, *tumefatione*, suggesting a localised and evident bulging of the shaft.[3] The *tumefatione* is drawn with a "circino" (compass) or an "asta longa"[4] (Fig. 12.3). The author also mentions the need to use a ruler as a guide to draw the required outline: "cum sia poi cosa necessaria a concordare epse conale linee e fare la lignea regula qual facia la perequata circumferentia, questa diligentia da te medemo la cognoscerai" ([16], book III, chap. II, fol. LVI r).

Four years later Albrecht Dürer analysed the subject in more detail. In his *Underweysung der Messung* [20], two alternative geometric constructions are proposed for a shaft profile (entasis is translated here

[3] The same term is found in a *codex* by Francesco di Giorgio Martini: "Entasis cioè tumefatione overo enflatione". See Francesco di Giorgio Martini, Codice Spencer 129 ("Opera di architectura"), New York, Public Library, copy from 1520–30 ca. The reference is found in Biffi ([9], p. 209).

[4] The table in Fig. 3, from Cesariano's *De architectura* [16], is numbered as fol. LXIII, but in fact it should be fol. LXII (wrong pagination in the original edition).

Fig. 12.3: Architectural orders and entasis in a plate of Cesare Cesariano's *De architectura* [16].

as *Bauch*, in line with the latin *Venter*). The first makes Cesariano's method more rigorous by reducing it to the problem of a circular arc passing through three fixed points.[5] The second proposes a point-by-point construction method (Fig. 12.4: [20], fol. Giiiiii verso) which in Sebastiano Serlio's later interpretation had a considerable following.

In his *Regole generali di architetura* [34], Serlio outlines a geometric construction which is based on a semi circumference traced on the shaft

[5] On this matter, Dürer refers to volume I, fig. 24 of the *Underweysung*. See ([20], Bvi, fig. 24).

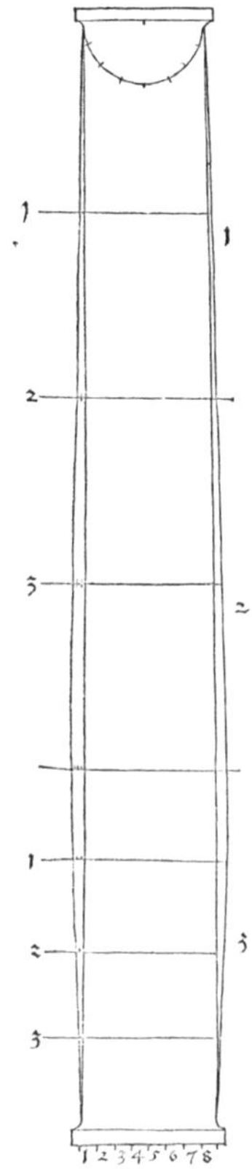

Fig. 12.4: Point-by-point method proposed by Albrecht Dürer in his *Underweysung der Messung* ([20], fol. Giiiiii verso).

(at the height at which the tapering starts), thus allowing points to be fixed to define its profile ([34], book IV, chap. V; Fig. 12.5): "Sotto il colarino, qui si menera due linee a piombo cascante sopra il mezzo circolo, & quella parte del circolo che restera da essa linea al estremo lato di la colonna sia divisa in altre tante parti equali quanto quelle de i dui terzi de la colonna, & cosi fatto da la destra & sinestra banda, sian tirate da li doi lati del mezzo circolo le sue linee a traverso & ad ogni linea postovi il suo numero per ordine venendo a basso, & cosi le linee che parteno la colonna posto li numeri col medesimo ordine, certa cosa è, che la prima linea del circolo si accordera con la linea sotto il colarino, poi si portera la linea seconda del circolo sopra la seconda linea de la colonna (...)."

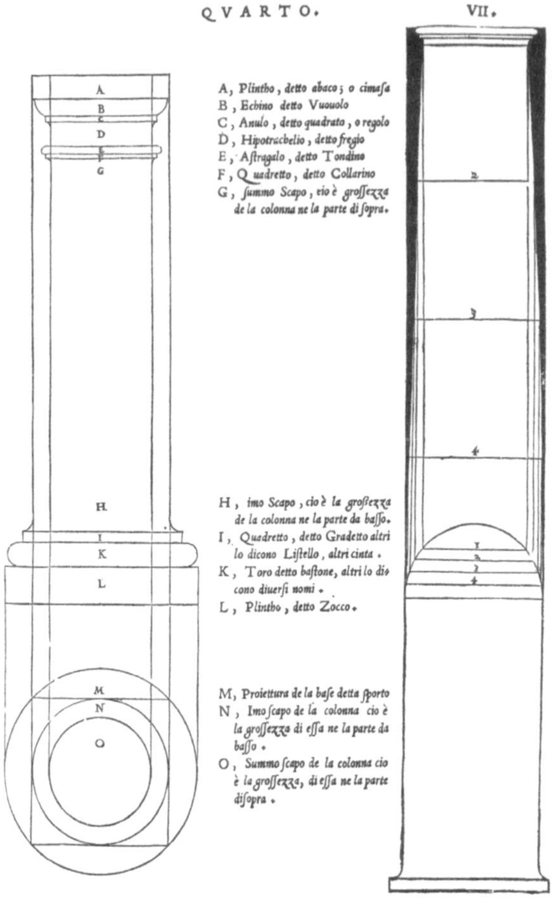

Fig. 12.5: The geometric construction proposed by Sebastiano Serlio in the *Regole generali di architetura* ([34], fol. VIII recto).

The 'Serlio method' gradually became a huge success and reappeared later, in an identical or almost identical form, in works by Coecke van Aelst [17] and Barbaro [4], among others. The extensive circulation of this method suggests that, after years of attempts and uncertainty, the question had finally been satisfactorily answered. The answer provided was, however, provisional and incomplete, as demonstrated in Vignola's *Regola delli cinque ordini* [36], which made a new, important contribution. Vignola describes two procedures (Fig. 12.6) to define the shaft's tapering profile (a third procedure concerns the design of twisting columns). The first corresponds to the method Serlio had already outlined,

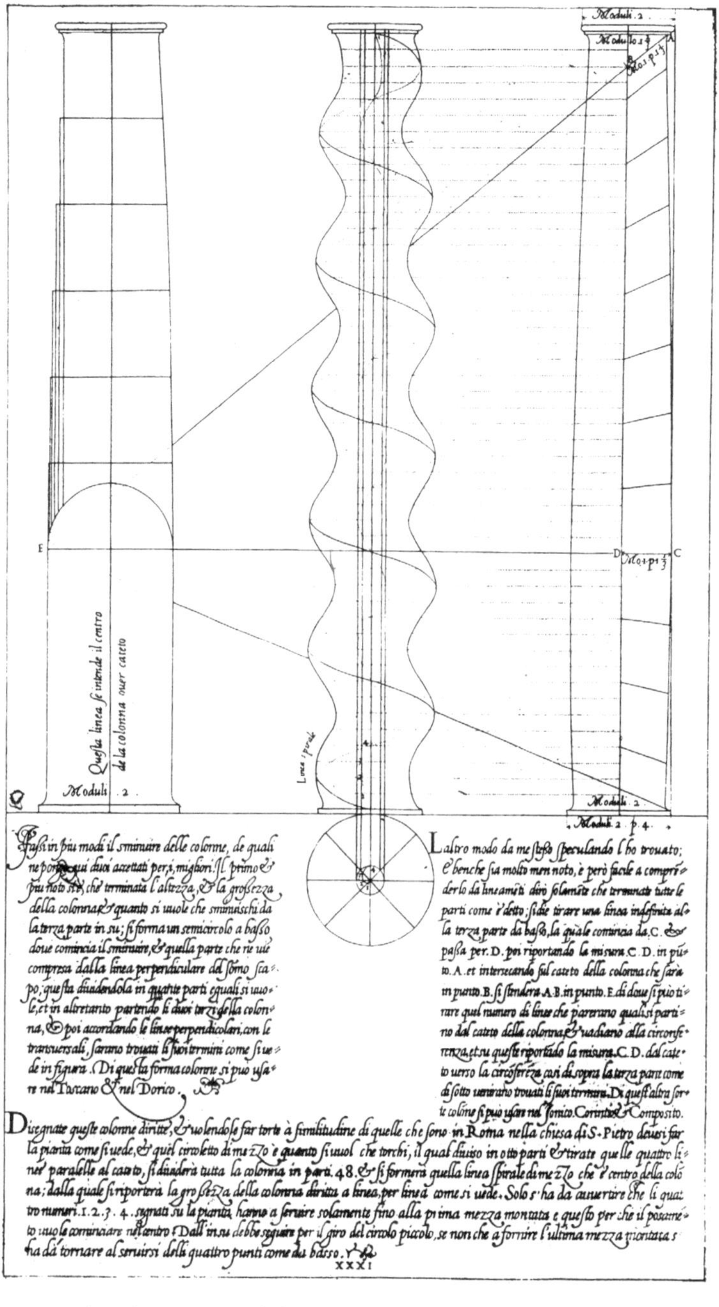

Fig. 12.6: The plate XXXI of the *Regola* by Jacopo Barozzi da Vignola [36]. On the right, the method he proposes for defining the shaft's tapering profile.

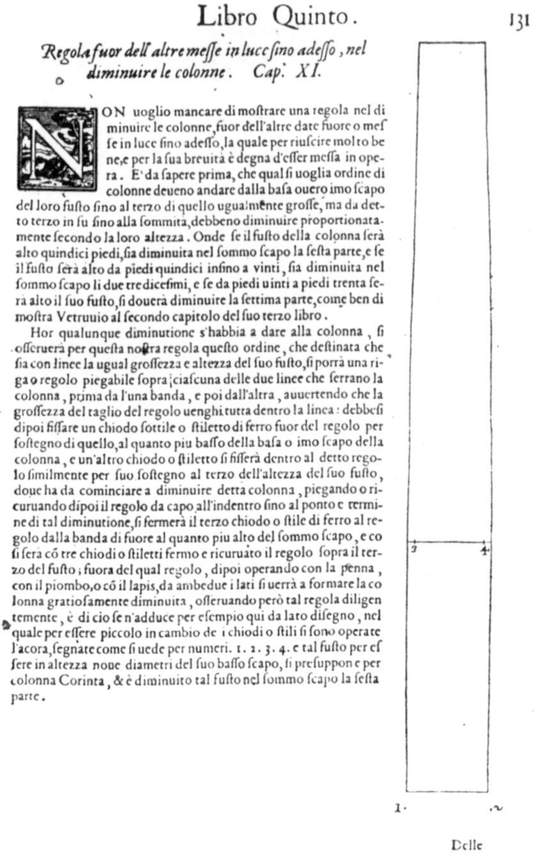
Libro Quinto. 131

Regola fuor dell'altre messe in luce fino adesso, nel diminuire le colonne. Cap. XI.

NON uoglio mancare di mostrare una regola nel diminuire le colonne, fuor dell'altre date fuore o messe in luce fino adesso, la quale per riuscire molto bene, e per la sua breuità è degna d'esser messa in opera. E' da sapere prima, che qual si uoglia ordine di colonne deueno andare dalla basa ouero imo scapo del loro fusto fino al terzo di quello ugualmente grosse, ma da detto terzo in su fino alla sommità, debbeno diminuire proportionatamente secondo la loro altezza. Onde se il fusto della colonna serà alto quindici piedi, sia diminuita nel sommo scapo la sesta parte, e se il fusto serà alto da piedi quindici insino a vinti, sia diminuita nel sommo scapo li due tredicesimi, e se da piedi uinti a piedi trenta serà alto il suo fusto, si douerà diminuire la settima parte, come ben dimostra Vetruuio al secondo capitolo del suo terzo libro.

Hor qualunque diminutione s'habbia a dare alla colonna, si osseruerà per questa nostra regola questo ordine, che destinata che sia con linee la ugual grossezza e altezza del suo fusto, si porrà una riga o regolo piegabile sopra ciascuna delle due linee che serrano la colonna, prima da l'una banda, e poi dall'altra, auuertendo che la grossezza del taglio del regolo uenghi tutta dentro la linea: debbesi dipoi fissare un chiodo sottile o stiletto di ferro fuor del regolo per sostegno di quello, al quanto piu basso della basa o imo scapo della colonna, e un'altro chiodo o stiletto si fisserà dentro al detto regolo similmente per suo sostegno al terzo dell'altezza del suo fusto, doue ha da cominciare a diminuire detta colonna, piegando o ricuruando dipoi il regolo da capo all'indentro fino al ponto e termine di tal diminutione, si fermerà il terzo chiodo o stile di ferro al regolo dalla banda di fuore al quanto piu alto del sommo scapo, e cosi serà cō tre chiodi o stiletti fermo e ricuruato il regolo sopra il terzo del fusto; fuora del qual regolo, dipoi operando con la penna, con il piombo, o cō il lapis, da ambedue i lati si uerrà a formare la colonna gratiosamente diminuita, osseruando però tal regola diligentemente, è di cio se n'adduce per esempio qui da lato disegno, nel quale per essere piccolo in cambio de i chiodi o stili si sono operate l'acora, segnate come si uede per numeri. 1. 2. 3. 4. e tal fusto per essere in altezza noue diametri del suo basso scapo, si presuppone per colonna Corinta, & è diminuito tal fusto nel sommo scapo la sesta parte.

Delle

Fig. 12.7: The method of the flexible rule described by Pietro Cataneo in his *L'architettura* ([15], p. 131).

while the second, which Vignola specifies "da me stesso speculando l'ho trovato", is described by the following ([36], plate XXXI): "Si die tirare una linea indefinita alla terza parte da basso, la quale comincia da C. & passa per D. poi riportando la misura C.D. in punto A. et intersecando sul cateto della colonna che sarà in punto B. si stenderà A.B. in punto E. di dove si può tirare quel numero di linee che parerano quali si partino dal cateto della colonna e vadiano alla circonferenza, et su queste riportando la misura C.D. dal cateto verso la circonferenza cosi di sopra la terza parte come di sotto (...)." A century later Francois Blondel [10] observes about the 'Vignola method' that the resulting curve corresponds to Nicomede's conchoid.

Although Vignola's proposal might at first seem particularly seductive another 'invention' appeared in Pietro Cataneo's *Architettura*[6] ([15], chap. V, p. 131; Fig. 12.7), introduced as *Regola fuor dell'altre messe in luce sino adesso, nel diminuire le colonne*: "Hor qualunque diminutione s'habbia a dare alla colonna, si osserverà per questa nostra regola questo ordine, che destinata che sia con le linee la ugual grossezza e altezza del suo fusto, si porrà una riga o regolo piegabile sopra ciascuna delle due linee che serrano la colonna, prima da l'una banda, e poi dall'altra, avvertendo che la grossezza del taglio del regolo venghi tutta dentro la linea: debbesi dipoi fissare un chiodo sottile o stiletto di ferro fuor del regolo per sostegno di quello, al quanto piu basso della basa o imo scapo della colonna, e un'altro chiodo o stiletto si fisserà dentro al detto regolo similmente per suo sostegno al terzo dell'altezza del suo fusto, dove ha da cominciare a diminuire detta colonna, piegando o ricurvando dipoi il regolo da capo all'indentro fino al ponto e termine di tal diminutione, si fermerà il terzo chiodo o stile di ferro al regolo della banda di fuore al quanto piu alto del sommo scapo (...); fuora del qual regolo, dipoi operando con la penna, con il piombo, o con il lapis, da ambedue i lati si verrà a formare la colonna gratiosamente diminuita, osservando però tal regola diligentemente (...)."

The way in which this banal method to trace the shaft's curved profile enters the history of entasis seems somewhat peculiar. Andrea Palladio resumes it in his *Quattro libri* [30] and credits himself with its *inventione*, by reminiscing that he himself had passed it on to Cataneo a few years earlier ([30], p. 15): "E benche io non mi abbia potuto imaginare altro modo più breue & espedito di questo, e che riesca meglio; mi son nondimeno maggiormente confermato in questa mia inuentione, poi che tanto è piaciuta à messer Pietro Cattaneo, hauendogliela io detta, che l'ha posta in una sua opera di Architettura, con la quale ha non poco illustrato questa professione."

The method that Cataneo and Palladio describe is reintroduced by Abraham Bosse [13] in a post-script added to Philippe de La Hire's memoir *De punctis contactuum* (...) *observationes* [28]. Recently, Anthony Gerbino took up the matter again, describing "Bosse's method for joining found points using a flexible rule" ([24], legend to fig. 17, p. 514). It seems that this time the historian's eye has been overly

[6] The method was not described in Cataneo's previous work, *I quattro primi libri di architettura* [14].

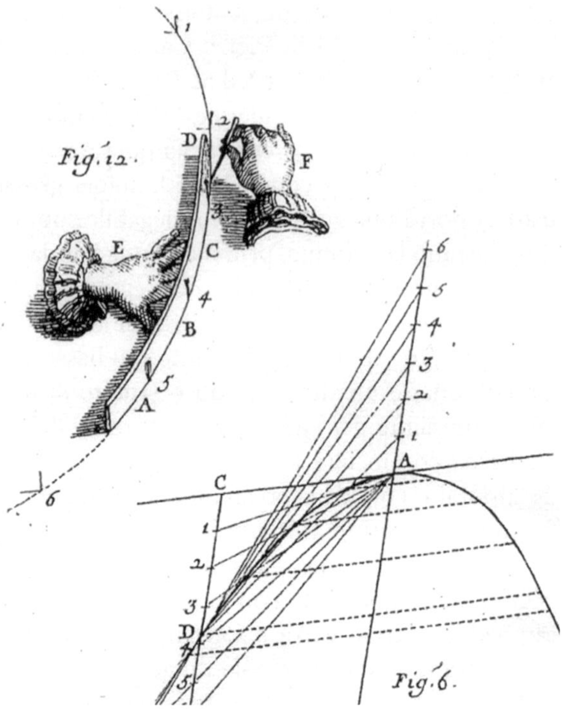

Fig. 12.8: Detail of a plate from Abraham Bosse's *Règle universelle* [13] showing the method of the flexible rule.

seduced by the fine illustration by Bosse (Fig. 12.8), renowned *graveur* of extraordinary talent.[7] Indeed, Gerbino neglects to mention that the "flexible rule method" had already been known for a while and was probably far older than Cataneo and Palladio intended to convey.[8] However, Bosse introduces the procedure specifically with reference to entasis [13]: "Dans mon Livre d'Architecture page XXXIII. je me suis servy de ce moyen pour décrire le fust de colomnes. Le bois de sapin est fort propre à faire ces regles, pour les ouvrages en grand, & la baleine pour les petits." Bosse describes the method on page 33 of the *Traité des manieres de dessiner les Ordres de l'architecture antique en toutes leurs parties* [12], whereas in his essay *Des Ordres des colonnes*

[7] The seduction this image exercised is confirmed by the choice of the *Journal of the Society of Architectural Historians*: to dedicate the front page of its issue containing Gerbino's essay [24] to it.

[8] See Heisel [27]. The "flexible rule method" is well known in Naval Architecture.

en l'Architecture [11] he confirms an explicit and direct inspiration to Palladio by quoting him on the same subject.

D'Aviler also commented on the Cataneo-Palladio method in his *Cours d'architecture qui comprend les ordres de Vignole* ([18], p. 103): "Vignole entend que sur les points donnez pour la diminution & le renflement de la Colonne on pose une regle mince d'une piece s'il se peut ou à deux ou trois reprises, & qui se courbe selon lesdits points, par laquelle on trace la ligne du contour; & cette operation est pour faire l'Epure (qui est le dessein au trait du Profil sur un mur enduit de plâtre) (...)." Two centuries later Gorham P. Stevens revived it, without providing any bibliographic reference ([35], p. 126): "The method employed in Italy today: the Italian architects of today take a long wooden straight edge, bend it to the desired entasis, and then draw the curve. The method cannot be used for large columns, for the inequality of a long strip of wood causes irregularities of curvature. At best, it is but a rough method and will not, therefore, be further considered in this paper."[9]

After Cataneo and Palladio, this series of proposals seems to come to an end. Philibert De l'Orme [19] and Hans Vredeman de Vries [38], for instance, resume geometric constructions already known. Giovanni Antonio Rusconi's illustrations of entasis ([32], p. 59) are neither explained in the text nor explicative of how to trace the curve. No detailed summaries of the *vexata quaestio* were available until Scamozzi's *Idea dell'architettura universale* [33].

Scamozzi illustrated two methods "di nostra inuentione, l'uno differente dall'altro, molto theoricali, e non però difficili" ([33], part II, book VI, chap. XXX, p. 142; Fig. 12.9, p. 145). Both turn out to be ingenious *variations on the theme* when compared to what Serlio (Scamozzi's 1st method) and Vignola (Scamozzi's 2nd method) had already illustrated. But unlike Vignola, Scamozzi does not taper the lower part of the shaft. There is also a third method ("un'altro poi assai più pratico, che theorico"), which corresponds to the one proposed by Cataneo and Palladio.

Scamozzi makes a clear distinction between the two phases we referred to at the opening (drawing the curve passing through entasis;

[9] The use of Nicomede's conchoid for the construction of entasis is known by archeologists principally from this Stevens's essay. The subject was taken up again by Gilbert Hallier [26].

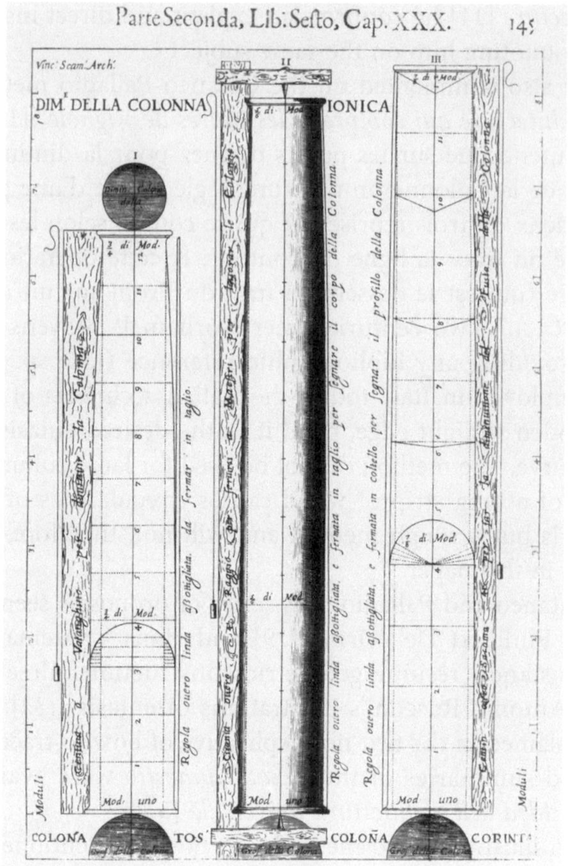

Fig. 12.9: Plate by Vincenzo Scamozzi ([33], part II, book VI, p. 145) summarizing the principal methods proposed in the Renaissance for drawing the entasis.

fabricating the template to use as lead while cutting the column) and he recommends the use of the "regola overo linda assottigliata da fermar in taglio" (that means a thin strip of wood) to trace the complete curve, which is defined by the points obtained from the first two procedures. In his closing lines, after marginally mentioning the names of Cataneo and Palladio, Scamozzi also briefly describes a fourth method, which is distinctly different from those proposed in previous work on the subject ([33], ibidem, p. 144): "se bene possono servire, à Pittori, e simili, i quali amano la facilità, e le cose loro non importano tanto: & alle volte possono fare tal diminutione con un lizzo bagnato di qualche tinta battendo alciatolo, e tiratolo all'infuori, in modo che poi fà una linea alquanto curva." No Italian dictionary, whether current or historical, quotes the term "lizzo", because it is a dialectal term (of the Veneto region). It refers to a string system used by weavers. A string, a rope, therefore, as "drawing instrument"—well-known to architects, but not only to architects. A few years earlier Guidobaldo del Monte [25] had studied the trajectory of bullets by using a rope or a chain; Galileo refers to the same procedure in the Seconda Giornata of the *Discorsi e dimostrazioni matematiche* [23] on the subject of drawing a parabola. François Blondel refers to Galileo in his *Quatre principaux problèmes d'architecture* [10], in the section dedicated to the "diminution des Colonnes".[10] His main issue is the resemblance between the methods used in architectural construction and those used in shipyards ([10], p. 383): "J'avertirai seulement les Ouvriers que Galilée leur en enseigne une [manière pour décrire les Paraboles] dans ses Méchaniques, que j'estime facile & ingénieuse, & que j'ai fait heureusement pratiquer par les Charpentiers du Roy, en la fabrique des Vaisseaux & Galeres, pour ce qu'ils appellent leur donner beau Galbe à la Pouppe."

Regrettably, the pages of Scamozzi's *Idea* (and its fine illustration) which refer to entasis had little following and a considerable number of authors never even read them: the circulation of Vignola's *Regola* was so widespread that it overshadowed Scamozzi's unique observations.[11]

The most well-known and widespread methods for defining and tracing entasis are therefore the following: Serlio and Vignola's geometric constructions, first published in 1537 and 1562 respectively and the Cataneo-Palladio tracing method (1567 and 1570). However, can one be sure that Serlio-Vignola and Cataneo-Palladio are the only noteworthy sources? Clearly not. For instance, a method based on the properties of the circumference can already be found in Francesco di Giorgio Martini's manuscripts and is resumed in a *folio* credited to Fra Giocondo (Fontana [21], fig. 15 and 16). Francesco di Giorgio also explains how to trace the whole profile of the column once the points have been defined by means of the above-described method ([22], vol. 2, p. 383): "Le quali sette particule overo linee per discrezione

[10] Bosse also mentioned "Cordeaux" when he described the methods for tracing entasis [11].

[11] On this matter, D'Aviler ([18], p. 103) wrote: "De tous les Architectes Anciens & Modernes Vignole est le premier qui ait donné des Règles du trait de Diminution & du Renflement des Colonnes."

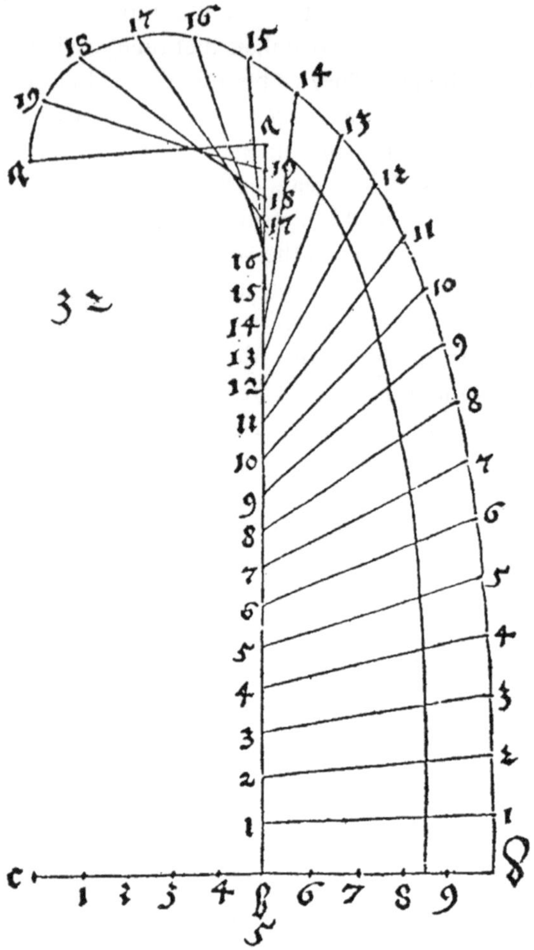

Fig. 12.10: Geometric construction proposed by Albrecht Dürer ([20], fol. Ciii recto, fig. 32) to define the tapering of a wall or a tower.

dello architetto dieno essare appropinquate alla natura della porzione d'uno circolo per non variare la diminuzione in uno punto ma successivamente, e con dolcezza così digradando sarà diminuita."

Francesco di Giorgio and Fra Giocondo's proposals suggest that the procedure might well be much older and perhaps very ancient, but certainly the elegance of Serlio's method has no affinity with the difficult construction Francesco di Giorgio describes.

Vignola's method also seems to have significant, though vaguer, precedents: a similar geometric construction is found in Dürer's *Under-*

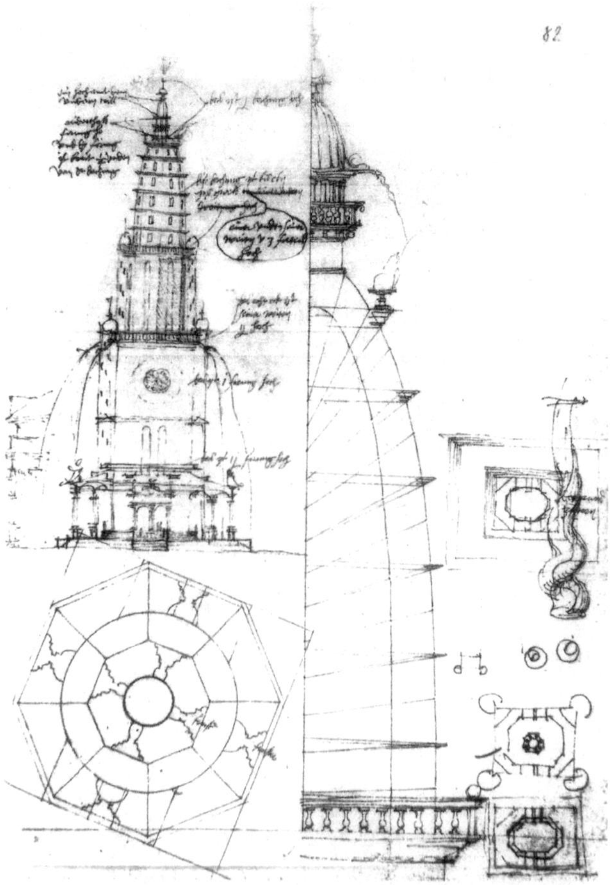

Fig. 12.11: A page of Dürer's *Dresdner Skizzenbuch* ([31], vol. III, fig. 253).

weysung, where it is recommended for some architectural applications[12] (for instance to define the tapering of a tower or a wall, Fig. 12.10; see also Fig. 12.11). Historians have more than once confirmed the link between Dürer's and Vignola's work, so we can justifiably presume that the method Dürer describes did in some way influence the Italian architect's *Regola*.

[12] Also other Dürer's drawings show similar constructions. See ([31], fig. 247–248, 251, 253).

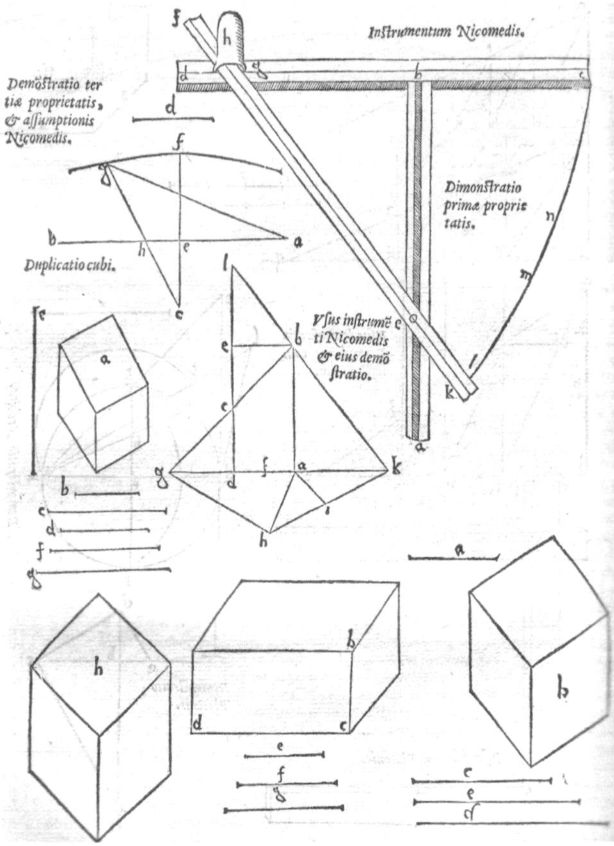

Fig. 12.12: The Nicomedes compass described by Daniele Barbaro in his commentary to Vitruvius' *De architectura* ([6], p. 280).

It is interesting to note that the interpretation of Vignola's method from a strictly geometric perspective, made by François Blondel a hundred years later [10], referred to a form which was not only well-known to mathematicians, but also to architects (some contemporaries criticised him for this, as if his compass for entasis were only a fruit of fraud. Actually Blondel confesses that clearly in his work). A conchoid is also described in the *Underweysung* and Nicomede's conchoid is accurately described, for instance, in Daniele Barbaro's editions of Vitruvius's *De architectura* ([5] and [6], Fig. 12.12). The 'mathematical' figure was therefore at hand, but it clearly did not arouse much interest with regard to entasis during the sixteenth century.

It was quite the opposite during the seventeenth century, when the solution's precision and the opportunity to use a bespoke instrument made Blondel's compass's fortune. Soon afterwards, entasis disappeared from architectural writings as a pretext for new *invenzioni*, but it became a key subject of discussion between physicists and mathematicians, who tried to lead it back to the principles of the *resistentia solidorum*. The works of Leonhard Euler, Louis Lagrange and many others, together with studies which over the years up to our times have referred to the subject in terms of structural optimisation, demonstrate that the Vitruvian word had in the meantime found other interlocutors. Entasis has shifted from the realms of what is possible to those of what is necessary, by turning into a very complex problem reserved for *calculatores*.[13]

References

1. Alberti, Leon Battista: "De re aedificatoria", Nicolò di Lorenzo Alemanno, Firenze 1485.
2. Alberti, Leon Battista: "L'architettura di Leonbatista Alberti, tradotta in lingua Fiorentina da Cosimo Bartoli", Lorenzo Torrentino, Firenze 1550.
3. Alberti, Leon Battista: "L'Architettura di Leonbatista Alberti, tradotta in lingua Fiorentina da Cosimo Bartoli", Francesco Franceschi, Venetia 1565.
4. Barbaro, Daniele: "I dieci libri dell'architettura di M. Vitruuio tradutti et commentati da monsignor Barbaro", Francesco Marcolini, Venetia 1556.
5. Barbaro, Daniele: "I dieci libri dell'architettura di M. Vitruuio. Tradotti & commentati da mons. Daniel Barbaro", Francesco de' Franceschi senese & Giovanni Chrieger, Venetia 1567.
6. Barbaro, Daniele: "M. Vitruuii Pollionis De architectura libri decem, cum commentariis Danielis Barbari", apud Franciscum Franciscium Senensem, & Ioan. Crugher Germanum, Venetiis 1567.
7. Becchi, Antonio: "Imaginer l'entasis. Constructions, définitions, malentendus dans les traités de la Renaissance", in Roberto Gargiani (ed.), La construction de la colonne, École Polytechnique Fédérale de Lausanne, Lausanne 2008 (in press).
8. Becchi, Antonio: "Les paradoxes (historiographiques) de la *vis columnarum*", in Roberto Gargiani (ed.), La construction de la colonne, École Polytechnique Fédérale de Lausanne, Lausanne 2008 (in press).
9. Biffi, Marco (ed.): "Francesco di Giorgio Martini. La traduzione del De Architectura di Vitruvio", Centro di Ricerche Informatiche per i Beni Culturali e dell'Accademia della Crusca, Pisa 2002.
10. Blondel, François: "Résolution des quatre principaux problèmes d'architecture", Imprimerie Royale, Paris 1673.
11. Bosse, Abraham: "Des ordres des colonnes en l'Architecture et plusieurs autres dépendances d'icelle", A. Bosse, Paris 1664.

[13] About this see [8].

12. Bosse, Abraham: "Traité des manières de dessiner les Ordres de l'architecture antique", chez l'auteur, Paris 1664.
13. Bosse, Abraham: "Règle universelle, pour décrire toutes sortes d'Arcs rampans dans toutes les sujections que l'on puisse proposer, sans se servir des Axces, des Foyers, ny du Cordeau", annex to Philippe de La Hire, De Punctis Contactum [...] Observationes, A. Bosse, Paris 1672.
14. Cataneo, Pietro: "I quattro primi libri di architettura", in casa de' figliuoli di Aldo [Manuzio], Venetia 1554.
15. Cataneo, Pietro: "L'architettura", [Paolo Manuzio], Venetia 1567.
16. Cesariano, Cesare: "Di Lucio Vitruuio Pollione De architectura libri dece traducti de latino in vulgare affigurati: commentati: & con mirando ordine insigniti", Gottardo da Ponte, Como 1521.
17. Coecke van Aelst, Pieter: "Generalen Reglen der architekturen", 1539.
18. D'Aviler, Augustin Charles: "Cours d'architecture qui comprend les ordres de Vignole", Nicolas Langlois, Paris 1691.
19. De l'Orme, Philibert: "Le premier tome de l'Architecture", Federic Morel, Paris 1567.
20. Dürer, Albrecht: "Underweysung der Messung", Nürnberg 1525.
21. Fontana, Vincenzo: "Fra' Giovanni Giocondo, architetto 1433c 1515", Neri Pozza, Vicenza 1988.
22. Francesco di Giorgio Martini: "Trattati di architettura ingegneria e arte militare", edited by Corrado Maltese, Il Polifilo, Milano 1967.
23. Galilei, Galileo: "Discorsi e dimostrazioni matematiche intorno a due nuove scienze attenenti alla mecanica et i movimenti locali", Elsevirii, Leida 1638.
24. Gerbino, Anthony: "François Blondel and the *Résolution des quatre principaux problèmes d'architecture* (1673)", Journal of the Society of Architectural Historians, vol. 64, n. 4, December 2005, pp. 498–521.
25. Del Monte, Guidobaldo: "Meditatiunculae Guidi Ubaldi [...]", Bibliothèque Nationale de France, Paris, MS Lat. 10246, 1587–1592.
26. Hallier, Gilbert: "Le premier Forum de Rougga", Bulletin archéologique du CTHS (Comité des Travaux Historiques et Scientifiques), n.s., issue 17B, 1984, pp. 101–113.
27. Heisel, Joachim: "Antike Bauzeichnungen", Wissenschaftliche Buchgesellschaft, Darmstadt 1993.
28. De La Hire, Philippe: "De Punctis Contactum [...] Observationes", A. Bosse, Paris 1672.
29. Pacioli, Luca: "De Divina Proportione", A. Paganius Paganinus, Venetia 1509.
30. Palladio, Andrea: "I quattro libri dell'architettura", Dominico de' Franceschi, Venetia 1570.
31. Rupprich, Hans (ed.): "Dürer. Schriftlicher Nachlass", Deutscher Verlag für Kunstwissenschaft, Berlin 1969, vol. III.
32. Rusconi, Giovanni Antonio: "Della Architettura, Con Centosessanta Figure Dissegnate dal Medesimo", Gioliti, Venetia 1590.
33. Scamozzi, Vincenzo: "Dell'idea della architettura universale", Venetiis 1615.
34. Serlio, Sebastiano: "Regole generali di architetura sopra le cinque maniere de gli edifici", Francesco Marcolini, Venetia 1537.
35. Stevens, Gorham P.: "Entasis of Roman columns", Memoirs of the American Academy in Rome, n. 4, 1924, pp. 121–152.
36. Barozzi da Vignola, Jacopo: "Regola delli cinque ordini d'architettura", [Roma] 1562.
37. Vitruvius, Marcus P.: "De architectura libri decem", Roma 1488 ca. See also Ingrid D. Rowland (ed.), "Vitruvius. Ten Books on Architecture. The Corsini Incunabulum with the annotations and autograph drawings of Giovanni Battista da Sangallo", Edizioni dell'Elefante, Roma 2003.
38. Vredeman de Vries, Hans: "Architectura", Typis Gerardi Smits, Antwerpiae 1577.

CHAPTER THIRTEEN

FROM ONE CURVE TO ANOTHER OR THE PROBLEM OF CHANGING COORDINATES IN STEREOTOMIC LAYOUTS

Joël Sakarovitch

The "sloping vault" is not *a priori* the most difficult mortarless piece of the stereotomic repertoire. Squinches, domes, spiral staircases in the style of the *vis de Saint-Gilles*, or vault penetrations are no doubt more difficult pieces to carve and, as is often the case, to draw.

However, the sloping vault poses greater representation problems, due to the choice of reference system, and allows one to grasp the methods authors use to approach a mortarless piece of architecture, if indeed they use a method. The sloping vault offers numerous solutions and, as we shall see later, many opportunities for making mistakes, which are

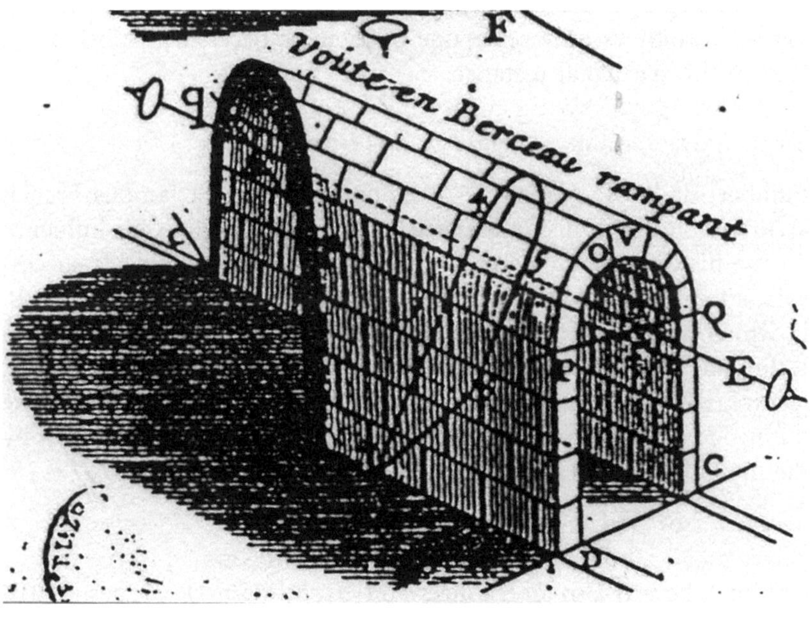

Fig. 13.1: A sloping vault, drawing from A. Bosse in [1], pl. 2.

always more revealing about the methods employed than correct drawings. It must be added that the choice of "sloping vaults" is not mine to begin with but Girard Desargues'. A 17th century mathematician and architect, this author is indeed partly characterized by having written a small opuscule entirely devoted to the study of this particular vault. If Desargues claims to present a "universal method" and only studies one vault, it is because, as I shall try to demonstrate, the corresponding blueprint indeed offers great geometrical wealth.

The authors

Before giving a detailed description of the mortarless architectural piece that will serve as a guide in this study, I would like to justify briefly why I chose the authors I wish to study, besides Desargues. The treatises of Philibert de l'Orme, Jousse, Derand and Frézier are, together with that of Jean-Baptiste de La Rue, the major French treatises on stereotomy. Hence it is not the presence of the first four that must be justified but rather the absence of the latter, and, to a lesser extent, that of Bosse or Millet de Challes. The reason for this is simple: the five authors chosen all study sloping vaults and present radically different solutions whereas the other authors only reuse one of their predecessors' solutions, at least in this particular instance.

Philibert de l'Orme

Philibert de l'Orme (1514–1570) is one of the most famous French architects, for both his built and written opus, as well as his influence on the history of architecture. He did not publish, as did his successors, a treatise on stone carving but a treatise on architecture [cf. 10], in which Books III and IV are devoted to stereotomy, a construction technique that is given its rightful place in his global theory on architecture. He thus inaugurated a new conception of treatises since, as he put it, "geometrical drawing [has] been used in architecture neither by the men of antiquity nor those of modern times" [10, Fol. 87].

Desargues

Although he was also an engineer and an architect, Desargues (1591–1661) has above all remained famous for his mathematical opus. His main contribution, the *Brouillon proiect d'atteinte aux événements des rencontres du Cône avec un Plan*, published in 1639, makes Desargues the father (or to be precise, the grand-father) of projective geometry. It is therefore to one of the greatest geometers of his time that we owe the small opuscule entitled the *Brouillon proiect d'exemples d'une manière universelle du S.G.D.L. touchant la pratique du trait à preuves pour la coupe des pierres en l'Architecture*, comprising four pages of text and four pages of figures.[1] Desargues did not write a treatise on stone carving and, as I said earlier, only studied sloping vaults in these few pages, by taking the most general point of view possible from the start.

Jousse

While Philibert de l'Orme or Desargues have, for various reasons, remained famous, Mathurin Jousse (1607–1650), master mason of La Flèche, is today quite unknown. The only construction attributed to him, the organ gallery of the Jesuit church in La Flèche, a true masterpiece of stereotomy, is now thought to be perhaps the work of another. A part from *Le secret d'architecture découvrant fidèlement les traits géométriques, coupes et dérobements nécessaires dans les bâtiments*..., published in La Flèche in 1642, we owe him *Le Théâtre de l'art du charpentier* (1650) and *La Fidèle ouverture de l'art du serrurier* (1627). Gabriel-Philippe de la Hire had the last two works reprinted in 1702, which goes to show the importance he gave them. While Philibert de l'Orme wrote for architects and Desargues for "excellent contemplative individuals", Jousse clearly wrote for master masons, stone carvers and "any builders who do not possess the broadest experience in this Science (namely Geometry)" [6, Introduction].

Derand

In 1643, François Derand (1588–1644) published *L'Architecture des voûtes ou l'Art des traits et coupe des voûtes*..., a treatise that enjoyed great success. Having entered the Jesuit order in 1611, Derand was both a mathematical teacher and an architect for his order. In Paris, he designed the Saint Paul-Saint Louis church and the Jesuit convent that has nowadays become the Lycée Charlemagne. The success of his architectural work is well deserved as it makes genuine progress

[1] A fifth figure page is part of the opuscule but pertains to the treatise on perspective published in 1636.

compared with his predecessors. Indeed, it displays almost all the different architectural devices used in the 17th Century. The blueprints are very carefully explained and far easier to read than those of the preceding treatises.

The 1640's were particularly fruitful for French treatises on stereotomy since the publications of Desargues, Jousse, Derand and Bosse are quasi contemporary. Such simultaneity is of course not fortuitous, the 17th Century being the golden age of mortarless architecture in France. One might mention for example the Louvre by Lemercier, the Sorbonne chapel (also by Lemercier), the Val de Grâce church (by Mansart and then Lemercier), the Paris Observatory built by Perrault or this masterpiece of French stereotomy, the hall of the Arles Town Hall, by Hardouin-Mansart. Among all the architectural works studied and listed by Pérouse de Montclos, half were built during the 17th Century [cf. 13].

Frézier

Amédée-François Frézier (1682–1773), a military engineer in charge of studying the means of defending the West Coast of Latin America against English attacks during the War of Spanish Succession, was nominated director of the Brittany fortifications in 1739. We owe him the great French stereotomy treatise that predates Monge's teaching course, *La Théorie et la pratique de la coupe des pierres et des bois pour la construction des voûtes…ou traité de stéréotomie à l'usage de l'architecture.* Published between 1737 and 1739, this work is very different from earlier treatises. The first tome, an initiation to three-dimensional geometry, opens on a Vitruve quotation boasting the merits of geometry, and starts with an ardent apology of the "usefulness of theory in the arts relating to architecture". Devoted to a theoretical and abstract study of spheres, cones, cylinders and their intersections, it is presented as a succession of theorems, corollaries and lemmas. But in spite of the theoretical nature of this study, Frézier takes great care to convince his readers that none of the problems studied should escape the attention of anyone interested in stereotomy. Each problem is immediately followed by a "practical application" that justifies the study.

As a mirror to his "practical applications", which follow general theorems in the first tome, the drawings proposed in the two other tomes are followed by "explicative demonstrations" that attempt to justify the graphical constructs with geometry. Although these attempts at demonstration are often not very convincing and quite incomplete, the approach is essentially novel. Frézier did not propose a geometry theory but inaugurated what Gino Loria called "scientific stereotomy" [cf. 9]. Indeed, after this publication, he began to be recognized as an authority on construction and stereotomy. René Taton [cf. 16] is nonetheless right in emphasizing that the wealth of the text is more easily perceived by readers with a knowledge of descriptive geometry than by the author's contemporaries for it is difficult to distinguish new methods in this imposing treatise of 1500 pages, drowned as they are in a sea of processes of very unequal interest.

Sloping vaults

The point of the blueprints studied here is to provide the layout necessary for the construction of a vault against a wall. In the simplest—and most common—situation, the wall is vertical, and the axis of the vault horizontal and perpendicular to the wall. In this case, elevation is sufficient and no blueprint is necessary to build the vault. The situation of interest here is, on the contrary, the most general possible. I know of no such example in architecture and it can be considered as a case study. However, one of the oculi of the Seville cathedral provides an example of a mortarless cylinder with any axis direction relative to the adjacent wall. The givens are the positions of the wall and the vault cylinder, their situation with respect to one another, and, since gravity plays a specific role in mortarless construction, their position in relation to the vertical of the place. In order to achieve the maximum degree of generality, one must choose a wall that is sloping rather than vertical, and a vault axis direction that is not horizontal, hence the term "sloping". Finally, the axis of the vault can have any direction relative to the wall except one that belongs to a vertical plane perpendicular to the wall, in which case the vault is no longer "sloping". In order for the vault to be fully determined, a directrix must also be given for the cylinder, which involves considering two situations. Either the curve on the face wall is given, if the architect wants the visible arch, or "face arch", to be semi-circular for the façade to be homogeneous; or, on the contrary, the cross-section of the cylinder (i.e. the section of the cylinder that is perpendicular to the axis of the vault) is fixed and the corresponding arch is said to be orthogonal. As Frézier wrote: "It is up to the architect to know whether he would rather have regularity on the outer face than inside, or whether he must cast the irregularity onto the face in order to make the inside of the vault more beautiful" [6, t. 2, p. 172].

The drawing is more or less difficult depending on whether the initial given is the orthogonal arch or the face arch. For the cutting of a voussoir, the true size of the orthogonal arch must first be determined as the initial block of stone will be pared down on this basis. Thus the bed joint surfaces will suffice to define the voussoir completely. Knowing the face arch is therefore not necessary for carving, and the tracing is very much simplified if we suppose that the orthogonal arch is given. Conversely, if the face arch is the initial given, the author must begin by indicating the construction of the orthogonal arch.

The specificity of the dry-stone piece under study is that everything is "skewed": the object itself has no horizontal or vertical plane on which one might "naturally" rely for drawing. Prior to any geometric construction, authors are therefore confronted with the problem of choosing a coordinate or reference system. More specifically, this drawing involves changing coordinates, i.e. going from one set of coordinates that express the givens of the problem (a horizontal and a vertical plane) to a set of coordinates allowing the panels to be drawn easily (a plane perpendicular to the vault axis and a plane parallel to it). Now the notion of changing coordinates is a complex notion, which Gaspard Monge, himself, did not explicitly develop in his descriptive geometry course.[2] Furthermore, the formalisation and use of coordinate change in stereotomy gave rise to measured opposition between the two main teachers of descriptive geometry in France in the 19th Century, Théodore Olivier and Jules Maillard de la Gournerie. It is therefore hardly surprising that the authors of stone carving treatises experienced some difficulty when faced with its use.

Geometrical Principles and Choice of Coordinates

Given the underlying difficulty, the various authors chose radically different geometrical principles for their graphical constructs. Philibert de l'Orme restricts the problem in two ways. First he supposes that the wall is vertical, draws the sloping vault starting from the orthogonal

[2] About the creation of the descriptive geometry by Gaspard Monge during the French Revolution, see [15].

arch (a semi-circle arch on his blueprint) and does not represent the face arch. The drawing is much simpler in this situation, as we saw previously, but is surprising for a "cellar slope", which is the title the author has given to this illustration. Indeed for a vault that covers a subsidiary space, priority should be given to the arch that appears on the face wall. The fact that Philibert de l'Orme makes such an "anti-architectural" choice certainly goes to show the difficulty he encountered in presenting (and possibly executing) the drawing of a sloping vault, starting from the face arch. The four other authors propose solutions both for the sloping wall and a regular face arch.

Philibert de l'Orme

Philibert de l'Orme and Frézier basically choose the same principle (though with different outcomes), consisting in bringing the problem back to one of drawing a horizontal vault adjoining a wall of any slope. But whether the wall is sloping and the vault horizontal, or the wall vertical and the vault sloping, is not equivalent. Rotating the entire vault suffices, it is true, to go from one position to the other. It is no doubt how Philibert de l'Orme reasoned although he does not say a word on the subject.

Judging the first situation more practical, he very naturally used this approach. In the case of an orthogonal slope, there is no disadvantage in supposing that we have a horizontal vault against a sloping wall. But obliqueness here creates an unavoidable difficulty and the reasoning becomes wrong. If one rotates Philibert de l'Orme's vault drawing from a sloping to a horizontal position, this rotation can only take place around an axis that is perpendicular to the vertical wall against which the vault is resting, i.e. an axis that is parallel to the ground line on the explicative drawing. But if the vault axis is made horizontal by performing such a rotation, the oblique section of the face wall will not be the same before and after the rotation. This problem does not arise in the case of an orthogonal slope because the rotation axis is parallel to the horizontal lines of the wall plane. Thus if we want to commute the problem from one involving a sloping vault to one involving a horizontal vault resting against a sloping wall (which is still possible), we must consider the modification of the oblique section, as a result of rotating the object [cf. Fig. 13.4]. This is in no way approached in the work of Philibert de l'Orme, who draws his layout using as reference planes the plane perpendicular to the frontal plane, which contains the vault axis, and a plane perpendicular to this axis. But the vault is only represented using this reference system, where neither the vertical, nor the slant angle with respect to a horizontal plane are given.

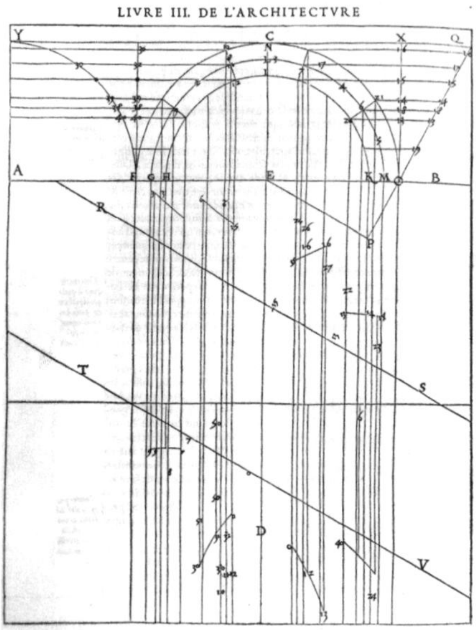

Fig. 13.2: De L'Orme's blueprint, in [10], Fol. 62, v°.

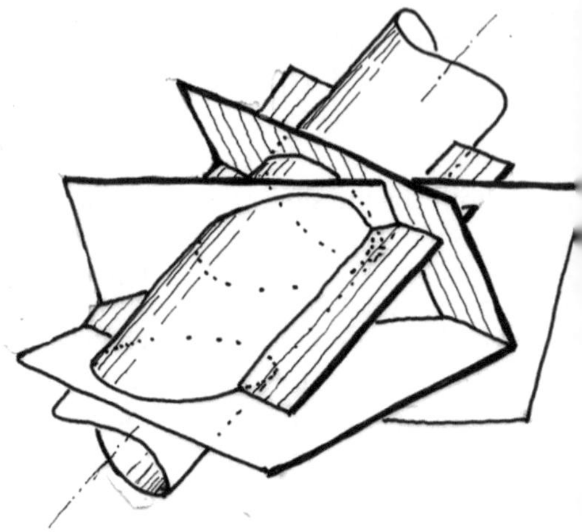

Fig. 13.3: De L'Orme's reference planes (Author drawing).

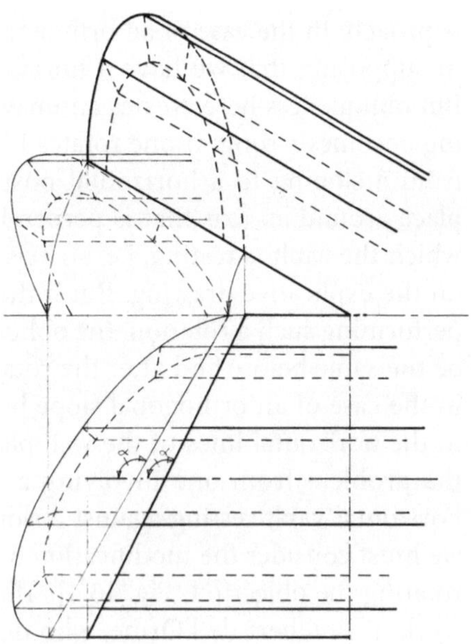

Fig. 13.4: De L'Orme's mistake (Author's drawing).

Frézier

Two centuries later, Frézier again adopts Philibert de l'Orme's principle but this time, without making the same mistake. With his usual thoroughness, Frézier considers all (or nearly all) the possible cases of vaults in his treatise: horizontal or sloping, orthogonal or oblique, adjoining a vertical or sloping wall, with a given orthogonal arch or face arch. Given the number of situations studied, and in order to derive a methodological approach to the problem rather than just display a juxtaposition of drawings, he systematically chooses to bring the problem back to the "previous problem". Frézier certainly exhibits great rigour and offers a very homogeneous presentation. Simpler cases can be deduced from more complex cases without changing the method in the slightest. In the blueprint of the orthogonal slope adjoining a sloping wall, for instance, if the slope angle is zero, we revert to the blueprint of the horizontal vault against a sloping wall. But, as we saw previously in the case of the layout given by Philibert de l'Orme, when the constraints accumulate—sloping vault, then a sloping vault against a sloping wall, keeping the initial principle and wanting to bring the problem back to the "previous problem" can present disadvantages.

In order to avoid the mistake of his illustrious predecessor, Frézier chooses other reference planes. He draws the profile view on a vertical plane that is parallel to the axis of the vault. This layout seems optimal given the problem treated because this view is essential. It is therefore natural, and more comfortable for the reader, to have this profile view appear as a main frontal view, i.e. horizontal on the page, and not by changing the frontal plane to a somewhat baroque position for the reader, as in Derand, de La Rue or… Hachette.[3] The second step is not so convincing: it consists in determining the projection of the vault onto what Frézier calls the "ramp plane", i.e. the plane (N on the fig. 13.6) from which the vault is erected, and folding it onto the profile plane of the vault in order for the projection to appear according to its true size. As mentioned earlier, when the given is the orthogonal arch, this plane is perpendicular to the frontal plane and the process involves performing what we now call a change in horizontal plane. In the new coordinate system, the layout is that of a horizontal vault against a sloping wall. The face arch is folded onto the new "horizontal" plane for it to appear according to its true size. But when the given is

[3] Jean Nicolas Pierre Hachette (1769–1834) succeeded to Monge, in 1797, in teaching descriptive geometry at the Ecole polytechnique.

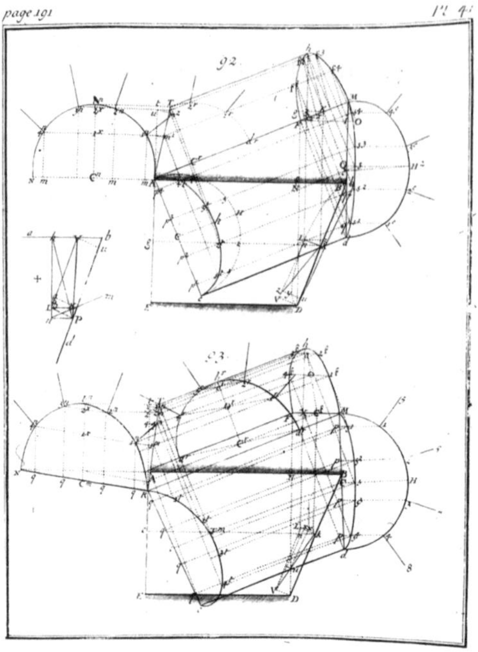

Fig. 13.5: Frézier's blueprint, in [6], pl. 41.

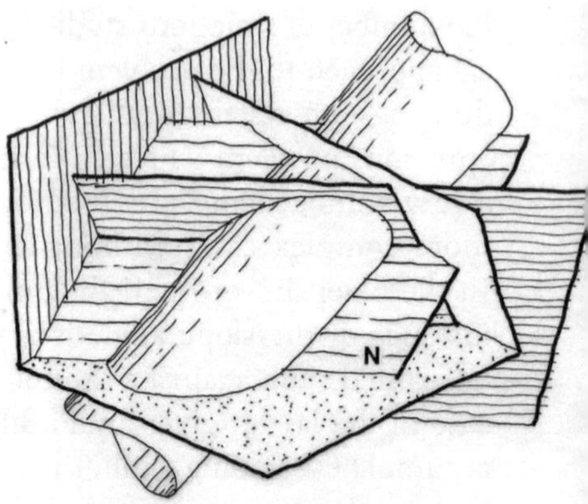

Fig. 13.6: Frézier's reference planes (Author drawing).

the face arch, the ramp plane can be any plane. Frézier explains this very well: "if we project the vault onto this plane, vertical lines will slope relative to it... Thus... we shall take the ramp plane as the horizontal and the vertical plane passing through the axis as the incline relative to the horizon". But we cannot think of this operation as a change in horizontal plane without losing the concept of verticality, and hence gravity.

Desargues

Desargues adopts another view point and the graphical construct he proposes, though very simple in principle, is rather confusing. Since the main problem is to derive a figure drawn on the plane perpendicular to the vault axis from a figure drawn on the face wall plane, the auxiliary plane that needs to be introduced to go back and forth easily between these two figures is a plane that is simultaneously perpendicular to the two preceding planes. Among this family of planes, it is then quite natural to choose the plane that contains "the axle", or, according to Arguesian terminology, the vault axis [Fig. 13.8]. This is the plane that the author calls the "sub-axle plane" and we can assume that Desargues' figures (apart from the first, which is a drawing of the jambs and not the voussoirs) are drawn in this plane. If we suppose the face arch is given, the first step in the construction, which is in fact the essential step, consists in deriving the orthogonal arch, that is to say the cross-section of the vault. With respect to the sub-axle plane, which is taken as a reference plane, going from the orthogonal arch to the face arch is an operation similar to the change in frontal plane in descriptive geometry. This coordinate system makes the construction of the orthogonal arch from the face arch and *vice versa* easier but requires an initial layout in order to present the assumptions of the problem.[4] The Arguesian choice is therefore very clever, but by taking a three-dimensional coordinate system that is intrinsically linked to the object, Desargues deprives his method of the universal quality he was aiming for. He also deprives himself of the possibility of treating architectural objects of a radically different nature from that of the object chosen, either because they comprise curved surface intersections, which would be the case if the opening were made in a cylindrical or conical wall, or because they present surfaces with a double-curvature radius such as Marseille *arrière-voussures*, squinches, domes or spiral-staircase strings. In addition, by choosing a coordinate system where all reference to gravity has disappeared, his opuscule becomes far less accessible.

As La Gournerie noted: "When considering an abstract system of lines and surfaces, one can suppose its transposition one way or another without any problem, but when dealing with an architectural work, such as a vault or a staircase, it is better to consider it in its natural position and study it on its plane and elevation. If it is made to rotate in space, if it is projected onto planes none of which are horizontal, the mind experiences some difficulty in imagining it, and the layouts, while just as simple from the geometrical point of view, become more difficult to grasp" [8, p. 45]. By (implicitly) taking the "sub-axle plane" as a reference plane, Desargues destabilises his readers, forcing them to reason on the basis of a plan that is neither vertical nor horizontal in space. He places his readers on a plane that rolls and pitches simultaneously, or thinks about the vault as if it were a satellite under zero gravity, which comes to the same thing [Fig. 13.8]. If, from the purely geometrical

[4] The component element of this "initial layout" is given by Bosse [1] in the form of 14 illustrations with comments.

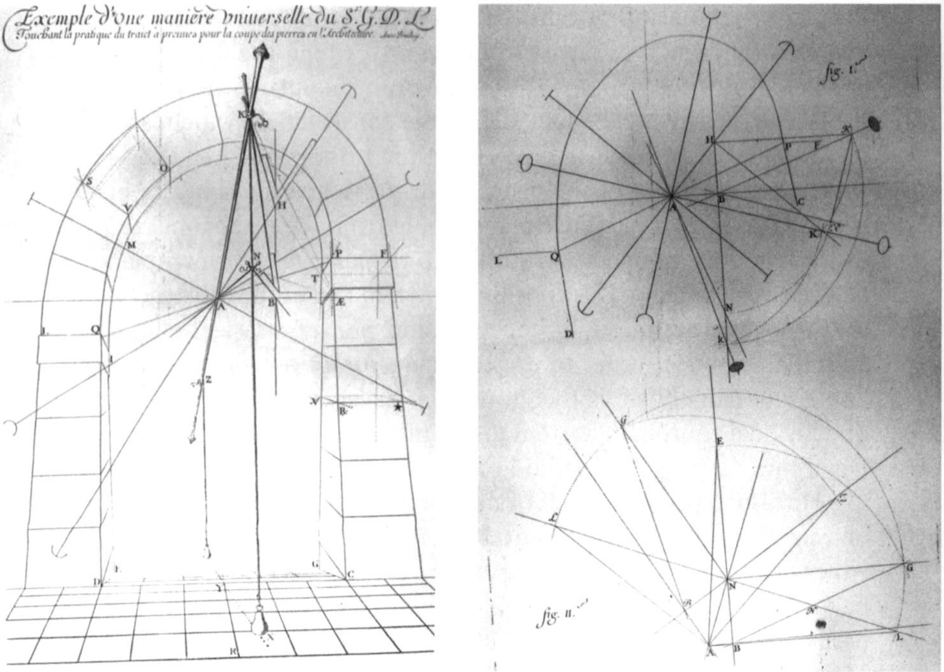

Fig. 13.7: Desargues' blueprints, in [5], pl. 1 and 2.

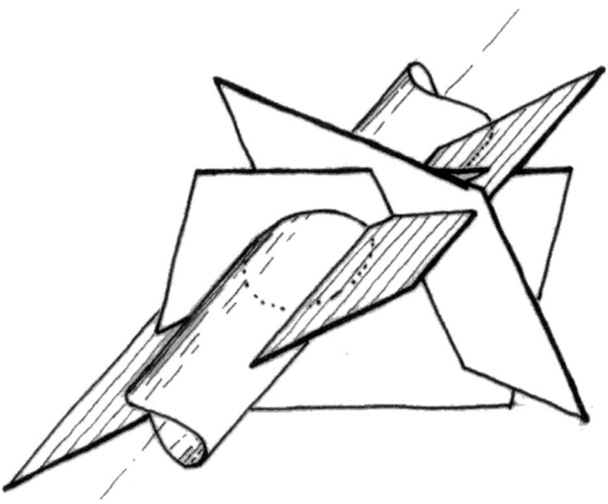

Fig. 13.8: Desargues' reference planes (Author's drawing).

point of view, when two planes, A and B, are superimposed, the result is the same whether one considers that plane A is fixed and plane B is folded over it or B is fixed and A is folded over it, in the case of an object with a mass, one of the two operations can be far more difficult to understand than the other. For example, if one of the planes, A shall we say, is horizontal in physical space and B has any direction, the first operation will be much easier to imagine than the second.

Now, in the case of Desargues' manual, the situation is complicated twice over. First of all because the game is played not with two planes but three or four planes. Secondly, because the "optimal" reading grid, the one that can reflect the geometrical operations in the simplest way possible, is not the same for the first figure, where Desargues determines the relative position of the four axes he has introduced, as for the following figures. Indeed, for the first figure, it is better to consider the plane of the face wall as "fixed", while for the following figures, it is simpler to suppose, as described above, that the sub-axle plane is the plane of reference. One can certainly object that these remarks are quite subjective, that each of us can find this or that situation more comfortable or "natural", and that the author, by giving no further details on this issue gives the reader the freedom to choose the reading grid that suits him best. In any event, being a good geometer, Desargues probably felt free of any constraints that were not strictly geometrical and one cannot doubt the ease with which Desargues went from one plane to the other or his mastery of the situation. But gravity remains the most shared of all things and, in a presentation that aims to be explicitly didactic, making total abstraction of it is not necessarily the best way to convince the reader.

Jousse

Jousse, like Derand or Frézier, clearly separates the various component steps and the case of the vertical wall from that of the sloping wall. Jousse chooses as a reference plane a horizontal plane and a vertical plane perpendicular to the horizontal projection of the vault axis. His layout consists in determining the orthogonal arch from the vertical section of the vault cylinder. If the face wall is vertical, this construct is satisfactory.

But, if the wall is sloping, it becomes arbitrary. Indeed, Jousse does not project the object on this vertical plane, but considers the intersection of the vault with this plane and uses this vertical arch to go from

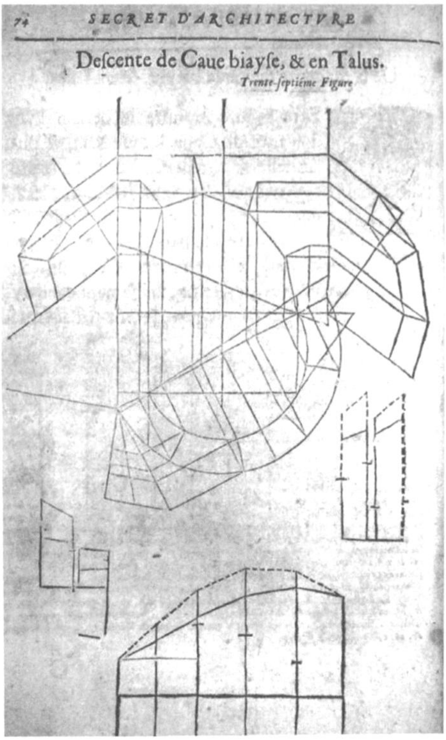

Fig. 13.9: Jousse's blueprint, in [7], p. 74.

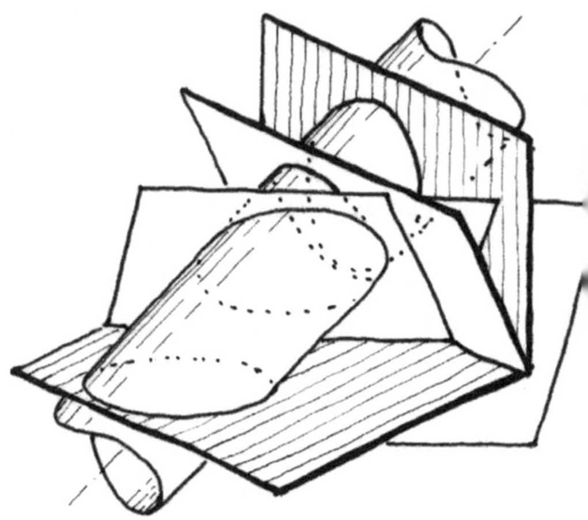

Fig. 13.10: Jousse's reference planes (Author's drawing).

the face arch to the orthogonal arch [Fig. 13.10]. The vault is seen as an unlimited cylinder with a directrix lying in a vertical plane including a horizontal line from the face wall. Although this layout presents the advantage of a tangential joining of the face arch on the edges of the jambs, this is probably not the reason for choosing this layout. While Jousse makes no comment to this effect, the construction he presents is rather general: it is valid for any arch of the face wall and is also "reversible". It can be read going from the orthogonal arch to the face arch or vice versa, which offers the architect a degree of freedom Philibert de l'Orme does not provide. Let us add also that this blueprint is coherent with the previous drawings. For example, if the slope is equal to zero, we obtain the blueprint given for the sloping horizontal vault.

Derand

Derand devotes eight chapters and thirty-five pages to the study of sloping vaults, "a type of vault among the most difficult that art displays"

[4, p. 48]. He proposes three different solutions for a semi-circle arch on the face wall and examines the case where the orthogonal arch is semi-circular. Derand then reverts to the three solutions previously presented, but assuming that the wall is a sloping wall, presents the construction of the sloping vault against a sloping wall for any rampant arch on the face wall, and finishes his study with the squaring method. Saying that Derand feels at ease with the subject, perceives all its wealth and happily juggles with all the different solutions would be an understatement. In his first method, Derand takes a coordinate system comprising a horizontal plane and a vertical plane parallel to the vault axis [Fig. 13.12]; in the second, he keeps the same coordinate system but projects the vault obliquely onto the vertical plane; in the third, he performs the same change in horizontal plane as Philibert de l'Orme. The new horizontal plane contains the vault axis and remains perpendicular to the given vertical plane. But, unlike with de l'Orme, the old horizontal plane of the vault is conserved, and both the old and new planes are superimposed in the same view.

While Derand's first illustration seems close to a descriptive geometry blueprint [Fig. 13.11], the other two are very different from it both in the form and spirit of their construction. In the second illustration, Derand replaces the profile view of the vault by a very special axonometry, where horizontal lines are represented by lines that are parallel to the vault generator lines and where the segments of lines parallel to the vault axis are projected according to their true size. The third method consists in separating the information supplied by the "axonometry" of the previous blueprint on two separate drawings. On an outlined profile, Derand only keeps the first part of the axonometry, which is very easy to draw and sufficient to deduce the orthogonal arch.

Continuing with the systematic study of the sloping vault, Derand provides the drawing of such a vault when the orthogonal arch is semicircular. This results, on the face wall, in a "depressed rampant arch, the effect of which can easily be visualized when it is well done" [4, p. 61]. Adopting, for this illustration, the principle previously followed in the first of the three solutions, he rapidly draws the blueprint by simply inverting the order of the operations. Finally, Derand considers the case of a sloping wall. Having previously shown his ability, we are therefore surprised, on this occasion, to see him make a blatant mistake. Indeed, he uses the same three methods previously developed for a vertical wall. He has no problem generalizing the first one.

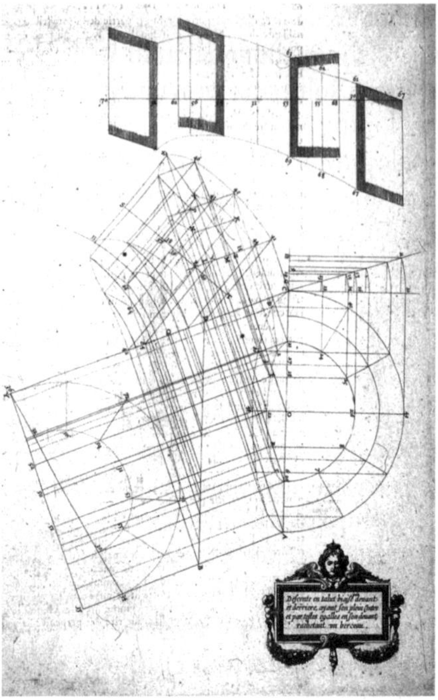

Fig. 13.11: Derand's blueprint, in [4], p. 66.

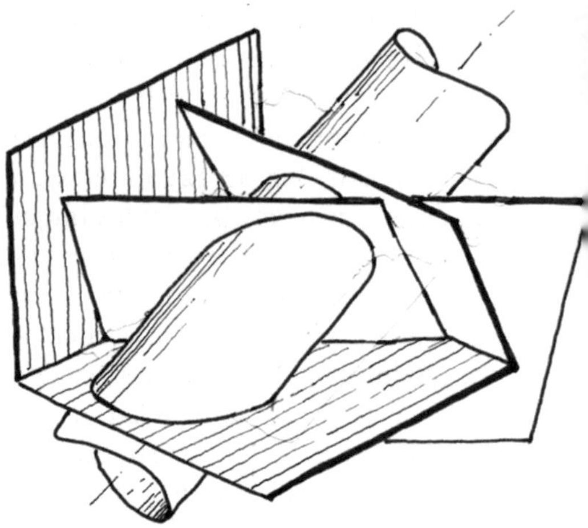

Fig. 13.12: Derand's reference planes (Author's drawing).

Derand adds a profile view, adjacent to the semi-circle arch folded over the face wall, on which he represents the angle of the wall. Unlike Jousse, he considers the semi-circle arch on the face wall. This profile view enables him to determine the horizontal projection of the face arch, to deduce its frontal view and finish the blueprint as for the vertical wall. But Derand is less successful when he attempts to apply the two other methods developed in the previous chapters. While, with some measure of good will, it is possible to interpret the diagram as an axonometry if the wall is vertical, when the wall is sloping, the construction given no longer corresponds to an oblique projection of the vault onto a plane, and the layout of the orthogonal arch is wrong. On Derand's blueprint, the error is minimal and even below the precision of the layout, but if the slant angle is increased, the error becomes significant. The rest of his reasoning is no less wrong, for the same reason. Derand indeed pursues his construction as he did in the case of the vertical wall. He performs a change in horizontal plane with the result that individual edges, but not the bed joint surfaces, are projected according to their true size onto the new reference plane.

"I must inform you...that these three methods for drawing sloping vaults adjoining a sloping wall...were performed using the same measurements, with a view that the resulting panels be equal in every way and that the same effect justify such good practices, which will therefore constitute proof for one another" [4, p. 74], concludes Derand. This justification is a little quick, first of all because the last two methods presented are coherent with respect to each other and simultaneously wrong, and secondly, because the error can only be noticed graphically, both in the case of the panels and the orthogonal arch, if the slant is substantial.

In the last chapter devoted to this subject (other than the squaring method), Derand studies the drawing of any rampant arch on the face wall and places himself exactly in the situation considered by Desargues. He rightly uses the first method presented for the semi-circle arch. Since this method never calls for the use of special properties of the face arch, it can be used without modification. It is, however, surprising that Derand adds a chapter to present a situation that is not really more general. The vault layout is correct because of this, but Derand makes a slight mistake in the simplest part of the layout, which concerns the jambs.[5]

There are as many methods of approaching the problem and different possible choices of reference planes as there are authors. This reveals the difficulty the authors experienced in approaching the problem. To draw the blueprint of a sloping vault, one of the reference planes must be parallel to the vault axis in order that the segments plotted from its generator lines appear according to their true size. The solutions listed above all consider the "reasonable" choices, given the premises of the problem. But this plurality of approaches shows that the basic principle, which consists in taking a horizontal plane and a vertical plane, was not inherently and automatically obvious before Monge. The latter nonetheless declared in his course at the Ecole Normale, that "artists...generally suppose that, of the two projection plans, one is horizontal and the other vertical" [11 p. 315]. Whatever Monge had to say, this custom was only well established before he began teaching when the object was itself naturally linked to such a coordinate system. The principle of the horizontal plane-vertical plane reference system

[5] Given the sloping wall, the vertical part of the passage wall is not, when projected horizontally, perpendicular to a horizontal line of the wall. Indeed, Frézier points this out in his work, [6, t. 2, pp. 187–188].

did not even prevail after Derand's treatise. Nor did it totally prevail for Derand himself, who somehow gives the impression of having used it by serendipity. What did he look for after having presented his first method? No doubt to present a blueprint with less lines to draw, which was a very essential preoccupation for masons. But this led Derand to make a blatant mistake, precisely because he ventured to draw layouts without any natural reference system. De la Rue makes use of the first two methods given by Derand, without particular preference for the one derived from a natural reference system. Finally, Frézier, who knew Derand's treatise thoroughly and presented the principle of the double projection in the first tome of his own treatise, proposes a different solution that is not quicker to draw and far more difficult to understand.

Indeed, Frézier writes in a paragraph of the first tome entitled *de l'Arangement des desseins dans l'Epure*: "Although it is more natural to separate each drawing, it is nonetheless true that this simplicity of object does not indicate so well the relationship between the lines, and that it is therefore not as convenient as gathering, and some times even mixing the layout, profile and elevation: the arrangement of their situation, close to, within, above, below or next to each other should however be viewed as arbitrary" [6, t. 1, p. 272].

Conclusion

A certain number of conclusions may be drawn from this comparative study.

The problem of two-dimensional representation of three-dimensional objects

One of the elements all these drawings and explanations have in common is the total—or quasi total—absence of an object in space. The constructions are drawn in the layout, without the points, curves or surfaces from space seeming to intervene. They are never named, except by Desargues, at the very beginning of his construction. Hachette distinguishes two steps in the resolution of a three-dimensional geometrical problem. The first concerns the resolution of the problem by three-dimensional geometry, and the second is the graphical rendering of the former. In the illustrations studied, and more generally in stone carving treatises, the first step is totally absent from explicative texts.

In descriptive geometry, on the contrary, accompanying texts give spatial geometrical constructions that are graphically translated into drawings. Descriptive geometry is precisely the language that is appropriate for describing space. As long as such a language has not been defined with sufficient precision by enunciating a few basic rules, certain mental operations remain untranslatable. This results in a distance, always surprising for present day readers, between the extraordinary ease the authors of such treatises seem to have for manipulating complex volumes and surfaces, and their capacity to clearly present their solutions. The explanations boil down to a succession of directions without justification: "trace this line, carry over this distance, etc..." without ever giving the reason for the tracing or the mental operation the reader ought to perform. The modern reading of these texts is summed up by the recurrent question "what is he doing"? Clearly, such questioning derives from a state of mind that is directly opposed to that prevailing among the various authors, who never sought to justify their drawings.

Transmission of know-how

In stone carving treatises, two different problems tend to interfere. The first is of a geometrical nature and involves determining the true size of the voussoir joint surfaces. The second is of a didactic nature: the reader must understand the drawing proposed in order to be able to reproduce it, or be in a position to produce a drawing to solve a precise problem he is faced with. Should a stone carving treatise be a collection of drawings for master masons? Philibert de l'Orme and Jousse appear to acquiesce. However, the real-size layout the master mason produces is one thing, and a drawing explaining how to make a drawing for a master mason is another. Confusion between these two steps may lead at worst to errors and at best to difficulties in reading and communication. Philibert de l'Orme seems to be aware of the problem he is facing. He explicitly admits it at the end of his commentary on the illustration of sloping vaults, when he writes that "other [drawings] might be made that are difficult to carry out...but apart from it being a head breaking exercise to reflect on them and represent them, I would also fear that few people in this world might sink their teeth into them and the simple demonstration I might make of them" [10, p. 62]. It is therefore not totally excluded that the drawing necessary for oblique vaults, when the orthogonal arch is to be determined, is beyond the pedagogical skills of the author, and even possibly his practical competence. According to Philibert de l'Orme, all cannot be presented in written form and a treatise cannot totally replace teaching by oral transmission and practical demonstration.

The drawings studied generally meet their main objective satisfactorily, in other words to give the exact geometrical construct of the joint surfaces. On the other hand, the second objective, the transmission of know-how, is only very partially attained. The main reason for this failure comes from the dichotomy, which we have just examined, between representation and geometrical construction. The graphical representation of the object is either totally absent, or entirely dependent on subsequent geometrical drawings. Descriptive geometry, on the other hand, makes it possible to unite these two functions on a single drawing, bringing together the technical drawing and the pedagogical tool.

Statical considerations

Curiously, statical considerations are totally absent from all the commentaries on sloping vault drawings. However, if the obliqueness is very pronounced, independently of the sloping vault or wall, the drawings proposed by the various authors, though they remain geometrically correct, become inapplicable in practice. Voussoirs with angles that are too acute might explode under the thrust of the vault, as Rondelet points out. The authors studied do not mention this limitation and remain strictly within the realm of purely geometrical considerations.

The strong dichotomy between the geometrical and statical approach to the problems inherent to dry-stone vault construction lasted a long time. One has to wait for Rondelet's *Traité théorique et pratique de l'art de bâtir*[6] to find a work that tackles the problems linked to mortarless architecture from a geometrical, statical and economical point of view, and expounding on the differences in the materials used, the quality of the cement and stones, etc.

[6] Jean-Baptiste Rondelet (1743–1829), a student of Blondel, became Soufflot's successor on the Sainte-Geneviève Church building site in Paris. He erected the dome and took care of its transformation into the Panthéon. Rondelet was a member of the *Commission des Travaux Publics*, who decided the creation of the Ecole polytechnique.

The issue of geometry

Placing geometry at the heart of architectural training was a wish Alberti had already formulated. But for Renaissance architects, the real need for geometrical knowledge in everyday practice remains very modest. Hence Alberti's wish only becomes truly meaningful and justified at very specific steps of architectural creation, drawing remaining the best case in point. Stereotomy, "that which is most refined and artistic in architecture", following a word of Claude Perrault, is the place where architecture meets geometry. From this meeting will arise a new geometrical theory, descriptive geometry, as well as the figure of the modern architect. Philibert de l'Orme clearly expresses the need for this marriage between geometry and architecture, which his work inaugurates to a certain extent. He insists on his goal, which is "to join the practice of architecture with the theory of the said Euclid" [10, p. 62]. But given the stakes at hand, the steps leading to this "wedding" were rather violent, jolted and conflict ridden.

The most symptomatic episode among these tensions was the quarrel opposing Curabelle, one of the most famous master masons of his time and Girard Desargues. The essence of this quarrel does not bear so much on the content of the stone cutting manual proposed by the Lyon geometer as on the manner by which one might be sure of its legitimacy. For Curabelle, feasibility is of course the criterion whereas for Desargues, the only thing that matters is the correctness of the geometrical reasoning. By virtue of this opposition, the entire status of the blueprint drawing is being questioned. If we accept, along with Curabelle, that the drawing can only be deemed worthy and validated by execution, the master mason remains the keystone of the building site. Preliminary drawings are certainly necessary but have no autonomy and cannot be dissociated from the construction they make possible. They are but a first phase, a first step of one and the same process of production.

If, on the other hand, a drawing can, as Desargues claims, have its very own legitimacy, if one can convince oneself of its correctness by purely theoretical considerations independent of any concrete actions, if geometrical reasoning rather than experience is what allows the finding of the optimal lines, then the status of the drawing itself is modified, and so is the status of the author and builder. The master mason is therefore deprived of part of his role and power. Desargues expresses this quite brutally: "just as Doctors…do not attend the school or receive the teachings of apothecaries... Geometers... neither attend the school nor receive the teachings of Masons; quite to the contrary, Masons... attend the school and receive the teachings of Geometers, which means that Geometers are the masters and Masons the disciples".[7] What clearly emerges from the polemic writing of Desargues is the claim that theory prevails over practice.

The camps are therefore well defined among stone carving aficionados. On one side, practitioners such as Jousse, Derand, Curabelle or de La Rue defend the master mason profession throughout their work, and on the other side, theoreticians such as Philibert de l'Orme, Desargues, Blondel, de la Hire or Frézier attempt, to various extents and with more or less success, to integrate academic geometry as a lever for transforming a construction technique and changing power relationships between the various social actors.

References

1. Bosse, Abraham, La pratique du trait à preuves de M. Desargues, pour la coupe des pierres en l'architecture, Paris, 1643.
2. Bosse, Abraham, Manière universelle de M. Desargues pour pratiquer la perspective par petit-pied, comme le géométral, Paris, 1647. (with the "Reconnaissance de Monsieur Desargues").
3. Coolidge, Julian Lowell, A History of Geometrical Methods, Dover Publications, New York, 1940.
4. Derand, François, L'architecture des voûtes ou l'art des traits et coupe des voûtes..., Paris, 1643.
5. Desargues, Girard, Brouillon-project d'exemple d'une manière universelle du S.G.D.L. touchant la practique du trait à preuves pour la coupe des pierres en l'Architecture..., Paris, 1640.
6. Frézier, Amédée-François, La théorie et la pratique de la coupe des pierres et de bois pour la construction des voûtes... ou traité de stéréotomie à l'usage des architectes, Strasbourg, 1737–39.
7. Jousse, Mathurin, Le Secret d'architecture, découvrant fidèlement les traits géométrique, couppes et dérobements nécessaires dans les bâtiments..., La Flèche, 1642.
8. La Gournerie, Jules Maillard de, Discours sur l'art du trait et la géométrie descriptive, Paris, 1855.
9. Loria, Gino, Storia della geometria descrittiva delle origini, sino ai giorni nostri, Hoepli, Milan, 1921.
10. L'Orme, Philibert de, Le premier tome de l'architecture, Paris, 1567, reprinted in Traités d'architec-ture, J. M. Perouse de Montclos ed., Léonce Laget, Paris, 1988.
11. Monge, Gaspard , Géométrie descriptive, in Les Séances des écoles normales recueillies par des sténographes et revues par des professeurs, Paris, 1795. reprinted in

[7] Desargues, "Reconnaissance de Monsieur Desargues" in [2].

L'Ecole normale de l'an III, Leçons de mathématiques, Laplace, Lagrange, Monge, J. Dhombres éd., Dunod, Paris, 1992, pp. 267–459.
12. Perrault, Charles, Parallèles des Anciens et des Modernes, Paris, 1688–1692.
13. Pérouse de Montclos, Jean-Marie, L'architecture à la Française, XVIe, XVIIe, XVIIIe siècles, Picard, Paris, 1982.
14. Rondelet, Jean-Baptiste, Traité théorique et pratique de l'art de bâtir, Paris, 1802–1817.
15. Sakarovitch, Joël, Epures d'architecture, de la coupe des pierres à la géométrie descriptive, XVIe–XIXe siècles, Birkhäuser, Basel, 1998.
16. Taton, René, L'Œuvre mathématique de G. Desargues, P.U.F., Paris, 1951; 2nd ed., Vrin, Paris, 1988.

CHAPTER FOURTEEN

THE SQUINCH OF ANET

Martina Lenz

In the 16th century Philibert De l'Orme was the first architect in France who occupied himself not only with building objects, but also with explaining his difficult work in detail, providing background information on stone masonry, stereotomy and a great deal of geometrical knowledge. In the fourth book of his treatise *Premier Tome de l'architecture* which appeared in 1567, he described a number of architectural objects, including the squinch of Anet. In 1988 Jean-Marie Perouse de Montclos prepared a new edition of De l'Orme's *Traité d'architecture* [1]. Volker Hoffmann was the first in Germany to concern himself with this treatise more intensively. De l'Orme's text and drawings address not only the world of scholars, but also the practical working craftsmen like the *"appareilleurs"* ([2] 7 and 235).[1]

The squinch of Château Anet was a point of great interest in architectural history because the geometrical construction drawings are presented and even explained by the architect. With this treatise De l'Orme created a work of great importance, since he showed us how to construct a difficult three-dimensional object. In this regard he had no predecessors, but a few interested followers like François Derand, Jean-Baptiste De la Rue and Amédée-François Frézier. With De l'Orme's treatise the French word *trompe*, in English squinch, was used for the first time to characterize a vault system that supports a part of a building. In his treatise De l'Orme described several different types of squinches for example the most popular as the *trompe quarrée* ([1] 100).

[1] The "appareilleurs" are the workers who choose the stones that are cut by the "tailleur de pierre."

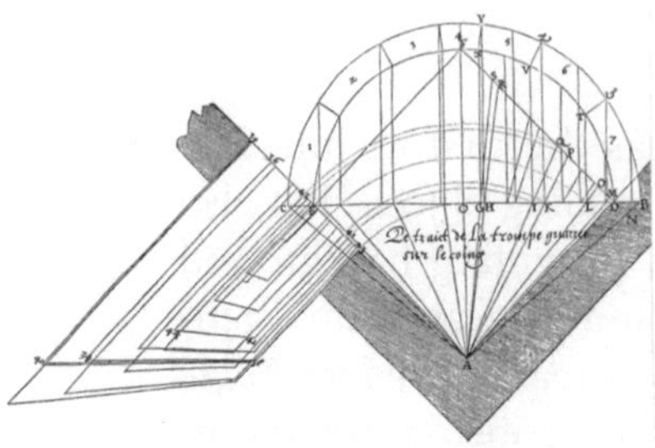

Fig. 14.1: Drawings for a trompe quarrée by Philibert De l'Orme 1567 in Perouse de Montclos: Ground plan.

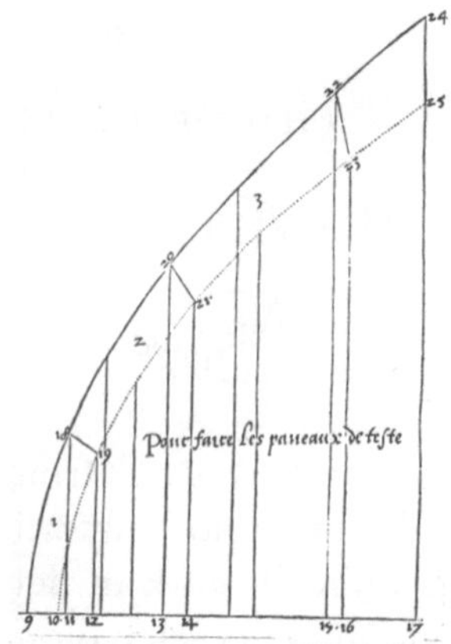

Fig. 14.2: Drawings for a trompe quarrée by Philibert De l'Orme 1567 in Perouse de Montclos: Panneaux de tête.

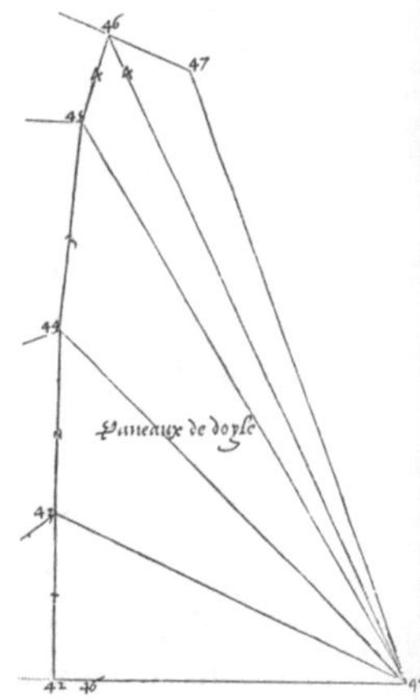

Fig. 14.3: Drawings for a trompe quarrée by Philibert De l'Orme 1567 in Perouse de Montclos: Panneaux de douelle.

Later theoreticians developed more than a dozen ingenious and extremely complicated variations on the theme, including for example, what was called the *trompe de Montpellier*.[2] The only type I would like to discuss here is the *trompe ondée et rampante*, the wave-like and inclined squinch of Château Anet. The technical function of a squinch in De l'Orme's sense is to create a self-supporting system to shore up an oriel above. For him the theoretical aspect of the stone masonry, which the French called "*stéréotomie*," was the most interesting thing. As he tells us, he was aware of several methods to support an oriel with iron or wooden construction elements inside the squinches, but for him stone masonry and the concentric commissures here constitute an ingenious construction method ([1] 90a). De l'Orme was one of the first who was no longer a craftsman in the medieval sense but an architect with considerable geometrical know-how.

Therefore he developed for the squinch of Anet four geometrical drawings that provide instructions for how to cut each individual stone in the complex piece of vault. These four drawings are connected to

[2] The "trompe de Montpellier" is the most common squinch built between two rectangular walls with a conical vault and a quarter circle ground plan explained for example in [3] or [4].

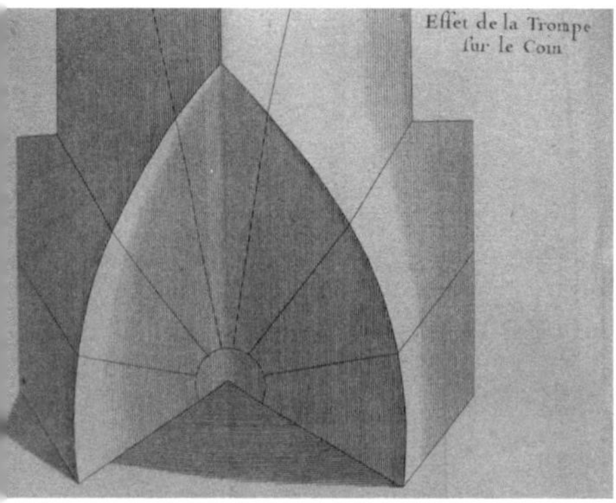

g. 14.4: Three-dimenional shape of a trompe quarrée by J. B. De la Rue, 1728.

Fig. 14.5: Three-dimensional shape of a trompe de Montpellier by J. B. De la Rue, 1728.

each other, with the measurements acquired in one drawing being used for the next. They belong to each other and work together. Using all of the sizes we obtain from these drawings and the written information De l'Orme provides in his treatise, it is possible to reconstruct the squinch ([1] 99a). And this is a particularly interesting subject for the very reason that the squinch of Anet disappeared, probably around 1800.[3]

There is only one three-dimensional view left, dating back to the period of construction between 1548 and 1552. This wood carving is very controversial, not only from today's point of view, but even from the perspective of De l'Orme, who criticized various of its aspects ([1] 106b).

Subsequent generations of architects and architectural theoreticians worked on the construction methods of the squinch, obtaining different results. François Derand ([3] 137f.), Jean Baptiste de La Rue ([3] 93f.)

[3] The exact date of the demolition is unknown. D'Aviler [2] talks about the year 1755. Roussel [5] tells us that the squinch disappeared between 1805 and 1810. Fact is, it is not there anymore.

Fig. 14.6: Wood carving from Le premier tôme de l'architecture de Philibert De l'Orme, 1567.

or Amédé-François Frézier ([6], livre 2, Charpitre IV, p. 305ff.) used technical drawings similar to those of De l'Orme, but none of them produced the three-dimensional object that results from these drawings. A three-dimensional squinch of Anet that is developed from these drawings can prove that they all fit together perfectly. Concerning the sources, the interest in such difficult geometrical refinements disappeared upon the publication of Gaspar Monge's *Géometrie descriptive* in 1788. This also means that there was no hope that researching different original sources would help to obtain fundamental results. So we had to approach the problem from a more interdisciplinary aspect, from the point of view of architects and stone masons.

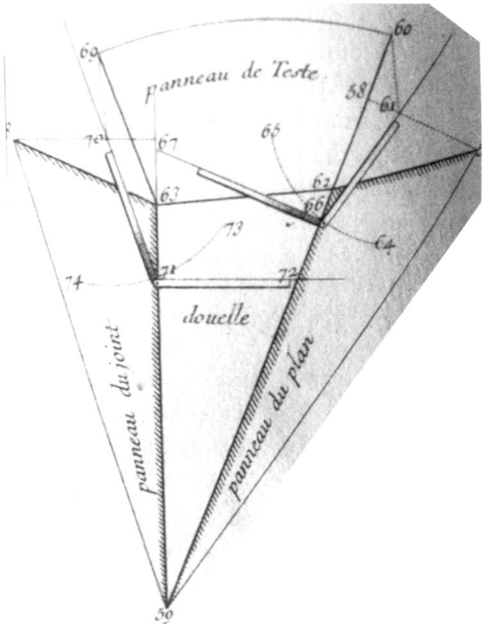

Fig. 14.7: Drawing that shows the different surfaces of the individual stones by J. B. De la Rue, Traité de la Coupe des Pierres, 1728.

Returning to De l'Orme's four drawings, these drawings are based upon the principles of geometrical orders ([1] 86f.). As apparent below De l'Orme used these principles to obtain true measurements of each individual stone from the different drawings. Today's architects and stone masons use the same rules to construct such a three-dimensional object on paper, in two dimensions. Thus it is clear that even today's stone masons could rebuild the squinch with the help of the drawings. They are sufficient. To clarify the following information, I will explain some of the technical vocabulary.

Every squinch consists of a certain number of stones. Each of these stones has different surfaces. In the case of squinches, we distinguish three surfaces for each stone, called the *panneaux de tête*, the *panneaux de douelle* and the *panneaux de joint*.

The *panneaux de tête* are the frontally visible surfaces of the vaulted stones. All *panneaux de tête* together form the contour line of the squinch. The *panneaux de douelle* are also among the visible surfaces. They form the soffit or intrados of a squinch. That means they both show the type of the vault of the whole squinch. The *panneaux de joint* are the non-visible surfaces between the stones.

For an exact construction of a squinch we need at least four different drawings. The first is a ground plan or horizontal projection that shows the shape of the oriel above the squinch. The second shows the heights of the *panneaux de tête* from the ground plan to the real contour line. The third shows the breadth of the *panneaux de tête*. The fourth shows the lengths of the *panneaux de douelle*. All drawings taken together represent the whole vault of the squinch.

1. De l'Orme's first drawing (untitled)

The first drawing by De l'Orme's hand shows the position of the squinch between the two walls G and H. Further we see the ground plan of the oriel that lies above our squinch, labeled CFDIB. It lies horizontally projected through the crown of the vault of the oriel. The thus created level is our main plane of projection for all subsequent drawings. Supplementing this first knowledge with the content of De l'Orme's treatise we get the following results: The squinch of Anet, above which was located the oriel with the *cabinet du roi*, was built between the two right-angled outside walls of the *corps de logis*. They are labeled G and H. The length of the distances from the crown of the vault A to the points C or B, where the oriel first touches the walls, are equal. The oriel is centered between the walls; it is not displaced to either of the sides G or H ([1] 92a).

Connecting the points C and B (where the oriel meets the walls) yields an isosceles triangle CAB. When we construct the bisecting line of CAB we locate the point labeled Z. De l'Orme himself used precisely this construction method for his t*rompe quarrée* ([1] 100b). Subsequent generations of architects such as François Derand and Jean-Baptiste De la Rue, designed squinches in exactly the same way. Extending the bisecting line reveals for the first time that the ground plan of the oriel is axially symmetrical with respect to the bisecting line. But soon it becomes apparent that the plan is not symmetrical in the vertical projection, but only in the horizontal. Using three concentric semicircles centred on Z, De l'Orme constructed the inner contour of the squinch, the wall thickness and the maximum overhang of the squinch, marked F, D and I. The contour of the middle window D is given by the outer semicircle. It follows this line before swinging back to the line that demarcates the wall thickness. The side windows F and G are symmetrical to the axis A.D. Like the squinch itself, they have

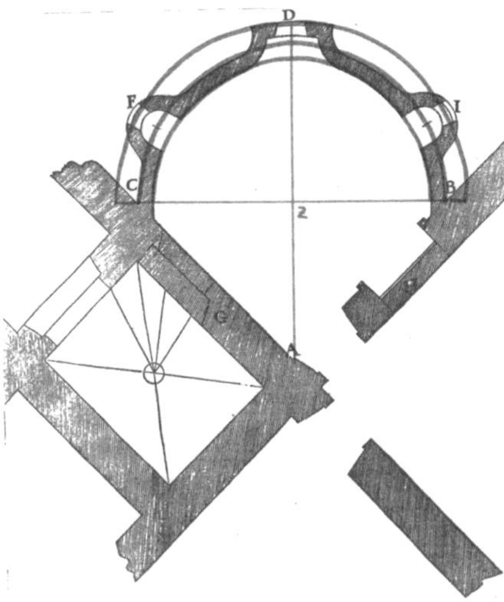

Fig. 14.8: Philibert De l'Ormes drawing of the ground plan of the squinch, 1567, with geometrical overlay by the author.

the appearance of a segment of a circle. All lines are constructed with compasses. But it is evident here, and in the subsequent drawings as well, that various of the construction lines are missing. De l'Orme criticized the quality of the print, not only because of this fault ([1] 106b).

This first plan is the simplest of the four drawings. It is the only one that gives geometrical information on its own. The others are much more complicated and can not be explained without the treatise.

2. De l'Orme's second drawing: "Le plan et le traict de la trompe d'Annet"

De l'Orme always described the squinch as a *trompe ondée et rampante*, which means a wave-like and inclined squinch. It is wave-like because of the extensions for the windows. It is inclined because it was not vertically symmetrical as were De l'Orme's *trompe quarrée* and the squinch of the Montpellier type.

This is because the imposts, the points where the vault begins, are not at the same height on both walls. Already in his treatise De l'Orme tells us why: Before he began with his design, there was a circular window in wall G. This window was supposed to admit light into the stairway behind the wall of the *corps de logis*. It was stipulated that this window remains unblocked ([1] 88ab).

The rooms of King Henry II. had already been completed in the *corps de logis* when the search began for a possible place to integrate a private study for the king. When this demand was expressed, there was no longer any room left. The idea was to built an oriel for this room. Because of the window the only possibility was to set one impost of the squinch higher than the other. Thus the construction of the oriel was predetermined by existing conditions ([1] 88a). The possibility of setting the entire oriel and squinch slightly higher to avoid the problem of the window was not discussed. Perhaps it was considered too inconvenient for the king to have to ascend stairs to reach his study.

The ground plan of the squinch as seen in De l'Orme's first drawing is integrated in this drawing identically. Further we see a flattened arch from F to B, the so called *arc droict Rampant*.[4] In the technical vocabulary of the stone masons this foreshadows a certain construction method to follow.[5]

As De l'Orme tells us in his treatise, the impost of the squinch was moved a distance of CF upward along side G ([1] 92a). Starting at B we follow the maximum height of the flattened arch to line CB. We obtain the point k, from which we construct a quarter circle with the radius kB on CB. That is the first half of our flattened arch. To get the center of the second half of the flattened arch we construct a vertical line through k, and then a parallel to CB through the point F. The intersecting point of these two lines yields point l, the center of the second quarter-circle.The second quarter-circle is constituted with the radius lF ([6] 184).

Now we know how to get this flattened arch, but what is it for? In three-dimensional reality the flattened arch is perpendicular to the ground plan and built on the line CB. Its function is to characterize the vault of the squinch. The second concentric flattened arch shows the thickness of the wall. Both arches are divided into seven segments ([1] 92a). The number of segments can be chosen by the architect at will. De la Rue limits the number of segments to five to ensure clarity in the drawings.

[4] Anthony Blunt [7] translates De l'Ormes "arc droit rampant" as "flattened arch". In the following, I use this term.

[5] The construction of flattened arches was known even in the Roman antiquity—see [8] 173, picture no. 11 ("Außenbau des Tepidariums der Thermen des Diokletian").

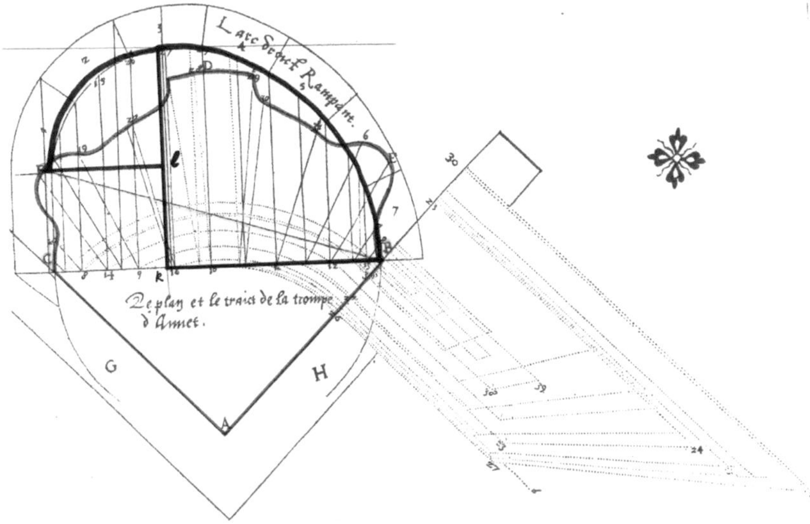

Fig. 14.9 [Col. Pl. XLVIII]: Philibert De l'Ormes drawing of Le plan et traict de la trompe d'Annet, 1567. Green Lines: Integrated ground plan. Yellow lines: Height of the inclination. Blue lines: Construction of the flattened arch.

After De l'Orme, this possibility of projection is later used by all known architects to characterize the vaulting systems of squinches. Derand, De la Rue and Frézier worked in the same way.

The main difference between the *trompe quarrée* or the *trompe de Montpellier* and the squinch of Anet is that the first are symmetrical and conical, and a semicircle can describe the vault to show three-dimensional contexts on a two-dimensional sheet of paper. There De l'Orme used geometrical methods for the first time. The ground plan, as mentioned above, was drawn before as a the main projection plane. By folding all points from the flattened arch 90 degrees into the main projection plane it was possible to obtain a two-dimensional picture of a three-dimensional object ([9] 15ff.).

Back to our drawing: On the right side of the drawing we see what are called the *lignes de pente*, the dotted lines.

De l'Orme explained the construction process for three different joints in detail ([1] 92a and 93b). Here only one commissure from segment two to segment three will be discussed: From the joints we construct the vertical lines on CB. With our vertical line we arrive at number 9 on CB. Positioning a pair of compasses at the angle point A, we move this point to wall AB, now it is designated 22. Through this point we

Fig. 14.10: Position of the ground plan and the flattened arch in the three-dimensional model.

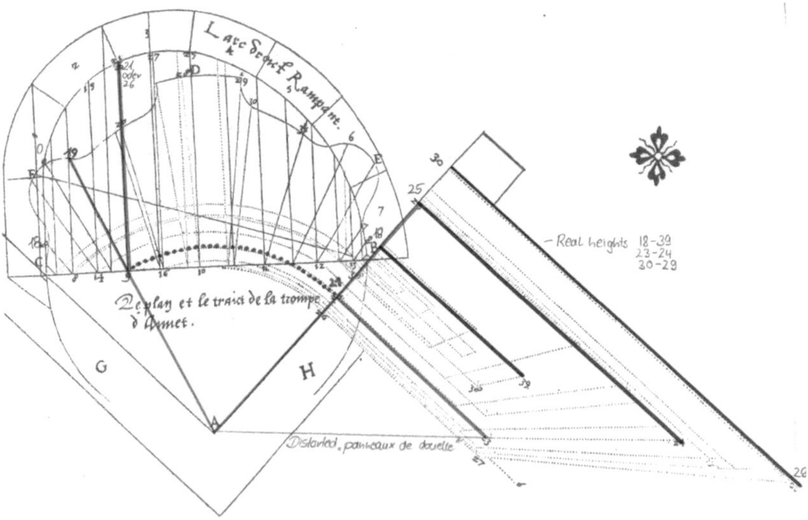

Fig. 14.11 [Col. Pl. XLIX]: Philibert De l'Ormes drawing of Le plan et traict de la trompe d'Annet, 1567. Coloured lines: Construction of the real heights of the panneaux de tête.

construct vertical dotted lines to AB. Their length is equivalent to the height from 9 to the flattened arch (9–21 or 26). This brings us to point 23. Now, from this projected point 22 on side AB, we construct the distance from CB to the real ground plan (9–19). The dotted lines show the result 22–25. This is what De l'Orme told us. But what is the point of this? With the dotted lines De l'Orme is able to obtain the true heights from the ground plan to the real contour of the squinch. He works with an orthogonal parallel projection.

All technical drawings are based on common fundamental geometrical principles ([9] 7ff.). One of them is the projection. Projection means the relationship between a point in the three-dimensional space and its two-dimensional picture on a sheet of paper. Technical drawing is a question of how to reproduce a three-dimensional object in one projection plane, so that it is clearly understandable on the one hand, and correct in its dimensions and proportions on the other. These two demands could not be both fulfilled at the same time in one drawing. With a method of drawing, using projection it is possible to obtain true dimensions. True dimensions are sizes that can be extracted from the different drawings without any distortions. This is important for the stone masons and the craftsmen—the practical working people.

True dimensions can be found in the plane that is parallel to the distance or level being sought ([11] 9f.; [9] 32ff.). Geometry knows three projection levels, each of which is perpendicular to the others. They are called main projection planes their position in space is fixed. We call them the "tripod" the ground plan level, the upright projection or vertical projection, and finally the lateral projection. These geometrical fundamentals are still necessary for architects and stone masons today.

If the projection lines are parallel and meet the projection plane at a right angle, we call this orthogonal parallel projection. With this method it is possible to bring three-dimensional objects to the main level. The best example to illustrate this is the drawing of a cube on paper. As to projections, we have to distinguish distances or planes, that are parallel to any of the three projection levels which have a special position, and those that are not parallel to any of the projection levels, which have a common, or general. position. We can take the lengths directly from the drawings where we find distances with a special position ([12] chapter "Projektionen"). Within all other drawings these distances appear distorted. This is much more difficult for distances with a common position. These distances are not parallel to any of the projection levels, meaning that they appear distorted in all

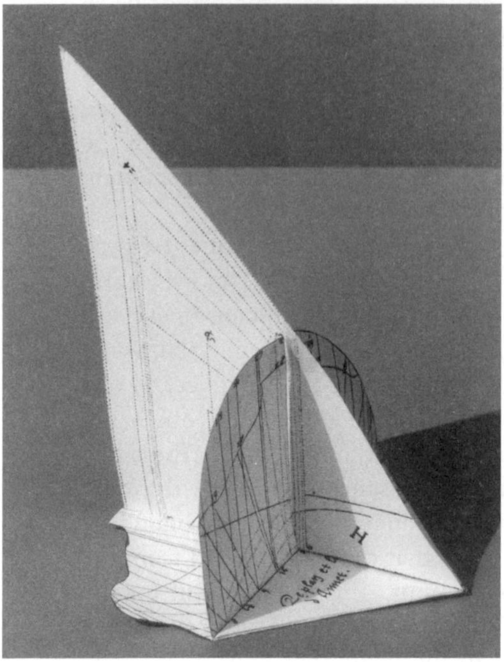

Fig. 14.12: Three-dimensional position of the ground plan, the flattened arch and the dotted lines.

Fig. 14.13: Three-dimensional position of the ground plan and the dotted lines integrated in the model.

of our geometrical drawings. There are two possible ways to find their true dimensions: First, by rotating the object, second, by revolving the object around an axis. With these methods the object can be brought into a special position where true dimensions can be obtained.

True dimensions are one thing, but the other is three-dimensional reality.

3. De l'Orme's third Drawing: "La Cherche ralongée & paneaux de teste par le devant de la trompe"

In the preceding drawings De l'Orme was able to calculate the heights of the *panneaux de tête*, but not their breadths. The breadth of each of the seven segments already had appeared in De l'Orme's second drawing, but because the segments are in a common position, the breadth

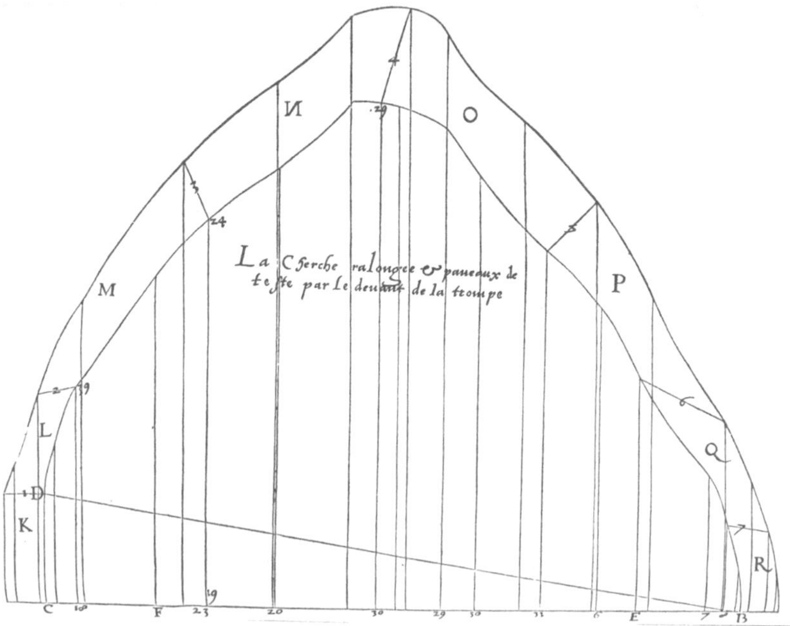

Fig. 14.14: Philibert De l'Ormes drawing of La Cherche ralongée & paneaux de teste par le devant de la trompe, 1567.

is unfortunately distorted. To get the true breadth of each segment he had to bring them into a special position. This is why De l'Orme's third drawing is necessary.

Remarks on the geometrical method: The method De l'Orme used is called unfolding of a geometrical object. This designates a special procedure important for structural plans in general, and for cases involving the construction of squinches especially. To unfold a circle, for example, you cut it once and then straighten it out to get a straight line. Therefore you cut the circle into small pieces using compasses, then, step by step, line up the pieces into a straight line. In this case the circumference of the circle is unfolded. For example, when you unfold a three-dimensional cone, you cut it along the generating line to bring it into two dimensions and obtain real measures.

But what did De l'Orme do to get the true breadth? First he unfolded the ground plan, and then above this he unfolded the real contour of the squinch.

1. *Unfolding of the ground plan*

In his treatise De l'Orme relates the steps of his work:

> Après ce il fault trouver avec le compas la ligne droicte ralongée de tout le devant de la trompe, par plusieurs petis rapports dudict compas. Ainsi que vous le voiez par exemple en la distance de C à 18, & de 18 à F, & de F à 0, & de 0 à 19, & de 19 à 20 & ainsi consequement iusques à ce que vous ayez trouvé toute la circonference du devant de ladicte trompe pour en faire une ligne droicte: comme vous le voiez à la figure ensuivant. ([1] 92a)

Then De l'Orme went back to the second drawing in "Le plan et le traict de la trompe d'Annet". From there he used the compasses to transfer the lines C-18, 18–F, F-O, 0–19, and 19–20 from the ground plan of the squinch to the line CB in his new third drawing, La Cherche ralongée et paneaux de teste par le devant de la trompe. In this manner he unfolded the ground plan onto CB.

2. *Unfolding the real contour of the squinch. Once again we look at the treatise*

> Et la hauteur que vous trouvez depuis ladicte ligne AB, iusques au point de 39, vous la portez en le figure de la cerche ralongée, & paneaux de teste cy apres descrits, & la mettez du poinct de 18, au point de 39. Cela vous monstre iustement la hauteur de doit avoir le paneau, & le devant de la trompe, au lieu marqué 2. Il fault ainsi trouver toutes les autres lignes que vous voyez designées en la figure ensuyant, qui vous montre les hauteurs du dessous de la trompe par le devant, suyvant lesquelles vous trouvez le pourfil de la voute de ladicte trompe. Et ainsi se trouvent tous les paneaux de teste marquez K L M N O P Q R, comme vous le pouvez cognoiste par la figure prochaine. ([1] 94a)

Over CB De l'Orme constructed the unfolding of the real contour line of his squinch, split into seven segments called K L M N O P Q and R. For this he returned to the *lignes de pente* in his second drawing. He transferred the real heights of the *panneaux de tête* from the second drawing to his third drawing. De l'Orme explained his construction method for three different joints. This concerns the lines 18–39, 25–24 and 30–29. But because he did not tell us what the drawing is for,it took some time to find this out.

The geometrical method De l'Orme used is the unfolding of the dark gray area between the ground plan and the real contour of the squinch.

What did De l'Orme do?

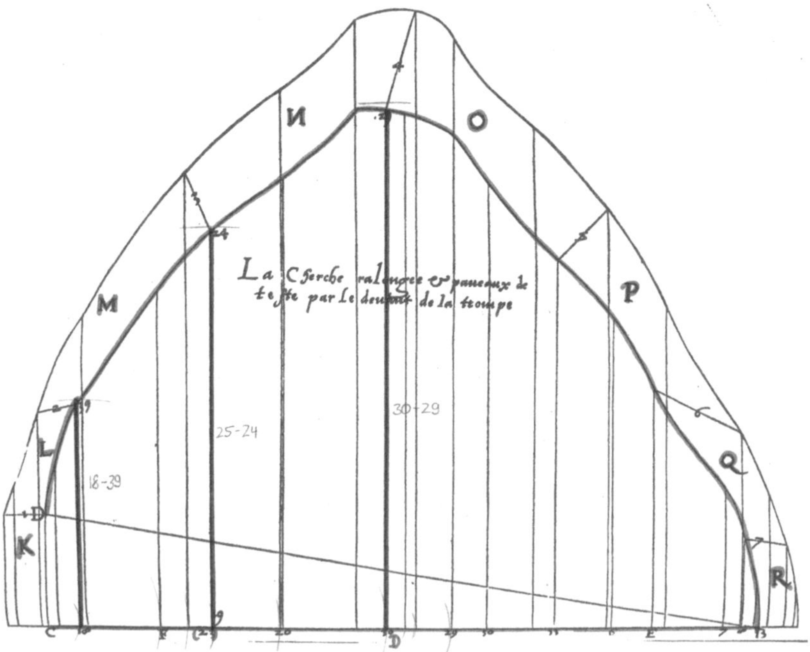

Fig. 14.15 [Col. Pl. L]: Philibert De l'Ormes drawing of La Cherche ralongée & paneaux de teste par le devant de la trompe, 1567. Green line: CB: Unfolding of the ground plan. Blue lines: Real heights from the second drawing. Red lines: Unfolded real contour of the squinch.

First he unfolded the contour line of the ground plan. Above the ground plan he constructed the true heights of the contour of the squinch that resulted from the preceding drawing. In this manner De l'Orme obtained the breadths of the *panneaux de tête*, one of the missing measures.

4. De l'Orme's fourth drawing: "Le paneaux de doyle par le dessoubs de Trompe"

The last measures missing are the true lengths of the commissures of the *panneaux de douelle*. The lengths already appeared in one of the drawings: we saw them as part of the dotted lines in the second drawing. But unfortunately, they are in a common position there, meaning they are distorted. So a new drawing was necessary that shows the lengths of the *panneaux de douelle* in their true dimensions.

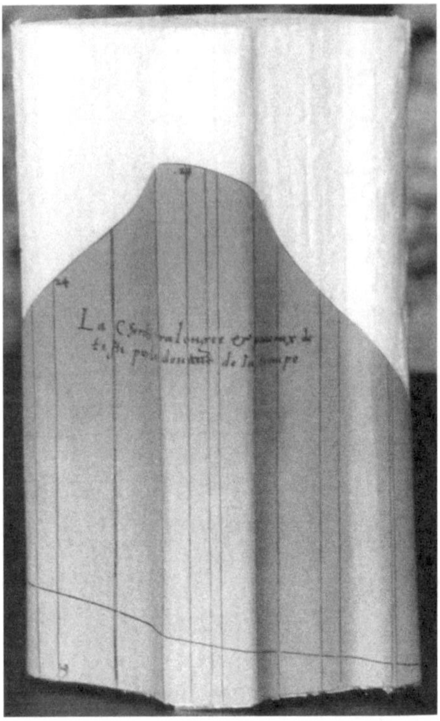

Fig. 14.16: Position of the drawing La Cherche ralongée…in the three-dimensional model.

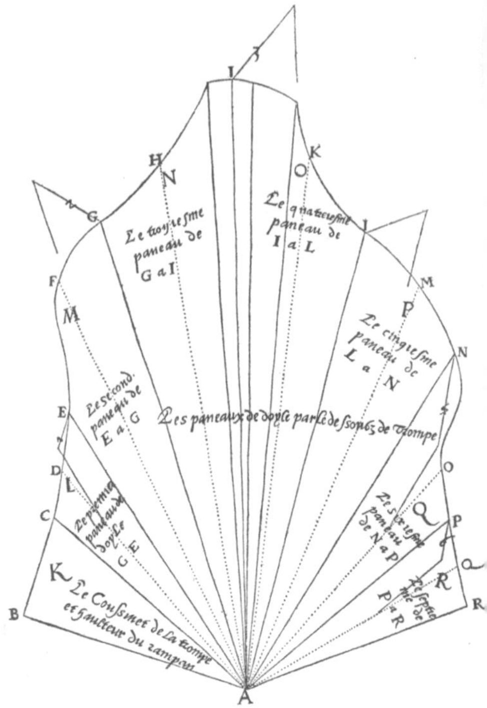

Fig. 14.17: Philibert De l'Ormes drawing of Les paneaux de doyle par le dessoubs de Trompe 1567. Red line: Contour of the impost.

That is the function of the final missing drawing. The right angled triangle CBA, labeled Le coussmet de la trompe et haulteur du rampant, shows the impost on the window side ([1] 95a). Its dimensions were to be taken from the second drawing, where they appear as true measures (as De l'Ormes tells us) ([1] 98b). In his treatise De l'Orme explains the complicated construction of the vaulted *panneaux de douelle*. Here we follow him just for one example, the "*panneau L*", the first vaulted stone:

> Le premier paneau marqué L, est fermé de trois lignes, comme vous voyez ACE, & ce faicte de ceste sorte: Vous retournez sur la figure du traicte de la trompe, & mettez la poicte du compas depuis le poinct de A, iusques à celuy de 18, & rapportez la distance en la figure precedete intitulée, La cherche ralongée & paneaux de teste & C. La colloquat sur la ligne CB, au point de 18, & tenant un pied du compas ferme sur ladicte ligne, vous ouvrez l'autre iusques à ce qu'il touche la commissure 2 au lieu de

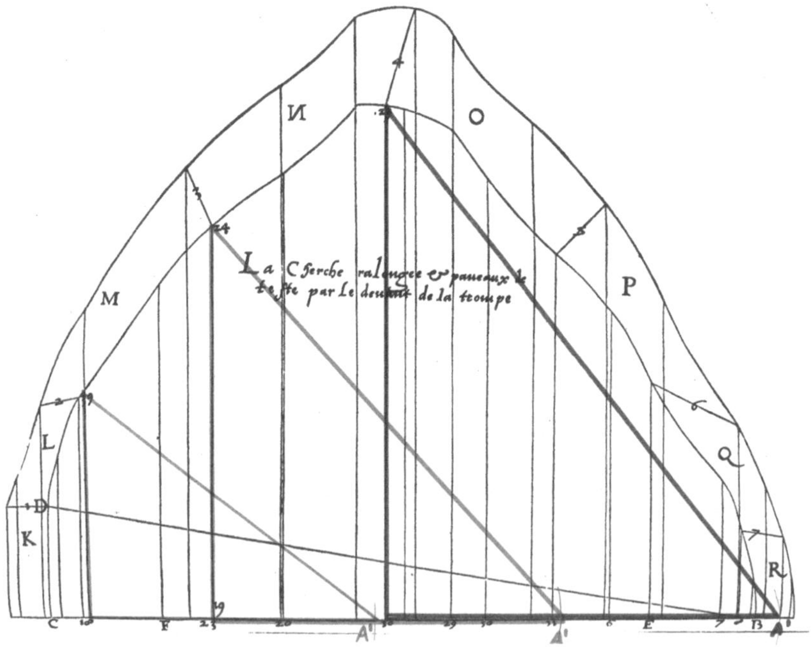

Fig. 14.18 [Col. Pl. LI]: Philibert De l'Ormes drawing of Les paneaux de doyle par le dessoubs de Trompe, 1567 with the integrated construction lines.

> 39. Puis vous rapportez la distance dudit 18 à 2 sur la figure qui est cy dessous inscrite, Les paneaux de doile par le dessoubs de la trompe: & la mettez du poinct de A, iusques à celui de E. En apres vous retournez à la susdicte figure des paneaux de teste, & prenez le distance du poinct de D, iusques à 39, laquelle vous rapportez en la figure cy apres ensuivant, & mettez depuis le poinct de C, à celuy de E, qui vous monstre le premier. ([1] 95a)

This sounds more complicated than it is. Here De l'Orme explains a geometrical projection method called rotation. He went back to the drawing of the Cherche ralongée to project the crown of the squinch labeled A to the same projection plane for every commissure. He rotates all necessary points. De l'Orme projects the true lengths of the *panneaux de douelle* to the same projection level as the ground plan and the position of the crown, where he obtained true sizes.

From the Plan et traict De l'Orme used compasses to record the distance A-18. He then transferred this distance to the drawing Cherche ralongée. He rotated the crown of the squinch labeled A to the same

Fig. 14.19 [Col. Pl. LII]: Three-dimensional squinch of Anet. Rebuilt in scale 1:1 to De l'Ormes drawings.

projection plane as the unfolded true ground plan of the squinch labeled CB. Then he took the distance from the crown A to the first commissure, labeled 39, and transferred it to the drawing Les paneaux de doyle. The breadth of the stones was taken directly from the Cherche ralongée.

This last drawing by De l'Orme's hand describes the three-dimensional shape of the squinch perfectly. Therefore it is possible to adjust De l'Orme's drawings to a three-dimensional model. There is no better proof for the three-dimensional shape:

5. Conclusion

In his treatise De l'Orme explains his constructions in detail, but usually only once. He frequently closes his descriptions with the words "*qui en entendra un les entendra tous*". If his difficult thoughts are not immediately comprehensible, there is no second chance for more clarification. To him all of these geometrical procedures seem pretty simple. Nevertheless he knew that his geometrical knowledge was not an easy subject for everyone, as he tells us in the treatise:

> Si est ce que telle chose ne se laisse pas traicter par maistres ignorants, car il fault qu'ils soient scavants en leur art, & ayent grande experience pour mettre telles choses en oeuvre. ([1] 90a)

And slightly later he tells us that we contemporaries are not the only ones who had to work hard to understand such difficult thinkings:

> Mais vous scavez que toutes nouvelles ecritures & inventions ne sont iamais sans grande difficulté et labeur. ([1] 87b)

We have no examples of the constructive virtuosity of geometrical methods in France earlier than De l'Orme. There are no predecessors for the construction of such a wave-like and inclined squinch, as De l'Orme himself declared ([1] 87b). Even later generations treated De l'Orme's theoretical work with great respect.[6] But he took these principles for granted and explained them just briefly. He seemed to regard the methods he used as fundamental. De l'Orme's treatise is the first source to show the state of knowledge of a few highly educated French theoreticians. Although we know of such difficult constructed objects, no earlier drawings seem to have survived in France.

References

1. Perouse de Montclos, Jean-Marie: "Philibert De l'Orme", Traité d'architecture 1561 and 1567, reed. Paris 1988.
2. D'Aviler, Augustin-Charles: "*Cours d'architecture*," Paris 1691.
3. Derand, R. P. François: "*L'architecture des voutes ou l'art des traits et coupes des voutes*", Paris 1643.
4. La Rue, Jean- Baptiste de: "*Traité de la coupe des Pierres, ou par une méthode facile et abrégé, l'on peut aisément se perfectionner en cette science*", Paris 1728.

[6] See [4] preface: "…cette science qu'on appelle ordinairement le traict est une invention très moderne. Philibert De l'Orme est le premier on ait donné des règles."

5. Roussel, Pierre Désiré: "*Histoire et description du château d'Anet*", Paris 1875.
6. Frézier, Amédée-François: "*Traité de la stéréotomie à l'usage de l'Architecture*", Paris 1738.
7. Blunt, Anthony: "*Philibert De l'Orme*, London 1958.
8. Hoffmann, Volker: "*Philibert De l'Orme und das Schloß Anet*", Architectura 1973, pp. 131–152.
9. Pechwitz, Franz: "*Darstellende Geometrie*", Nürnberg 1958.
10. Appel, Gerhard: "*Projektionszeichnen und darstellende Geometrie*", Kassel, 1960.
11. Pechwitz, Franz: "*Axonometrie*", Nürnberg 1950.
12. Frieder, Bernhard (ed.): "*Der Steinmetz und Steinbildhauer*", Munich 1996. Chapter Projektionen.

Additional literature and sources

A. *Literature*

Brauner, H. and Kickinger, Wolfgang: "*Baugeometrie*", vols.1 and 2, Wiesbaden, Berlin 1982.

Brion-Guerry, Liliane: "*Philibert De l'Orme*", Milano, Paris 1955.

Brissac, Philippe Cossé de: "*Châteaux de France disparus*", Paris 1947.

Gébelin, François: "*Les châteaux de la Renaissance*", Paris 1927.

Gloton, Jean-Jacques: "*L'architecture française de la Ren*aissance", Etat de la question, L'information d'histoire de l'art, Paris 1959, 4e année. pp. 133–143.

Hautecouer, Louis: "*Histoire de l'architecture classique en France*", Paris 1963.

Heliot, Pierre M: "*Documents inédits sur le châteaux d'Anet*", Mémoires de la société Nationale des Antiquaires de France, Paris 1951, Serie 9, pp. 258–269.

Monduit, Louis and Denis, Alexandre:. "*Stéréotomie du point de vue de la coupe des pierres*", Paris 1889.

Monduit, Louis: "*Traité théoretique et pratique de la stéréotomie au point de vue de la coupe des Pierres*", Dourdan 1980.

Noel, Pierre, Ingenieur E. T. P.: "*Technologie de la pierre de Taille*", Paris 1965.

Perouse de Montclos, Jean-Marie: "*Vocabulaire de l'architecture*", Paris 1972.

Potié, Philippe: "*Le projet constructif de Philibert De l'Orme*", l'idée constructive en architecture. Actes du colloque tenu à Grenoble du 28 au 30 novembre 1984, Paris 1987, pp. 21–31.

Prevost, Jean: "*Philibert De l'Orme*", Paris 1948.

Recht, Roland: "Théorie *et Traité d'architecture au Moyen Age*", Les Traités d'Architecture de la Renaissance. Actes du colloque tenu à Tours du 1 er au 11 juillet 1981, Paris 1988.

Roy, Maurice: "*Artistes et monuments de la Renaissance en France*", Paris 1929, vol. 1.

Sellenriek, Jörg: "*Zirkel und Lineal*", Kulturgeschichte des konstruktiven Zeichens, Munich 1987.

Vachon, Marius: "*La* Renaissance française", l'architecture nationale, Les grands maître maçons. Paris 1910.

B. *Sources*

Avaux, André Félibien des: "*Des principes de l'architecture de la peinture et des autres arts qui en dependant*," Avec un dictionnaire des termes propres à chacun des arts, Paris 1676.

Berty, Adolphe: "*Les grands architects français de la Renaissance*", Paris 1860.

Berty, Adolphe and Legrand, H.: "*Instructions de Monsieur d'Ivry, dict De l'Orme, abbé de Sainct Sierge, et cestui architecteur du Roy,*" Topographie historique du Vieux Paris, Région du Louvre et des Tuileries, Paris 1868.
Bosse, Abraham: "*Manière universelle de Monsieur Desargues pour pratiquer la perspective par petit pied,*" Paris 1648.
Chabat, Pierre: "*Dictionnaire des termes employés dans la construction*", Paris 1881.
Doulot, J.-P.: "*Traité special de coupe des pierres*", Paris 1825.
Frézier, Amédée-François: "*Eléments de stéréotomie à l'usage de l'architecture pour la coupe des pierres*", Paris 1760.
Laborde, Marquis Leon de: "*Les comptes du bâtiments du roi*", Paris 1877–1880.
Monge, Gaspard: "*Géométrie descriptive*", Paris 1788.
Pfnor, Rodolphe: "*Monographie du château d'Anet*", construit par Philibert De l'Orme en 1548, Paris 1867.
Riquet, Compte Adolphe De Caraman: "*Anet, son passé, son état actuel*", Paris 1860.
Rondelet, Jean: "*Traité théorétique et pratique de l'art de bâtir*", Paris 1812.
Vachon, Marius: "*Philibert De l'Orme*", Paris 1887.

CONCLUSIONS

The theme of this book and the preceding workshop addressed the history of shape creation knowledge in two related disciplines, civil and naval architecture. This topic appeared to be original and unique since we are not aware of any earlier attempts to apply an approach of *cross-disicplinary comparison* to the knowledge history of two distinct disciplines. Our preface has described the objectives we pursued in offering a discussion forum to scientists with expert background in both disciplines. The harvest of information was rich in substance and complex in structure. It is too early to draw any systematic conclusions from this wealth of information. But in retrospect we are convinced that it was worthwhile to propose such an ambitious goal and to make an opening move toward a deeper analysis of the parallel knowledge histories in these two disciplines. Let us examine some of the essential results achieved by these presentations and discussions.

The study of a new historical subject usually begins with the collection of source material on the pertinent developments to be investigated. It appears that in our joint efforts we have collected and compiled a substantial critical mass of such evidence on the subject of the knowledge history of shape design in civil and naval architecture. This evidence is documented in this book and in some precursor articles in the related MPIWG Preprint. It cannot be claimed that this material in any sense is exhaustive, and its analysis has just begun. But the information is of such scope and substance that it merits deeper scrutiny and continuing research. This is a valuable result.

Methodically the agenda of our meeting and the ensuing report in this book provided an opportunity to develop a new approach that we may call: "*Comparative Knowledge History in Two Related Disciplines*". We were particularly interested in those elements of knowledge genesis in shape design which allowed us to judge the interactions and interdependencies of shape design knowledge in the two applications, both, at a given time and place, and across different periods and regions. Both the presence and absence of such interdependencies shed new light on the knowledge genesis processes in the individual disciplines. This approach in the workshop has yielded some first interesting results and deserves to be further pursued.

Shape design, a characteristic subset of professional knowledge in both disciplines, has had its own knowledge history, which must have begun very early with human creative activities in art and technology. But in contrast to other branches of science and technology relatively little is known and well documented on this subject in historical sources. Archaeological artifacts are often the best evidence of creative methods in design. Much of this knowledge used to be informal, incompletely communicated and publicized, difficult to capture in words, though often quite rational and effective. Yet today the methodologies of creative shape design are reasonably well understood in fields like constructive geometry, geometric synthesis, geometric design, and more recently computational geometry. How did this knowledge once originate? To what extent was it consciously used in historical design applications? The discussions leading to our book have posed these questions and offered a few initial insights into this chapter of knowledge history. A more complete account on the history of shape creation methodology remains to be written. The results presented in this book offer a helpful fresh starting point for such continuing research.

One of those initial insights is the particular attention that must be paid to the tools and means of shape creation with respect to the historical development of design methods in the two disciplines. Which kind of tools and means of shape creation were employed in different periods and regions? We observed during the workshop that certain tools like drawing instruments, templates, patterns, molds, and plans played an important role in both disciplines for certain periods and appear to have been of crucial significance for the design methodology. Other methods such as canons of proportions and tools and procedures of the construction process seem to have supported the shape design activities procedurally. Did a shared "tool box" of such means exist in a certain region in a certain period and how did such means develop in both disciplines—independently of each other or by way of exchange? Attention to the kinds of tools, means, and specific procedures employed for the definition, production, and control of shapes also paves the way toward the distinction of which design methods can be subjected to meaningful comparisons and which ones cannot.

Conscious awareness of the commonalities between the two architectures has always varied and still does. We observed during our meeting that even among contemporary specialists there are those who recognize similitudes and analogies between the two fields and there are also others who are more impressed by the differences in requirements, functionalities and methodologies. Contrasting viewpoints do exist, but this does not detract from the value of a comparative analysis.

In order to concentrate on those paramount arguments in favor of a closer proximity, at least for limited periods when similar approaches were clearly perceived in the two professions, we would propose the following evidence, also underscored in the discussions of the meeting:

- The *naming* of civil and naval architecture has common roots in Greek antiquity (ἀρχιτέκτων and ἀρχιτέκτων ἐπὶ τὰς ναῦς). This is no coincidence. In fact the two related professions had similar roles and responsibilities in the urban society of Athens. They are likely to have used a coherent platform of technological knowledge in shape design and construction.
- In late antiquity and in the Middle Ages one can detect continuing commonalities in the tools and methods of the crafts involved in terrestrial and naval construction projects. For shape creation this centers in the concepts and practices of marking and measuring parts to be assembled. The practices of *moulding* (v. "modulus") and *lofting* in cathedral building lots (chantiers) and shipyards (chantiers navals) evolved during this early age.
- During the Renaissance the awareness of a close relationship gained new recognition and expressed itself also in literary activities describing the history and practices in the two architectures. Alberti, Lavanha, Furttenbach and later humanists, who consciously denoted the two professions as *architectura civilis* and *architectura navalis* and described them in close analogies, are witnesses of a consciousness of closely related identities.
- The body of knowledge in shape creation and geometric design has continued to grow steadily. Today it is almost a trivial statement that designers and constructors in the two architectures have access to a shared knowledge basis whose roots lie in their distinct experiences, but also in a common scientific platform for shape design. This knowledge basis has been available for at least the last two centuries.

The book and the meeting on which it is based have helped to substantiate these observations and claims. We have likewise identified many open questions which require further study. This book describes a challenging beginning and invites to continuing scientific efforts.

Horst Nowacki
Wolfgang Lefèvre

APPENDICES

APPENDIX A

FROM WORDS TO TECHNICAL PRACTICES: MOULDS AND NAVAL ARCHITECTURE IN THE MIDDLE AGES*

Eric Rieth
Translation: Richard Barker

1. Introduction

In the field of the history of Mediterranean naval architecture, the end of antiquity to the beginning of the Early Middle Ages corresponds to a period of profound changes, from which numbers of techno-economic aspects still raise questions. These modifications will touch on what can be called a sort of "architectural revolution". The Atlantic system of construction by the shell-first principle [20], will be substituted, in effect, by the new system of construction by the frame-first principle proper to the Mediterranean in the Middle Ages. It is in this architectural context, well defined historically and geographically, that this study is directed.

In a first part, the principal characteristics of this architecture will be recounted, on the basis of archaeological facts. A particular regard will bear on the determinant role held by transverse carpentry, as much from the theoretical point of view as practical. In the second part the manner in which the mediaeval written sources testify to this privileged function of the transverse carpentry will be examined. More precisely, we will seek to establish the list, brief in the event, of the most significant terms for this role of the frames, and to discuss their meaning. Finally, in the third part, the fashion in which one word particularly revealing of this vocabulary of mediaeval shipbuilding is found in that of terrestrial architecture of the same period will be considered. Does this technical term indicate a relationship other than purely formal between the terminology of the two architectural worlds? That is the central question studied at the end of this third part.

* "Des mots aux pratiques techniques: gabarits et architecture navale au Moyen Age", Chronique d'Histoire Maritime, n° 56, Paris, pp. 13–34, 2004.

2. The archaeological facts

With regard to the present state of archaeological research, it is between the 6th and 7th centuries A.D. that the great changes appear, intervening in the Mediterranean shipbuilding yards according to two different processes, the one of transition or mixed methods, the other of rupture.

The first, which is characterised schematically by the use of processes of shell-first construction for the hull, and frame-first for the part situated above the waterline, was long ago attested by a single wreck with badly preserved architectural vestiges. It is a question of the Yassi Ada I wreck from the first half of the 7th century, in Turkey [27], which is that of a modest Byzantine coaster loaded with amphorae.[1] Of the other wrecks for the architecture of transition, the most recently excavated is that of Dor D, situated in the waters of the Tantura Lagoon (Israel), and dated to the middle of the 7th century [16]. Unfortunately the very poor state of preservation of this wreck lying at shallow depth makes all restitution of the forms of this ship impossible, even hypothetically.

The second process, that of rupture, was similarly attested long ago by a single wreck, that of Saint-Gervais II, located in the gulf of Fos-sur-Mer (Rhône estuary), and dated to the 7th century [14, 15]. After that underwater excavation, others have taken place of which one of the most interesting, on account, notably, of the dating of the wreck studied, is that of Tantura Lagoon A, in Israel. This wreck, which would date from the end of the 5th–beginning of the 6th centuries [17], presents all the basic characteristics of a hull built frame-first. In that wreck, as in that of Saint-Gervais II, the integrity of the hull rests henceforth on a frame-first architecture. Furthermore, no active[2] architectural element is present from the ancient system of shell-first construction, characterised archaeologically by the assembly of carvel planks by means of a network, more or less dense according to the period considered, of keys chased in mortices arranged in the edges of the planks and blocked by small treenails. It is principally the frames that ensure the coherence of the structure of the whole of the hull. Before returning, shortly, to this function occupied by the frames, it is

[1] We recall that the edges of the bottom of the Yassi Ada I wreck are assembled by means of tenons simply embedded (without treenails) in the mortices. These points of assembly, widely spaced, seem to have been principally intended to guide the placing in position of the strakes of the hull.

[2] This is the case for the Saint-Germain II wreck which preserves some essentially passive vestiges of assembly by mortices and tenons.

perhaps worth asking about the limits of the archaeological evidence in so far as it is like an historical source.

The first element to envisage is that of the representativity, in terms of the history of naval architecture, of these wrecks. It appears weak, on account of the small number of sites discovered, excavated and wholly published. The study of this period of profound evolution of the techniques of shipbuilding is, in truth, presently based on a dozen sites, at the very most. The second element to take into consideration is that of the place of construction of the ship, an origin difficult to establish archaeologically, despite the recent and promising support of dendrochronology.[3] Effectively, the location of the site of the shipwreck, the most evident, is not to be confused with that of the place of loading of the cargo, nor with that of the building of the ship, nor even with that of the zone or zones of supply of the timber for the work. There can exist relationships of proximity between that zone and the shipyard, but these are not systematic. The timber can be imported from a place distant from the construction site. Prudence is necessary, consequently, even if there is a tendency, and good reason, it seems, in these two specific cases, to associate the origin of the Yassi Ada I wreck, as a ship, to the Eastern part of the Mediterranean basin, and that of the Saint-Gervais II, also as a ship, to the Western sector of this basin. The distinction between the two geographic poles of the Mediterranean represents a factor of the highest importance for historical interpretation. In the case of a vessel (like the wreck of Saint-Gervais II), produced by a shipyard situated in the Western part of the Mediterranean basin the hypothesis can thus be envisaged, with all the indispensible precautions, of a possible Western influence, in constructive practices of Romano-Celtic or Gallo-Roman tradition, characterised, notably, by a privileged role held structurally by the frames and, above all, by the flat floors of heavy scantling which are nailed to the strakes of the flat bottom. Another illustration of the importance of the identification of the construction site concerns the two paths of evolution. In other words, would the process of transition or mixed method be found limited only to the Eastern part of the Mediterranean, as the origin of the wrecks presently known would seem to indicate? This first question

[3] It is necessary however to qualify these contributions because of the difficulties encountered in establishment of reference dendrochronologies of certain Mediterranean species. It is a matter of long-term research which requires the analysis of a very great number of samples.

opens a second question. Would the passage of the shell-first to the frame-first architectural system imply, necessarily, an intermediate phase, to which the wrecks of Yassi Ada I and Dor D amongst other examples would testify, or, according to the regions considered, could it have been effected directly, according to a process of rupture, as the wrecks of Saint-Gervais II and of Tantura Lagoon A would seem to be able to illustrate? The debate remains open.

Returning now to the new characteristics of this mediaeval architectural system that the Anglo-Saxons qualify as skeleton-first or frame-first, and which we generally call *membrure première*, or sometimes *sur couples* [on frames]. Whatever the expression chosen, the major fact to retain concerns the henceforth central position of the transverse carpentry, to the detriment of that of the shell planking in the structural architecture. The consequences have been perfectly summarised by J. R. Steffy, when he writes: "…the greatest difference between the mentalities of ancient and later shipwrights—the difference in their structural philosophies—was that one visualised his hulls in longitudinal bands, the other in athwartships configurations" [28, p. 419]. In that transversal perspective of the forms which, as J. R. Steffy terms it with much justice, corresponds to a structural philosophy, the frames, or more or less, some amongst them, have acquired an *active* function, according to the qualification used by L. Basch [2, p. 39], who defines an *active* frame as "…one which determines the side planking", in contrast with a *passive* frame, "one that takes its own shape from either the side planking or the provisional battens". In this second case, these are notably the curvatures and the inclinations of the strakes of the side or eventually those of the temporary ribbands, returning to the notion of the *longitudinal bands* evoked by J. R. Steffy, which have an *active* function from the point of view of the conception of the forms.

How is this fundamental dimension of the frame-first architectural system found to be expressed in the archaeological sources? Two wrecks provide a certain number of elements to reply. The first is that of Serçe Liman, in Turkey, dated to the first half of the 11th century [29]. The analysis of the vestiges of this coaster originating from the Eastern basin of the Mediterranean has led to the restitution of a system of conception of the forms of the hull based on a predetermination, geometrically elementary, of some elements of frames having the function of moulds. Once these elements of the transverse carpentry are predetermined, they are followed by particular strakes or ribbands which intervene in the processes of the definition of the form of the coaster of Serçe Liman.

The second wreck is that of Culip VI, in Catalonia (Spain), dated to the end of the 13th–beginning of the 14th centuries [22]. The study of the transverse carpentry of this coaster, built in a shipyard of the Western basin of the Mediterranean has permitted us to place in evidence a predetermination of the forms, much more developed than in the preceding case, of a great part of the floor timbers, the only members of the frames perfectly preserved, according to the Mediterranean mediaeval method of the master-mould and the staff (*tablette*) [23].

One first question is posed. Would these two archaeological attestations of a predetermination of the forms of the hull starting with the frames be revelatory of an evolution of the methods of conception between the 11th and the end of the 13th–beginning of the 14th centuries? A second question appears. Would this evolution, as a sign of progress, be marked, then, by the acquisition of a mastery, more and more affirmed, of the predetermination of the geometric outline of the frames? And in this hypothesis, would Muslim Spain have been able to have a role in the diffusion of geometric knowledge towards the environment of the shipyards? A third question can be formulated, finally. Would the Eastern origin of the construction of the coaster of Serçe Liman, chronologically earlier than that of Culip VI, be the indication of a technical advance in the shipyards of the Eastern Mediterranean, through the eventual influence of Byzantium on those of the Western Mediterranean?

It will be hazardous to reply, even with infinite precautions, to these three questions, from the fact of the still very partial character of the archaeological documentation which we currently have available.

In conclusion of this sketch of the contribution of the facts from archaeology to the knowledge of mediaeval Mediterranean naval architecture, it remains to recall the fundamental role acquired by the frames as much at the theoretical level—that of the conception of the forms—as the practical—that of the construction of the hull according to the predefined forms of the frames. The most notable consequences of this active participation of the frames in the architectural system are of diverse orders. From the point of view of the supply for the shipyard, for example, the frame-first architectural system implies a growth of the selection of curved timber, most particularly as a function of the geometry of the future frames. In relation to the predetermination of the form of the frames, this architectural system supposes that there was recourse to specialised instruments, permitting the carrying of the predetermined outline onto the pieces of timber to be fashioned into frames. In what way do the written sources express these different aspects of mediaeval Mediterranean naval architecture? Such is the object of the second part of this study.

3. The contribution of the written sources

Our documentary enquiry has concerned essentially the published sources, which are a question of lexicographical works like the *Trésor du langage des galères* [8] and the *Glossaire nautique de la langue d'Oc* [11], or the edition of unpublished documents like the account of the construction of galleys at Narbonne between 1318–20 [26, 10, 6], of which the 116 folios drawn up in Latin are of exceptional interest.

This account mentions, without any ambiguity, the recourse to moulds, these specialised instruments that we come to evoke, for the creation of the frames, more exactly the *madiers* corresponding in the Mediterranean vocabulary to the classic floor timbers (*varangues*) of the Western language.

Thus wood is bought to make the moulds for the *madiers* at the opening of the shipyard: "Item solvit Arnaldus Figerie pro fusta empta per Petrum Videlli in primo escordio fabricationis galearum, pro faciendis *modulis* madaire dictarum galearum" [26, p. 93, f°5]. These moulds, occurring here in the context of "pro faciendis modulis", are attested in the same document under two other forms: those of *galbe* and of *mensura*. Thus it is that small nails are bought to nail the moulds called *galbes*: "Item pro parvis clavellis qui dicuntur tachie, ad clavendum *modulos* qui dicuntur *galbes*" [26, p. 268, f°93v]. At the time of a purchase of timber by Petrus de Berra, "Grand-Master of the galleys", it is specified that this material is going to serve to make moulds: "...pro faciendis galbis seu mensuris galearum" [26, p. 95, f°6]. These moulds for the floor timbers, designated under the form of *galbe, mensura* or *modulus*, correspond, in all likelihood, to the commonest definition of the mould taken in the sense of a pattern reproducing full size the form of a piece of carpentry. Blaise Ollivier, in his treatise for construction in the form of a dictionary, of 1736 [18], a work which we will appeal to on different occasions on account of the quality of his definitions, indicates that the term mould (*gabarit*) "is still spoken of for the model in timber of different parts of the vessel...One makes the moulds in timber with common planks, or re-sawn planks" [18, p. 181]. Concerning these moulds, he adds that: "...when they are composed of several pieces, one overlaps these pieces one over the other around 15 *pouces* [406 mm] and attaches them with nails" [18, p. 181]. One can ask whether these compound moulds are not comparable to those mentioned in the account for construction of the galleys of Narbonne in relation to the purchase of small nails: "…clavellis parvulis ad clavendum mensuras tebues que vocantur galbe pro madayris galearum mensurandis" [26, p. 208, f°57v]. In this hypothesis, these moulds consisting of several elements nailed together could correspond to the whole of a frame, the word *madier* being found used besides in a broad sense, and not in that limited and strict sense, of synonym for floor timber.

The meaning of these different attestations of the term mould (*gabarit*) corresponds perfectly to the active role of the frames, proper to the architecture of frame-first principle, characteristic of mediaeval Mediterranean naval architecture. Unfortunately the documents do not permit us to go beyond this relationship of a very general nature between the mould (*gabarit*) and its use in the chain of operations of a hull built frame-first, except in the context of the earliest operations of the shipyard, that of the selection and acquisition of straight and above all curved timbers, these last being more especially reserved for the creation of the frames.

We make a new appeal to the construction treatise of Ollivier. At the entry *bois de gabarit*, one can read the following definition: "it is curved oak timber, which from this form is suitable to make the pieces of the stem, the frames, the riders (*porques*), the winding-timber and the knees of a vessel. One also calls them twisted timber (*bois tort*) and *bois de courbeau*. This last term is used in the ports of the Mediterranean (…). The floors of the bottom that one fashions in the forests for vessels of a known size have the length and the figure determined by their mould (*gabarit*) or some few inches more in length if the oak permits it" [18, pp. 44–45].

Two points are worth underlining here. On the one hand, the selection of the timbers, practised in the forests, supposes a more or less definitive prior knowledge of the form of the pieces to be worked, in fact of the frames here. Logically, this knowledge implies a predetermination of the form of one part, more or less important, of the transverse carpentry. On the other hand, the pre-fashioning of the frames in the forest, for the reason of the ease of transport of the pieces, is effected by making appeal to the moulds that reproduce the form of the frames in correspondence with the definition given to the terms *galbe, mensura* and *modulus*.

This practice is attested in the Middle Ages. One of the oldest illustrations is provided, once again, by the account of the construction of the galleys of Narbonne. It is thus that the moulds of the floor timbers are transported in the forest of Mossans, one of the timber supply areas for the shipyard: "...homini qui portavit modulos madaire cum ferramentos magistri Raymundi Bonaventura in nemore de Mossans cum uno animali" [26, p. 293, f°105v].

The usage of the *modulus* in the forest for the selection and the prefashioning of the floor timbers raises the question of a hypothetical control of curvature of the trees with a view to artificially obtaining, after a certain number of years, pieces of wood whose curvatures would be adapted to the needs of mediaeval shipyards [25]. If this hypothesis is to be considered with some reserve, from the fact of the difficulty of interpreting the documents, it does not seem technically illogical. In the 18th century, for example, Duhamel du Monceau has evoked this practice of forced growth in his treatise *Des semis et des plantations des arbres et de leur culture* (1760), emphasising that "...as in the environs of certain ports one lacks curved timbers for the Navy, some persons have advised pruning the stems (*tiges*) of certain trees to make them take suitable curvature" [1, p. 75]. He considers always that this twisting of the trees (*torture*), according to the expression imagined by Ballu, tended to somewhat indecisive results [1, p. 75] to the extent where "...the branches are developed in the upper part of the curve and rob nourishment from the rest of the stem", with, consequently, the taking up of vertical growth again.

We return to the context of the shipyard to examine another term used to designate a particular type of mould (*gabarit*). It is a matter, this time, of no more than a mould as pattern or model serving to carry the form of a piece of carpentry and, notably, of a frame, but very probably, of the master-frame defined as much as a privileged instrument of a method of conception based on the master frame properly called, reproducing *at natural size* the outline of the master-frame, the rising staff (*tablette*), and eventually the breadth staff (*trébuchet*). These three wooden instruments permitted the predetermination, starting from the figure, in a geometric sense of the term, of the master-frame, those of the frames called moulded comprising the space between the two quarter frames (*couples de balancement*). It is at the entry for "*tablette*: method of construction by the *tablette* and the *trébuchet*" of Ollivier's construction treatise that one finds a very complete definition of the method of conception resting on the use of the master mould (*maître-*

gabarit): "It is the method by which one determines the figure of all the moulds of the vessel comprised between the last floor timbers fore and aft, by means of the master-mould, the *tablette* and the *trébuche*t" [18, p. 319]. Ollivier adds that "... this method is practised in all the ports of the Mediterranean by most of the constructors and it is the oldest of all those that are in use today to determine the figure of sea-going vessels" [18, p. 319]. We recall that the treatise of Ollivier makes reference to a method of mediaeval and Mediterranean origin: a good illustration in truth, of the weight of tradition in the field of techniques, which, nonetheless, does not exclude innovations and progress. At the same period, indeed, the use of plans which would be followed, some years after, by recourse to the first hydrostatic calculations, was generalised in the space of the military arsenals of the great maritime nations.

The classical Latin term designating the master-mould, in the sense defined by Ollivier, is *sextus*, though it is rendered in the mediaeval documents as "*pro sextis*". One of the most ancient attestations of this word dates from the year 1273. It is extracted from the contract specification (*devis*) for the construction of a *huissier* (horse-transport) by a Neapolitan master carpenter named Gratia [9]. The ship was intended to be incorporated into the fleet of King Charles I of Anjou, then in conflict with the Genoese. The *devis* enumerates, the quantity and the cost of the materials necessary to the creation of the carpentry of the *huissier:* "Extimatio pertinencie et rei necessarii corpore unius usseri nigri. Videlicit faccta per magistrum Gratia de Neapoli... Curris 12 de altano (pro taccis battaliis fornimento puppis et prore formis et sextis) ad rationem de tari octo et medio pro quolibet" [9, p. 265]. A second *devis*, a little different from the preceding, establishes however a better distinction between the terms *sextis* on the one hand, and *formis* on the other. The *devis* specifies: "... item pro faciendo fieri taccis battaliis pro fornimento puppis et prore et pro formis et sextis curris lignamini 12 de altano de tari 8-1/2 pro quolibet curru" [9, p. 266]. The mediaeval Latin form "*[pro] sextis*" corresponds to the Venetian term *sesto* or *sexto* which is defined as ".... a particular mould which permits the whole series of frames progressively different one from another, by means of marks inscribed on the surface".[4] One finds again there a definition

[4] Free translation of the definition given in G. Penzo [19, p. 250]: "Il sèsto è una particolare sàgoma che permete—per mezzo di segni riportati nella sua superficie—di tracciare tutta una seie di pezzi progressivamenté differenti tra di loro".

identical to that of the master-mould as an instrument participating in a method of predetermination of the forms of the hull.[5] The frames defined by the method of partition (*partison*) are qualified as moulded frames, or again, in the Venetian dialect, *di sesto*. One of the most ancient mentions of this expression is contained in a message coming from a shipyard of Brindisi with the date of 17 February 1275 for King Charles I of Anjou concerning the description of the famous *red galley* of Provence.[6] It is specified that the galley ought to have 96 moulded frames: "…habet corvas de sexto nonginta" [3, p. 259].

To summarise, it seems that two principal meanings of the term mould (*gabarit*) can be distinguished. The first, under the forms of *galbe*, *mensura* or again *modulus*, corresponds to the definition of the mould as a pattern reproducing at full size the form of a frame. The second sense, under the form of *sextis*, corresponds to the definition of the master-mould in as much as the instrument of conception, always created at full size, is associated with a method of predetermination of the frames. The oldest written attestations are situated in a chronological bracket between the years 1275 and 1320. An important fact to emphasise: it is to this same period that the wreck of the coaster Culip VI is attached, which, we recall, constitutes, in the present state of research, the oldest archaeological testimony for a ship of frame-first conception, of which a great part of the floor timbers have been moulded by means of the method of the master-mould and the rising staff (*tablette*).

The last point on which we ought to pause concerns the Latin term *formis*, attested in the *devis* for construction, already cited, of the *huissier* for King Charles I of Anjou: "…et pro formis et sextis". By reference to the definition the *Vocabolario marino e militare* published in 1889 by Guglielmotti, Fourquin establishes an equivalence between the word *forma* and the term *gabarit* [9, p. 272]. It would seem that another definition could be proposed in correspondence with the second sense given by Fennis to the substantive feminine *forme*: "…a long, temporary, piece of timber serving to find the exact form of the frames" [8, p. 949]. The modern and Western synonym of this *forme* is

[5] In the Venetian context of the Middle Ages, S. Bellabarba [4, p. 274] qualifies this method as the method of the *partison*, of which, from its definition, the essential characteristic is to assure the predetermination of the curved forms of the hull by means of *geometrical aids*, according to his expression.

[6] It concerns a particularly successful flag-galley of 27 banks which, from this fact, had been chosen to serve as model for several galleys intended for the fleet of Charles I of Anjou.

that of the ribband (*lisse de construction*). The oldest attestation cited by Fennis dates from 1494: "two pieces of *leigna* that he has sold to make the *formes* for the said galleys" [8, p. 949]. A more recent attestation (1691) puts the function of conception of the *forme* in evidence well, which arises essentially in the shipyard, in the course of construction. It is not a question, then, like the master-mould, of an instrument for predetermination of the frames: "one cannot find their proportions (of the crooks) by the theory as those of the floor timbers...; one is obliged to search for them in the construction by means of *formes*" [8, p. 949]. With regard to this other definition of the word, one can ask, it seems to us, whether in the *devis* for construction of the *huissier* dated from 1273, the distinction susceptible to be made between the words *sextis* and *formis* would not indicate a difference of architectural practises between the the use of the master-mould for the predetermination of the moulded frames and the use of ribbands for the determination, in the course of construction, of the other frames.

Under what form is this precise vocabulary of mediaeval naval architecture to be found in that of terrestrial architecture? Does there exist, besides, other documentary data permitting the establishment of an eventual connection, and other than purely formal, between the two architectural worlds? Such are the two questions discussed in the third and last part of this reflection.

4. Naval architecture, terrestrial architecture

An idea presented often enough in the studies devoted to mediaeval terrestrial carpenters is that of a certain convergence between the technical practices of the ship-carpenters and those of the carpenters working on land. In his *Dictionnaire raisonné de l'architecture française*, Viollet-le-Duc thus emphasises that: "...the Normans, a sea-people, seem to be in these regions the first to have made a considerable step in carpentry...What characterises the Anglo-Norman carpentry, is its analogy with the means of assembly employed at all times in naval carpentry" [30, t.3, p. 3]. Still today, it is not rare to find this supposed influence put in evidence by art historians comparing the vaulted carpentry of a church to an inverted ship (relying on nave = nef). Beyond an apparent similitude of form of the structure, certainly, in certain cases, the constructive logic of the two families of carpentry, naval and terrestrial, appears fundamentally different. The conception and the construction of a hull of a boat cannot be confused, it seems to us, with those of church carpentry.

It would be very ambitious to engage in a discussion on this subject. Our intention is much more limited. In the prolongation of the remarks of the second part of this study, the first aspect that we wish to examine is that of the moulds (*gabarits*).

A first examination of the documentation relating principally to Western France and to England[7] shows that amongst different words of the mediaeval terminology of the Mediterranean shipyards, one alone seems to be attested in a frequent fashion in the context of terrestrial architecture: that of mould (*molde*) which recalls the term *modulus* used by ship-carpenters as a synonym for *gabarit*. The use of the *moulds* or *moles* seems circumscribed to the domain of the stone-cutters. It is a question of moulds reproducing, at full size, the contour of the mouldings to be created.[8] After the definition of these moulds given by Recht, " One of the first tasks of the architect, after the conception of the project as a whole is the preparation of the moulds (*gabarits*). These are the sheets (*panneaux*) of wood or of metal to which are given the form of the different mouldings that will be used in the building. The stone-cutter then places them on the squared stone and follows the contour" [21, p. 87]. In 1282 for example, a certain Stephen Jeignur was paid "...for various planks (*tabulas*) for moulds for the work of the masons" [24, p. 21]. It is a question in this case, of wooden moulds. An account dated 1445 specifies that the planks are cut out by the carpenters according to the pre-established outlines: "...and made the carpenters to carve them according to the draughts" [24, pp. 20–21]. The question that arises then is that of knowing what these draughts could be. An excellent illustration of these moulds (*moldes* or *moles*), in wood or metal, is furnished by Villard de Honnecourt in his celebrated notebook of designs, dated to the first half of the 13th century.[9] One of the oldest written attestations for these moulds particular to terrestrial workshops dates from 1174. It concerns the sending to Caen, by Guillaume de Sens, of *moles* intended for the stone-cutters. These moulds (*gabarits*) reproduce the profile of the mouldings to [be] cut

[7] This geographic localisation is connected to the documentation consulted.

[8] According to L. F. Salzman [24, p. 93], the moulding intended to decorate the periphery of the openings of the buildings would generally be called *forma*: "....The tracery is called forms, or form pieces, and occasionally molds".

[9] Reproduction in R. Recht, [21, p. 87].

in the stones extracted from the Caen quarries, and which serve in the reconstruction of Canterbury cathedral [7, p. 25].

These *moldes* or *moles* in wood or metal are endowed with a function analagous to that of the wooden moulds utilised by ship-carpenters in the restricted sense of "...model in wood of different parts of the vessel" [18, p. 181], that is to say of patterns reproducing the contour of a piece of carpentry. But there, from our point of view, the relationship of form and of meaning between the two categories of moulds ends. In the framework of terrestrial construction, these moulds, intended essentially (exclusively?) to working stone, are inscribed in an architectural perspective often more decorative than structural. They are attached, moreover, to a body of workmen, that of the stone-cutters, fundamentally different to that of the carpenters. In the case of shipbuilding, in summary, the moulds, in the limited sense of models, are associated in a direct and active fashion to the constructive system. In these conditions, the identity of the form of the word does not at all infer an identity of usage and of functional context.

What other categories of data are susceptible to allowing the establishment of a link between the practices of terrestrial workshops and those of the shipyards of the Middle Ages? It is this second aspect that we wish to touch on.

In examining our documentation, one problem seems particularly interesting to envisage: that of the marking of the pieces of terrestrial carpentry and of the frames which rests on the ciphers of analogous form and a numeration of similar principle.

In terrestrial carpentry, the marks based on numbers are established according to a numeric progression corresponding to the order of placing the pieces during the phase of raising and definitive assembly of the carpentry. The individual marking of the pieces takes account of the whole carpentry structure. Previously, a first stage has consisted of temporarily assembling the pieces of carpentry on the ground. This is the trial assembly at the end of which the marking of timbers is effected. According to the most recent research,[10] the most current mode of numeration seems to make use, from the middle of the 13th century,[11] of Roman numbers, sometimes contracted. One of the most recent precisely dated examples is, in the present state of research, that

[10] Compare notably: Hoffsummer P. [12], J.-Y. Hunot, J.-Y. [13].

[11] This period corresponds, we note, to that of the great Gothic workshops.

of the carpentry of the North transept of the church of Notre-Dame, at Etampes (Essonne), which goes back to the years 1207–1217 [12, p. 63]. Moreover, a lateralisation mark is superposed to the numeration serving to distinguish, in relation to the observer and the axis of the roof, the pieces disposed to the left of the axis of symmetry from those situated to the right. Generally, the lateralisation mark, of which the most ancient attestations, in Northern France and in Belgium more or less, seem to date from the beginning of the 13th century, operating by the bias of a contra-mark integrated to the mark [12, p. 66] properly called.

In the field of mediaeval ship-carpentry, a similar system of marking with numbers is found at the level of the frames associated with an architecture of frame-first principle. The objective is comparable to that preceding. It is a question of identifying, following a numeric progression, the predetermined (moulded) frames comprised between the two balance frames, before their mounting and assembly on the keel. The wreck of Culip VI (Spain), dated to the end of the 13th–beginning of the 14th centuries, constitutes the most ancient and most complete archaeological evidence of this system of marking [22, pp. 169–180], at present. The numbers cut in the moulded floors of this wreck present three characteristics absolutely identical to those that are found associated to the numbers incised in the trusses (*fermes*) of terrestrial carpentry: Roman numbers, with contraction in the case of composed numbers, and lateralisation of the marking [22, p. 170, fig. 85]. The only notable difference concerns the spatial reference of the marking. In the case of terrestrial carpentry, the axis of symmetry of the marking is that, longitudinal, of the roof. In the case of the wreck of Culip VI, the axis of symmetry of the marking is that, transversal, of the master-frame. This transverse axis, which testifies to the fundamental role occupied by the *sextus* (according to the classical Latin form) the master-mould of the mediaeval written sources, defines two numeric progressions, the one going from the master-frame towards the forward balance frame, the other directed from the master frame towards the after balance frame.

How can one interpret these convergences which, in a first analysis, appear much more significant from the point of view of architectural practises than those evoked with respect to the usage of the same term—*molde* or *mole*—to designate a type of mould utilised in a fashion probably very different in terrestrial architecture and in naval architecture? There would seem to exist a certain kinship of usage of a numeric language between the house-carpenters and ship-carpenters. In a workshop, faced with the necessity of identifying the pieces of carpentry, trusses of terrestrial carpentry or frames, with a view to their raising and assembly, any of them could effectively have made appeal to analogous ways of working. At this level, there could very well have existed a part of know-how shared between the carpenters, those who constructed church carpentry or the hull of a ship. But it will be totally hazardous to go beyond this possible common denominator of know-how, which, moreover, it is necessary to recall, does not at all mean a professional identity. House-carpenters and ship-carpenters are holders, it seems to us, of a technical culture which is their own. The constructive logics of the two categories of carpentry, like the whole chain of command, of finance, of organisation... the terrestrial workshops and the shipyards rely well and truly on two totally different technical and socio-economic worlds.

5. Conclusion

At the outcome of this study, two new axes of research are delineated. The first concerns the growth of mediaeval terms associated with the frame-first system of construction proper to the Mediterranean area and totally absent from the Western nautical vocabulary of the Middle Ages.[12] In the case of England, for example, there is hardly any doubt that this absence is connected to the techniques of clinker construction in use in the shipyards of this country in the Middle Ages. The shell-first clinker architecture, indeed, does not know of the use of the master-frame as an instrument of conception. In summary, in what way can the English vocabulary of the end of the 15th and the first half of the 16th centuries, the period of introduction and development of carvel construction in England, express these profound modifications of architectural systems? In this respect it will perhaps be interesting to follow the linguistic path of a term like that of *modulus*, from the Mediterranean area towards the Channel. Could the most ancient attestations of the Spanish equivalents—*modelo*, Portuguese—*moldes*, Basque—*molde*, English—*mould*, German—*Mall*, not correspond to a

[12] Compare for example, B. Sandahl [25].

chronology, more or less accurate, of the introduction and development of carvel construction in these territories?

The second axis for research concerns the worlds, in all appearance very distinct, of terrestrial and naval architectures. One enquiry, without doubt very difficult to conduct because of the complex nature of the written sources, to present an abstract on the activities of the house-carpenters and those of ship-carpenters could perhaps allow us to follow the possible passages of the two categories of carpenters from one place of work—a workshop for religious buildings—to another—a shipyard—and vice versa.

These two paths, amongst others, show how vast and fascinating this field of the history of techniques remains which, in the precise context of this study, ought to incite, moreover, the reinforcement of exchanges between specialist researchers (historians and archaeologists) of terrestrial and naval architectures.

Acknowledgements

We thank Odette Chapelot, assistant professor at EHESS, Paris, and Marie-Hélène Joly (†), librarian at the Musée national de la Marine, for their valued collaboration and Richard Barker for his translation and commentary.

References

1. Ballu, Jean-Marie: "Bois de marine. Les bateaux naissent en forêt", Editions du Gerfaut, Paris, 2000.
2. Basch, Lucien: "Ancient wrecks and the archaeology of ships", International Journal of Nautical Archaeology, vol. 1, pp. 1–58, 1972.
3. Bellabarba, Sergio: "The origins of the ancient methods of designing hulls: a hypothesis", Mariner's Mirror, vol. 82, pp. 259–268, 1996.
4. Bellabarba, Sergio: "The ancient methods of designing hulls", Mariner's Mirror, vol. 79, pp. 274-292, 1993.
5. Ciciliot, Furio: "Il legname da garbo (secoli XIII–XVIII)", in Marzari, Mario (ed.), Navis, vol. 1, pp. 77–86, 1999.
6. Dennebouy, Julie: "Un compte de construction de galères à Narbonne (1318–1320). Traduction, commentaire et approvisionnements du chantier naval en matériaux ligneux", thesis for Master's degree in mediaeval history under the direction of Monique Bourin and Eric Rieth, Université de Paris I (Panthéon-Sorbonne), Paris, 2001.
7. Du Colombier, Pierre: "Les chantiers des cathédrales", Editions A. et J. Picard, Paris, 1973.

8. Fennis, Jan: "Le trésor du langage des galères", Max Niemeyer Verlag, Tübingen, 3 vols, 1995.
9. Fourquin, Noël: "Un devis de construction navale de c.1273", in Villain-Gandossi, Christiane, and Rieth, Eric (eds), "Pour une histoire du "fait maritime". Sources et champ de recherche", Editions du CTHS, Paris, pp. 263–278, 2001.
10. Fourquin, Noël: "Navires marseillais au Moyen-Age", in Miège, Jean-Louis (ed.), "Navigation et migrations en Méditerranée. De la préhistoire à nos jours", Editions du CNRS, Paris, pp. 181–250, 1990.
11. Fourquin, Noël, Rigaud Philippe: "De la nave au pointu. Glossaire nautique de la langue d'Oc", Editions Objectifs Mer et Capian Méditerranée, Saint-Tropez-Toulon, 1993.
12. Hoffsummer, Patrick (ed.): "Les charpentes du XIe au XIXe siècle. Typologie et évolution en France du nord et en Belgique", Monum, Cahiers du Patrimoine, n° 62, Editions du Patrimoine, Paris, 2002.
13. Hunot, Jean-Yves: "L'évolution de la charpente de comble en Anjou, du XIIIe au XVIIIe siècle", Patrimoine d'Anjou : études et travaux, vol. 1, Angers, 2001.
14. Jézégou, Marie-Pierre: "L'épave II de l'anse Saint-Gervais à Fos-sur-Mer (Bouches-du-Rhône): un navire du haut Moyen Age construit sur squelette", in Tzalas, Harry (ed.), Tropis I, Athens, pp. 139–146, 1985.
15. Jézégou, Marie-Pierre: "Eléments de construction sur couples observés sur une épave du haut Moyen Age découverte à Fos-sur-Mer (Bouches-du-Rhône)", in VI Congreso Internacional de Arqueologia Subamarina, Cartagena 1982, Madrid, pp. 351–356, 1985.
16. Kahanov,Yaacov, Royal, Jeffrey: "Analysis of hull remains of the Dor D vessel, Tantura Lagoon, Israel", International Journal of Nautical Archaeology, vol. 30, pp. 257–265, 2001.
17. Kahanov, Yaacov, Royal, Jeffrey: "The 1995 INA/CMS Tantura A Byzantine Shipwreck Excavation-Hull Construction Report", CMS News, University of Haifa, Center for Maritime Studies, n° 23, pp. 21–23, 1996.
18. Ollivier, Blaise: "Traité de construction contenant par ordre alphabétique la description des vaisseaux de tout rang, galères, frégates, 1736", (Vincennes, Service historique de la Marine, ms SH 310), Editions Oméga, Nice, 1992.
19. Penzo, Gilberto: "La gondola. Storia, progettazione e costruzione della piu straordinaria imbarcazione tradizionale di Venezia", Istituzione per la conservazione della gondola e la tutela del gondoliere, Venice, 1999.
20. Pomey, Patrice: "Principes et méthodes de construction en architecture navale antique", Navires et commerces de la Méditerranée antique. Hommage à Jean Rougé, Cahiers d'Histoire, vol. 33, pp. 397–412, 1988.
21. Recht, Roland: "Glossaire des termes d'architecture médiévale", Dossiers Histoire et Archéologie, n° 47, pp. 84–90, 1980.
22. Rieth, Eric: "L'arquitectura naval", in Palou, Hug, Rieth, Eric, Izaguirre, M. *et al.*, "Excavacions arqueològiques subaquàtiques a Cala Culip. 2. Culip VI", Monografies del CASC, 1, Girona, pp. 115–117 and 137–201, 1998.
23. Rieth, Eric: "Le maître-gabarit, la tablette et le trébuchet. Essai sur la conception non-graphique des carènes du Moyen Age au XIXe siècle", Editions du CTHS, Paris, 1996.
24. Salzman, L. F.: "Building in England down to 1540. A documentary history", Clarendon Press, Oxford, 1967 (first edition 1952 under the title of Building in England down to 1540. Documentary evidence).
25. Sandahl, Bertil: "Middle English Sea terms. I. The Ship's Hull", The English Institute in the University of Uppsala, Uppsala, 1951, (Essays and Studies on English Language and Literature, VIII).
26. Sosson, Jean-Pierre: "Un compte inédit de construction de galères à Narbonne (1318–1320)", Bulletin de l'Institut historique belge de Rome, vol. 34, pp. 57–318, 1962.

27. Steffy, J. Richard: "Reconstructing the hull", in Bass George F. and Doorninck, Frederick H. van (eds), "Yassi Ada, vol. I. A seventh-century shipwreck", Texas A & M University Press, College Station, pp. 65–86, 1982.
28. Steffy, J. Richard: "Ancient scantlings: the projection and control of Mediterranean hull shapes", in Tzalas, Harry (ed.), Tropis III, Hellenic Institute for the Preservation of Nautical Tradition, Athens, pp. 417–428, 1995.
29. Steffy, J. Richard: "The Mediterranean shell to skeleton transition: a Northwest European parallel?", in Reinders, Reinder and Paul, Kees (eds.), "Carvel construction technique. VIth International Symposium on Boat and Ship Archaeology", Oxford, Oxbow Monograph 12, pp. 1–9, 1991.
30. Viollet-le-Duc: "Dictionnaire raisonné de l'architecture française du XIe au XVIe siècle", Editions Ve A. Morel et Cie, Paris, vol. 3, 1875.

APPENDIX B

THE RARE TRACES OF CONSTRUCTIONAL PROCEDURES IN "PRACTICAL GEOMETRIES"

Jens Høyrup

In a customary dichotomy, geometry (like many other fields, mathematical as well as nonmathematical) falls into "theoretical" and "practical". In full agreement with this, Stephen K. Victor [28: ix] writes about his Ph.D. project that

> My first assumption, and that of most of the people I have spoken to about the topic, was that "practical geometry" must relate somehow to architecture, surveying and city planning, to those areas, in other words, where geometry plays a central role in the exercise of other professions. The study of medieval buildings, fields and towns from extant physical evidence was not a fruitful approach for me, and I have left it to those better trained in the methods of archaeology and art history. Since I was working as a historian of science, I chose to concentrate on the written tradition of treatises called "practical geometry".

The treatises he chose to work on—the Latin late-twelfth-century *Artis cuiuslibet consummatio* and a vernacular (Picardian) *Pratike de geometrie* from the late thirteenth century which is largely a translation of the former work—led him to a different view,

> namely that practical geometry has its greatest importance as a popularization of mathematics. The treatises on practical geometry were a way of teaching some basic principles to those who would not remain in school or university long enough to become philosophers or theologians and would not necessarily exercise a mathematical profession, but who might want, or even need, some mathematics in their everyday lives. The sampling of arithmetic and astronomy in ACC and of commercial arithmetic and metrology in the Pratike argues for the generally pedagogic, rather than scholastic, purpose of the treatises. The development of a vernacular version of ACC is further evidence that the practical geometries sought their homes outside of the universities, perhaps in the bureaucratic and commercial milieus. Nonetheless, as the Introduction shows, the formalized structures of university education had an influence even on the nonscholastic tradition of practical geometry. As the tradition developed, the practical geometries acquired an increasingly theoretical underpinning, to the point where they are sometimes considered works on measurement rather than simply practical geometries.

Part of this conclusion depends critically on the choice of a Latin treatise and a vernacular treatise which in the main was derived from it. Other features, however, are shared not only with the Italian vernacular *Pratiche di geometria* and with Fibonacci's *Pratica geometrie* but also with most Arabic,[1] Sanskrit, Chinese, Greek, Babylonian and ancient Egyptian writings on the subject matter. They deal, not at all or not much with *measurement*, as Victor states euphemistically, but rather with *how to calculate* something on the basis of measurements that have already been performed or which are presupposed to have been performed, either on preexisting objects or on configurations which are supposed to have been already constructed.[2] In general terms, they belong within the "scribal" sphere, or in Victor's words, the "bureaucratic and commercial milieus".

Nonetheless, a few traces of constructional procedures hide within these texts. I shall present two instances, one fairly certain and the other not much more than suggestive.

Constructing the circular diameter

In *Metrica* I.xxx [24: 74] Hero explains that "the ancients"—οἱ ἀρχαῖοι—in their formula for the area of a circular segment seem to have "followed those who took the perimeter to encompass the triple (τριπλάσιος) of the diameter", whereas I.xxxi [24: 74] states that "those who made more precise investigations" must have followed the course according to which the perimeter is the triple diameter and in addition $^1/_7$ of the diameter.

Hero himself teaches (I.xxvi, [24: 66]) to multiply the perimeter by 22 (using the construction "22 ἐπὶ") and then to take the seventh, but in the pseudo-Heronian *Geometrica*[3] [14]—throughout using the "more precise" variant—the diameter is invariably taken "thrice" or "tripled",

[1] An Arabic exception to this rule is Abū'l-Wafā''s *Book on What is Necessary for Artisans in Geometrical Construction* [ed., Russian trans. Krasnova 1966].

[2] Actually, the genre studied by Victor—Latin practical geometries such as *Geometria incerti auctoris* and Hugh of Saint Victor's *Practica*—deals to some extent with mensuration, namely the determination of (e.g.) inaccessible heights by means of equilateral right triangles. This also had a slight (very slight!) impact on Italian abbacus geometries.

[3] Definitely *not* Heronian, and actually a composite created by the modern editor from two rather incompatible manuscript groups, respectively A+C and S+V, as Heiberg [15: xxi] points out.

and this triple calculated explicitly, after which a supplementary seventh is added. The terms for tripling are without exception τρισσάκις or τριπλάσιον even when neighbouring multiplications are ἐπὶ *n*.[4]

The same terminological distinction between tripling and multiplication is found already in Old Babylonian geometry (c. 1700 B.C.E.). Here, the perimeter is always found as the diameter "repeated until three" (*ana* 3 *esēpum*), or it is "tripled" (*šalāšum*). It is not calculated by means of the normal multiplication (*našûm*, "to raise") used, e.g., when the area of the circle is found as 0;5 (= $^5/_{60}$ = $^1/_{12}$) times the square on the perimeter.

The explanation of this linguistic puzzle is found in two texts from the fourteenth and the fifteenth century (C.E.). One is Mathes Roriczer's *Geometria deutsch* from c. 1488 [26: 121]:

> If anyone wishes to make a circular line straight, so that the straight line and the circular are the same length, then make three circles next to one another, and divide the first circle into seven equal parts,

one of which is marked out in continuation of the three circles—see Figure B.1.[5]

The other is the old Icelandic manuscript A.M. 415 4to from the early fourteenth century, according to which (fol. 9^v) "the measure around the circle is three times longer as its width, and a seventh of the fourth width",[6] obviously a reference to a similar construction.

---

[4] Thus mss AC, **17**.8, between **17**.7 and **17**.9, and ms. S, **17**.6, after **17**.6 [14: 336, 334].

[5] Shelby [26: 182] observes "some resemblance between [Roriczer's procedure] and one of the theorems in a brief *Tractatus de quadratura circuli*—traditionally attributed to Campanus de Novara, but authorship and date uncertain". The passage in question [6: 591] deals with how to "give a straight line equal to a circularly drawn line", and runs as follows:

> Using mathematical knowledge and physical truth, a circle is divided into 22 equal parts, and with one part subtracted, that is, the 22nd part, a third of the remainder, namely, 7, is the diameter of the circle. Therefore, let the diameter be tripled and let there be added a seventh of the diameter, and let these parts be ordered in a straight line. We shall have a straight line equal to a circular line, as is apparent in the figure.

This could well be an attempted "theoretical" explanation of Roriczer's construction, but since the diagram shows a circle divided into 22 parts (with a diameter prolonged indefinitely toward the right) it could at least as well be a justification of the *calculation* found in the *Geometrica* and writings of the same kind, like that fifteenth century *De inquisicione capacitatis figurarum* to which Shelby [26: 6–65] refers in his introduction.

[6] "Ummæling hrings hvers þrimr lutum lengri en bréidd hans ok sjaundungr of enni fiorðo breidd" [1: 231*f*]. I am grateful to Peter Springborg for localizing a passage

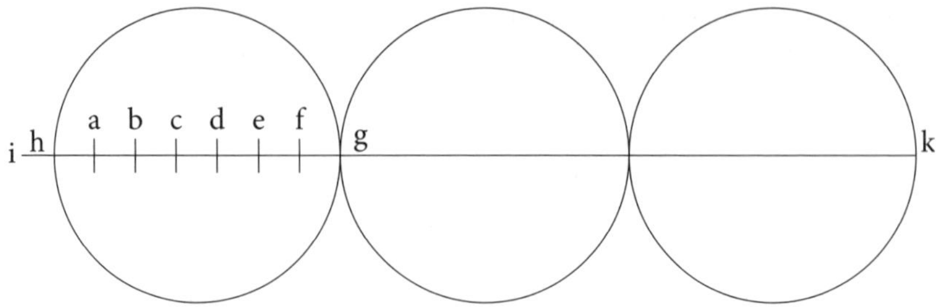

Fig. B.1: Roriczers construction of the circular perimeter. After [26: 121].

Roriczer was a Gothic master builder; what he tells is the way to find by means of a drawing, *without calculation*, the perimeter corresponding to a given circular diameter. The Icelandic text confirms that the method was widespread; there seems to be little doubt that it offers the explanation why both the Greek and the Old Babylonian text refer to a tripling, a material repetition, and not to a mere numerical multiplication. This trick had thus been known for more than three thousand years in the late Middle Ages, first as a simple tripling, after the acceptance of the Archimedean improvement with an addition of an extra seventh—still a separate supplement, and still to be provided in physical space.

The regular octagon and the side and diagonal numbers

The other example is differently balanced, in the sense that the traces in the calculational material are fewer but those in other sources more copious.

One trace is constituted by the Old Babylonian approximations to the ratio between the diagonal and the side of a square. One, already quite good, is 1;25 = $^{17}/_{12}$; the other, excellent, is 1;24,51,10. The former may have been found by iteration of a procedure also known from elsewhere in the Old Babylonian record, corresponding to the formula

$$\sqrt{a^2+d} = n + \frac{d}{2n}$$

which is quoted withour reference by Menninger [21: I, 91] and for providing me with a photocopy from the microfilm in the Arnamagnean collection, Copenhagen.

—actually, the text VAT 6598 contains what may be a failed attempt at such iteration [17: 271*f*]. The latter can be found *by us* by further iteration, but hardly by the Babylonians: as pointed out by David Fowler and Eleanor Robson [8], the calculations have to pass through repeated divisions by very unpleasant sexagesimally irregular numbers; if we try to approximate by regular divisors (in agreement with what we know about Babylonian computational techniques), the reconstruction no longer yields the approximation it should but either one which is too rough or one which is even better.

Neugebauer and Sachs [23: 43] propose a different way to the same approximations, namely through alternating arithmetical and harmonic means. Algebraically, this gives the same results—and computationally it runs into the same problem.

A third possibility—also algebraically equivalent—is the use of the "sideanddiagonalnumber algorithm",

$$s_1 = d_1 = 1, \quad s_{n+1} = s_n+d_n\,, \quad d_{n+1} = 2s_n+d_n.$$

The value of $2s^2-d^2$ oscillates between -1 (for odd n) and $+1$ (for even n). Since s and d increase exponentially, the ratio $d{:}s$ therefore converges rapidly toward $\sqrt{2}$.

The procedure is first described by Theon of Smyrna (*Expositio* I.XXXI, [7: 70–74]), but according to his own statement in agreement with Pythagorean traditions without any addition whatsoever (book II, the introduction). It is also habitually assumed that Plato's reference to "a hundred numbers determined by the rational diameters of the pempad lacking one in each case" (*Republic* 546C, [27: II, 247]) shows him to be familiar with the same algorithm. Actually, all it shows for certain is that he was familiar with the use of 7 as an (approximate) value for the diagonal in a square with side 5.[7] In any case, another discussion of the algorithm is found in Proclos's commentary to the passage in question from the *Republic*.[8] Finally, Proclos's commentary

[7] Heath [13: I, 399] and others read the "lacking one" as a reference to the fact that 7^2 is lacking 1 compared to the square on the true (irrational) diameter in the square with side 5, which corresponds to an essential feature of the sequence of approximations produced by the algorithm. Actually, as pointed out to me by Marinus Taisbak (personal communication), Plato's point is rather that the number 48 (the number which is required) is lacking one with regard to the "number on the rational diameter 7" (and 2 with regard to that on the irrational diameter *dynamei*, as Plato goes on). This is indeed also Proclos's explanation, cf. Hultsch in [20: II, 407].

[8] [20: II, 24*f*]; cf. discussion in [29: 351*f*].

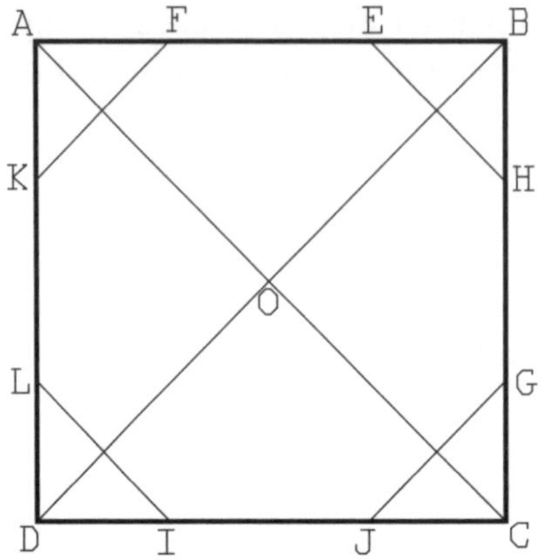

Fig. B.2: The construction of a regular octagon according to *De mensuris*.

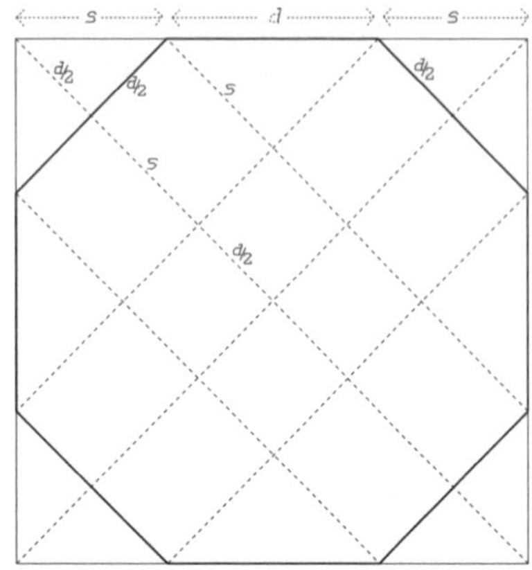

Fig. B.3: A diagram showing why the *De mensuris* construction works.

to *Elements* I contains an oblique but unmistakeable reference to the topic[9] and speaks of it as σύνεγγυζ, "proximate".

Though moderately to quite competent in mathematics, both Theon and Proclos have affinities to the environment which took mathematics as a way to or a kind of *gnosis*—in very loose terms, the Neopythagorean Platonizing ambience. As I have discussed elsewhere [16], this ambience, being unable to follow mathematics at the Euclidean or Archimedean level, borrowed from the practitioners' level. Since no word about the algorithm has reached us from the ancient Greek highlevel mathematicians, it seems reasonable to look for the roots of the procedure in some practitioners' environment.

The algorithm does not turn up as such in "mensuration" treatises, but the pseudoHeronian *De mensuris* [15: 206] prescribes a construction of a regular octagon (under the misleading heading "mensuration of an octagon") which suggests the reasoning that may have led to its invention. In a square *ABCD*, the corners of the octagon *FEHGJILK* are found by making *AE* = *BF* = *BG* = *CH* = … = *AO*—see Figure B.2.

Figure B.3 explains the correctness of the construction; the very same argument shows what we might call the "sideanddiagonal rule": namely that if *s* and *d* are the side and diagonal of a square, so will *s*+*d* and 2*s*+*d* be.

9 [9: 427^{21-23}], trans. [22: 339].

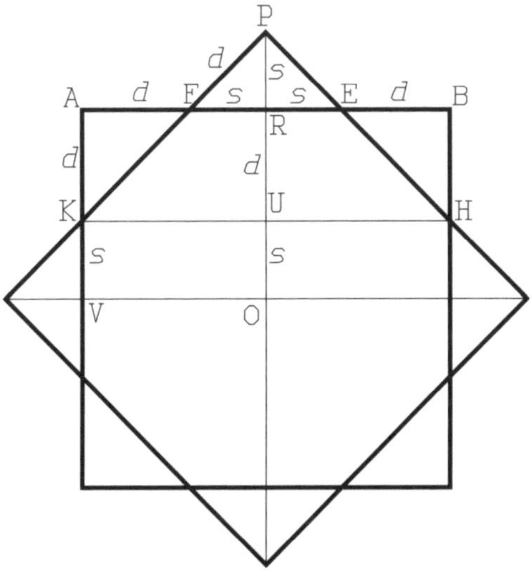

Fig. B.4: The completed version of Figure B.2.

The same construction is found in several other sources: in Abū'l-Wafāʾ's *Book on What is Necessary from Geometric Construction for the Artisan* as problem VII.xxii [19: 93]; in the *Geometria incerti auctoris* no. 55 [2: 360*f*]; in Roriczer's fifteenthcentury *Geometria deutsch* [26: 119*f*]; and in Serlio's *Primo libro di geometria* [25: C2ʳ]. However, it is difficult to believe that anyone would get the idea to draw this diagram if the construction was not known already; and indeed, a much more intuitive diagram can be drawn, of which Figure B.2 is simply a reduced version—namely the one shown in Figure B.4. For symmetry reasons it is intuitively obvious that the superposition of two identical squares of which one is tilted 45° produces a regular octagon; but if we look at the diagram we also observe that $FR = RE = RP = KV = UO$; this length we may call s; then the corresponding diagonal is $d = PF = AF = AK = RU$. Therefore, the semidiagonal PO is $s+d+s = 2s+d$, thus equal to AE. Furthermore, since $KF = FE = 2s$, $UP = s+d$ and $KP = 2s+d$ are, respectively, the side and diagonal of a square—that is, the argument that shows the correctness of the *De mensuris* construction from this diagram also leads to the sideanddiagonal rule.

This construction was employed in actual architecture at least in Classical Antiquity: according to Hermann Kienast (personal communication) it can be seen to have been used in the ground plan of the Athenian "Tower of the Winds" from the first century B.C.E. (outside the octagon itself, the point *P* is marked).[10] The superimposed squares producing the regular octagon are also found as an illustration to the determination of its area in Epaphroditus & Vitruvius Rufus [3: 212, Fig. 40].[11] Since the area is found from the octagonal number, this (as well as any other) geometrical construction is irrelevant to the calculation; it can only be there because it was familiar. Finally, Roriczer's *Wimpergbüchlein* [26: 108*f*] makes use of the configuration.[12]

The conclusions to be drawn from this are somewhat shaky. It appears that the construction of the octagon, both by means of superimposed squares and via the simpler diagram of Figure 2, was known in Classical Antiquity and by late medieval Gothic masterbuilders; it is near at hand to assume some kind of continuity. In the absence of better explanation it is also tempting to presume that the sideanddiagonal algorithm was inspired by one or the other of these constructions. Equally in the absence of better explanations, it is tempting to conjecture that the same algorithm was used by Old Babylonian calculators, and that even they had come to know it in this way (nothing neither excluding nor guaranteeing that the Classical knowledge of the algorithm was due to independent discovery).

Concluding observations

Fairly broad reading of writings on practical "mensuration" from a variety of preModern cultures have thus permitted me to locate one rather certain instance of inspiration from a (very simple) *construction*, and one more dubious case. Even in this field it is confirmed that "practitioners' knowledge" was not unspecific "folk" but specialists'

[10] The construction described by Vitruvius in *De architectura* I.vi.6–7 [10: I, 58–61] is thus a (mistaken) reconstruction, explaining only how Vitruvius thought the construction *could* be made.

[11] The text is also in Bubnov's edition [2: 539], but the diagram is omitted.

[12] Cantor [4: 108] refers to the superimposed squares as common in Pharaonic wall painting, but this can hardly be considered as evidence, neither for use in actual architecture nor for mathematical reasoning based on it. But at least is shows the idea to be near at hand.

The several apparently regular octagons in Villard de Honnecourt's sketchbook [12: Taf. 18, 63] are not accompanied by verbal or geometric indications as to how they were constructed. Only familiarity with Roriczer's description allows us to surmise that Villard's specimens were made in the same way; they cannot count as independent evidence.

knowledge, and that specialists belonged to distinct cultures with little mutual communication.

As an illustration of the rarity of such communication I shall mention one instance, albeit rather of communication between "scribes" and surveyors than between "scribes" and constructors. In the introductory remarks I mentioned that the vernacular *Pratike de geometrie* was largely a translation of the Latin *Artis cuiuslibet consummatio*. However, on one point it is not (in fact on several points, but only this one concerns us here). The *Artis cuiuslibet consummatio* I.15 [28: 158–160] finds the area of an equilateral pentagon as the corresponding pentagonal number (in agreement with the agrimensorial tradition, and in spite of Gerbert's explanation of the fallacy in the triangular case [2: 45], even though this same explanation is reported in chapter I.2 [28: 130]). In contrast, the "*Pratike* [28: 1979: 489] suggests to multiply each side by half the height (which must be supposed to be measured, since no value is told) and to add the five partial areas afterwards.

A very similar procedure is proposed in the treatise *Geometrie due sunt partes principales* [11: 155], whose earliest manuscript also dates from the thirteenth century. Here, for any regular polygon it is proposed to construct the perpendicular bisector of each side, to see where they meet, and measure the heights—etc. Finally, the *Trattato di tutta l'arte dell'abacho*, written in 1334 in Tuscan language but in Montpellier and under obvious Provençal influence,[13] gives an alternative "by geometry" to a corrupt version of the "arithmetical" computation by means of the pentagonal number. This alternative looks as a mixture of the two Latin prescriptions—which can only mean that all three texts share a common background where scribal "mensuration" had contact with real mensuration, probably in French vernacular culture.

I know of no evidence beyond these three passages for the character of this point of contact, and it is much of an accident that I noticed them. Other evidence for interaction between different geometrical cultures of the time may be hidden in odd corners of manuscripts and wait for detection. On the other hand, the very possibility of hiding shows that such contacts were exceptions: on the whole, the preModern geometrical cultures of scribal administrators, surveyors and master builders were

[13] Florence, Biblioteca Nazionale Centrale, fond. prin. II,IX.57, fol. 133^{v}. For the date and place where the treatise was written, see [5]. For the Provençal origin of much of the material, see [18, *passim*].

as isolated from each other as, say, dentists, air traffic controllers and public relation experts nowadays.

References

1. Beckman, N., & Kr. Kålund (eds): "Alfræði íslenzk. Islandsk encycklopædisk Litteratur". Vol. II. "Rímtol". Samfund til Udgivelse af Gammel Nordisk Litteratur / S. L. Møller, København, 1914–16.
2. Bubnov, Nicolaus (ed.): "Gerberti postea Silvestri II papae Opera mathematica (972–1003)". Friedländer, Berlin, 1899.
3. Cantor, Moritz: "Die römischen Agrimensoren und ihre Stellung in der Geschichte der Feldmesskunst. Eine historischmathematische Untersuchung". Teubner, Leipzig, 1875.
4. Cantor, Moritz: "Vorlesungen über Geschichte der Mathematik. Erster Band, von den ältesten Zeiten bis zum Jahre 1200 n. Chr." Dritte Auflage. Teubner, Leipzig, 1907.
5. Cassinet, Jean: "Une arithmétique toscane en 1334 en Avignon dans la cité des papes et de leurs banquiers florentins", pp. 105–128 in "Commerce et mathématiques du moyen âge à la renaissance, autour de la Méditerranée". Actes du Colloque International du Centre International d'Histoire des Sciences Occitanes (Beaumont de Lomagne, 13–16 mai 1999). Éditions du C.I.H.S.O., Toulouse, 2001.
6. Clagett, Marshall: "Archimedes in the Middle Ages". Vol. I. "The Arabo Latin Tradition". University of Wisconsin Press, Madison (Wisconsin), 1964.
7. Dupuis, J. (ed., trans.): "Théon de Smyrne, philosophe platonicien, Exposition des connaissances mathématiques utiles pour la lecture de Platon". Hachette, Paris, 1892.
8. Fowler, David H., & Eleanor Robson: "Square Root Approximations in Old Babylonian Mathematics: YBC 7289 in Context", Historia Mathematica, vol. 25, pp. 366–378, 1998.
9. Friedlein, Gottfried (ed.): "Procli Diadochi in primum Euclidis Elementorum librum commentarii". Teubner, Berlin, 1873.
10. Granger, Frank (ed., trans.): "Vitruvius, De Architectura". 2 vols. (Loeb Classical Library 251, 280). Heinemann, London / Harvard University Press, Cambridge (Mass.), 1970.
11. Hahn, Nan L. (ed): "Medieval Mensuration: QUADRANS VETUS and GEOMETRIE DUE SUNT PARTES PRINCIPALES..." (Transactions of the American Philosophical Society, vol. 72, Part 8). The American Philosophical Society, Philadelphia, 1982.
12. Hahnloser, H. R. (ed.): "Villard de Honnecourt. Kritische Gesamtausgabe des Bauhüttenbuches ms. fr 19093 der Pariser Nationalbibliothek". Anton Schroll, Wien, 1935.
13. Heath, Thomas L. (ed., trans.): "The Thirteen Books of Euclid's Elements", 2nd revised edition. 3 vols. Cambridge University Press, Cambridge/Macmillan, New York, 1926.
14. Heiberg, J. L. (ed., trans.): "Heronis Definitiones cum variis collectionibus. Heronis quae feruntur Geometrica". (Heronis Alexandrini Opera quae supersunt omnia, IV). Teubner, Leipzig, 1912.
15. Heiberg, J. L. (ed., trans.): "Heronis quae feruntur Stereometrica et De mensuris". (Heronis Alexandrini Opera quae supersunt omnia, V). Teubner, Leipzig, 1914.
16. Høyrup, Jens: "Alchemy and Mathematics: Technical Knowledge Subservient to Ancient γνωσις", pp. 38–54 in Vincent F. Hendricks & Jesper Ryberg (eds),

"Readings in Philosophy and Science Studies", vol. I. Department of Philosophy and Science Studies, Roskilde, 2001. (Marred by numerous computer conversion errors.)

17. Høyrup, Jens: "Lengths, Widths, Surfaces: A Portrait of Old Babylonian Algebra and Its Kin". (Studies and Sources in the History of Mathematics and Physical Sciences). Springer, New York, 2002.
18. Høyrup, Jens: "Jacopo da Firenze's Tractatus Algorismi and Early Italian Abbacus Culture". (Science Networks. Historical Studies, 34). Birkhäuser, Basel etc., 2007.
19. Krasnova, S. A. (ed., trans.): "Abu-l-Vafa al-Buzdžani, Kniga o tom, [1]to neobxodimo remeslenniku iz geometri[1]eskix postroenij", pp. 42–140 in A. T. Grigor'jan & A. P. Juškevi[1] (eds), "Fizikomatemati[1]eskie nauki v stranax vostoka". (Sbornik statej i publikacij. Vypusk I (IV)). Izdatel'stvo "Nauka", Moskva, 1966.
20. Kroll, Wilhelm (ed.): "Procli Diadochi In Platonis Rem publicam commentarii". 2 vols. Teubner, Leipzig, 1899, 1901.
21. Menninger, Karl: "Zahlwort und Ziffer. Eine Kulturgeschichte der Zahl". I. "Zählreihe und Zahlsprache". II. "Zahlschrift und Rechnen". 2. neubearbeitete und erweiterte Auflage. Vandenhoeck & Ruprecht, Göttingen, 1957–58.
22. Morrow, Glenn R. (ed., trans.): "Proclus, A Commentary on the First Book of Euclid's Elements". Princeton University Press, Princeton (N.J.), 1970.
23. Neugebauer, O., & A. Sachs: "Mathematical Cuneiform Texts". (American Oriental Series, vol. 29). American Oriental Society, New Haven (Conn.), 1945.
24. Schöne, Hermann (ed., trans.): "Herons von Alexandria Vermessungslehre und Dioptra". (Heronis Alexandrini Opera quae supersunt omnia, vol. III). Teubner, Leipzig, 1903.
25. Serlio, Sebastiano: "Tutte l'Opere d'Architettura". Francesco de' Francesci, Venezia, 1584.
26. Shelby, Lon R. (ed.): "Gothic Design Techniques. The Fifteenth Century Design Booklets of Mathes Roriczer and Hanns Schmuttermayer". Southern Illinois University Press, Carbondale & Edwardsville, 1977.
27. Shorey, Paul (ed., trans.): "Plato, The Republic". 2 vols. (Loeb Classical Library 237, 276). Heinemann, London / Harvard University Press, Cambridge (Mass.), 1930, 1935.
28. Victor, Stephen K. (ed., trans.): "Practical Geometry in the Middle Ages. Artis cuiuslibet consummatio and the Pratike de geometrie". (Memoirs of the American Philosophical Society, vol. 134). The American Philosophical Society, Philadelphia, 1979.
29. Vitrac, Bernard (ed., trans.): "Euclide d'Alexandrie, Les Éléments". Traduits du texte de Heiberg. Vol. I. Presses Universitaires de France, Paris, 1990.

APPENDIX C

DRAUGHTING CURVES USED IN SHIP DESIGN

Jobst Lessenich

Ship lines plans as orthogonal view drawings were first used routinely in naval architecture since the early 18th century. Draughting tools for such drawings in addition to the traditional compass and rulers evolved gradually to deal with increasingly complex and free form shapes. The shipbuilder's spline and the draughting curve are the most prominent and unique tools developed to produce ship lines plans accurately and neatly. The development of splines is described in Horst Nowacki's paper Splines im Schiffbau [1].

This paper will present a documentation of the early history of draughting curves which came into wider use during the last few decades of the 18th century. Their shapes were appropriate to draw waterlines, diagonals, buttock and bow lines etc. and also the sections of the body plan after the formal compass-and-ruler design of the midship section had been given up.

Source material on draughting curves is scarce and unsystematic. For this article the available information was collected from primary and secondary sources and is presented in synoptical form. This material may serve as a reference for later comparisons with similar developments in other disciplines.

Ch. G. D. Müller

In his translation of Duhamel du Monceau's Élémens de l'Architecture Navale [2] the translator Müller added also his own knowledge and ideas to the subjects of the various chapters. In a footnote to Chapter 4 in [2], the drawing of the body plan, he suggested the use of self-made templates to obtain a neat drawing in ink of the body plan. To make these templates the naval architect should copy the sections of the body plan separately on stiff paper, cut them out and check the contours. If the result was satisfactory the contours were transferred to thick card board and cut out carefully. Running a fingertip along the edge was the final test for the smoothness of the template.

Fig. C.1: The draughting curve under Chapman's arm, adapted by the author from contemporary engraving.

From this special solution for one lines plan it is not far to the idea of having a more universal set of templates or curves of various shapes to draw the sections of the body plan already during the process of designing the ship.

It may be assumed that the early examples of curves were cut individually by the master shipwright or naval architect following his own ideas, and that he kept his curves as concealed from the public as the other special skills of his trade. But someone must have had the idea to combine a number of curves of various size and different shapes to a standardized set, with which all contours in a lines plan of a ship could be drawn, and he manufactured such sets and they were traded like other drawing tools.

Portrait of F. H. af Chapman

A contemporary engraving with the title "F. H. af Chapman" shows him holding a pair of dividers in his right hand and a book and a curve under his right arm. As Chapman had been ennobled in 1772 the engraving must be dated to a later year. The curve is approximately 500 mm long and has a small to moderate curvature, a shape that would serve well to draw waterlines and diagonals. This particular feature of the portrait is shown in figure C.1. The role of Chapman in the development of ship design methodology is well described in a book by Harris [3].

Gerhard Timmermann

In an essay about ship design in former times, Skibskonstruktionstegningen [4], Timmermann informs the reader that the Maritime Museum in Bergen, Norway (Sjøfartsmuseum Bergen) possesses thirteen drawing templates or ship curves, which are signed "RR3". These curves were once owned by the shipwright Rasmus Rolfsen and are dated from 1779. Nine of these thirteen templates resemble the curves of the Copenhagen Set (see Fig. C.2) with the numbers 8, 12, 16, 20, 24, 30, 34, 35 and 37. With the exception of N° 8 the curves have shapes very appropriate for drawing the sections in a body plan. Curve N° 8 has some similarity with the one Chapman is holding under his arm (see Fig. C.1).

Copenhagen Set

Gerhard Timmermann writes in 1962 [4] that the oldest and still existing Copenhagen Set (Københavner-sæt) is in possession of the Danish Maritime Museum at Kronborg (Handels- og Sjøfartsmuseet på Kronborg) and had been manufactured in 1817. It once belonged to the master shipwright Rasmus Møller of Troense.

A photograph in [4] gives the impression that this set already contains nearly all the curves which will be found in the present Copenhagen Set (see Fig. C.2). Therefore it may be supposed that the first set of this kind appeared close to the year 1800.

Another set, owned by the Naval Shipyard in Copenhagen (Orlogsværftet i København), is dated to 1835 and is still used (in 1962) as pattern for new sets [4]. The Copenhagen Set of the 20th century represents its status of 1835, according to Timmermann [4]. Some manufacturers

Fig. C.2: Copenhagen Set [9].

added to the Copenhagen Set the French Curve N° 4, others the French Curves N° 4 and N° 21 (see Figs. C.2 and C.7).

David Steel

In the chapter "Explanation of the terms used in shipbuilding" in The Shipwright's Vade-mecum [5, p. 118] Steel explained the term "moulds" in the familiar way as fullsize templates to cut the timbers for frames, stem etc. But then he continued that the term mould is also given to "thin flexible pieces of peartree or box, used in constructing the draughts and plans of ships, which are made in various shapes; viz. to the segments of circles from one foot to 22 feet radius, increasing six inches on each edge, and numerous elliptical curves with other figures". (There is no further description of "other figures"). In a footnote to this explanation the name of a dealer was given where "moulds, &c. of every sort requisite for marine drawing may be had". This advertisment indicates that in 1805 these moulds were already known and demanded tools and must have come into use some years earlier.

Steel never mentioned the use of these "moulds" in his description of constructing the body and the half-breadth plan (page 186 and following in [5]). He described only the old fashion of drawing the

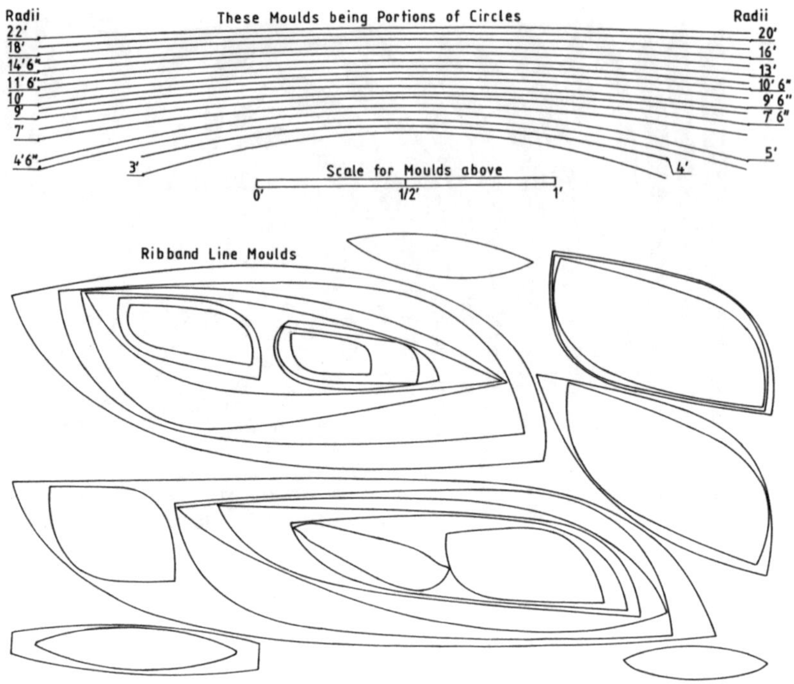

Fig. C.3: Moulds and sweeps used in ship draughting, adapted by the author from Schnellsegler [8].

frames using "arches of circles called sweeps" (see page 106 in [5]), but the drawings of a West Indiaman of 300 tons (plate XXI), a Collier Brig of 170 tons (plate XXII) and a sloop of 60 tons (plate XXVII), all reedited in Éléments et pratique de l'architecture navale 1805 [6], show that their body plans have been drawn using "moulds" of elliptical or similar shapes.

Peter Hedderwick

A plate with the title "Moulds and Sweeps used in Ship Draughting" in Hedderwick's book "*A Treatise on Marine Architecture*", published 1830 [7], illustrates David Steel's verbal description of moulds. The plate shows 17 "*Moulds being Portions of Circles*" with radii from 3 to 22 feet (not all radii mentioned by Steel are presented here) and 26 other moulds, many of their contours set into one another. Most of them show a hyperbolic shape, only few of them seem to be parts of elliptical curves.

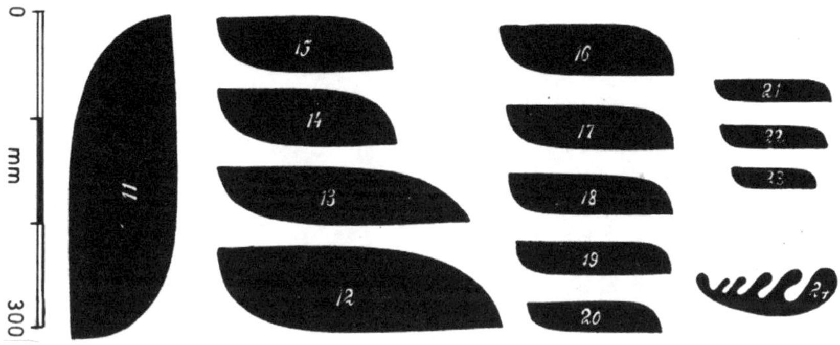

Fig. C.4: English Set [9].

The plate (see Fig. C.3) has been redrawn from a small reproduction in Schnellsegler 1775–1875 [8] and a scale in feet has been added to the circular moulds. There is no scale given for the other 26 moulds nor is it known if all of them are of the same scale.

Other sets of the 20th century

The various sets of curves for the special purpose of drawing ship lines shown here (Figs. C.4, C.5 and C.6) are taken from a catalogue of a dealer and manufacturer of drawing tools [9]. They represent the status of these traditional drawing tools which they had reached by the beginning of the 20th century and which lasted as long as these curves were manufactured. The numbering of the curves is arbitrary and serves only as identification for ordering single curves out of a set. A scale in millimeters is added at the side to give the approximate size of the original curves. These curves were originally cut out of thin bords of steamed wood of the pear-tree, the box-tree or white beach. By steaming the wood warping was avoided. Later in the 20th century various plastic materials, transparent or coloured brown, were used.

English Set

The English Set (Englischer Satz, see Fig. C.4) looks like a subset of the 26 moulds Peter Hedderwick presented in his book [7] (see Fig. C.3) and which Steel [5] described as “numerous elliptical curves with other figures”.

Fig. C.5: Hamburg Set [9].

Hamburg Set

The English Set and part of the Copenhagen Set were obviously combined to form the Hamburg Set (Hamburger Satz, see Fig. C.5). At least one "new" curve N° 26 was created for this set by reducing in scale curve N° 25.

German Set

The youngest set is the German Set (Deutscher Satz, see Fig. C.6). It seems that this set was a combination of several curves from the Copenhagen Set and the Hamburg Set to which about twelve new curves were added (see lower part of Fig. C.6). Timmermann wrote [4] that these new curves had been adapted to the change in geometry of iron and steel hulls.

French Curves

"French Curves", in the narrower sense of the word as it is used in Britain, indicates that these drawing tools originated in France where these curves were called "pistolets à dessin". Both names are linked to

Fig. C.6: German Set [9].

the same type of curves, a selection of which is shown here in Fig. C.7 [9]. Other French Curves are shown in Drawing Instruments 1580–1980 [10] on page 103.

The definition of "French Curve" in The Random House Dictionary [11] shows a curve very similar to N° 15 of Fig. C.7 and the Dictionnaire usuel illustré [12] gives as examples for "pistolets à dessin" three curves similar to N°s 4, 7, and 17 of Fig. C.7.

French Curves were first manufactured in the early 19th century in France and sold in sets. The material was pearwood, lime or other similar stable wood. Later in the 20th century vulcanite, celluloid or acrylic were used [10].

The outlines of French Curves were not derived from geometrical curves as ellipses, parabolas or hyperbolas but cut, as it is stated in the Grand Larousse [13], following curves of fantasy ("découpé suivant des courbes de fantaisie"), ending in spirals, curls or corner points. French Curves were used in architecture and any other design or decorative work; their use in naval architecture, specially to draw the body plan of a ship, had been described by d'Étroyat in Traité élémentaire d'Architecture Navale [14]: The French Curve (he called it simply "pistolet"), cut out of thin bord, served to draw lines with a strong curvature which could not be achieved with an ordinary spline. He pictured only one sample of a French Curve (planche I, Fig. C.5 in [14], similar

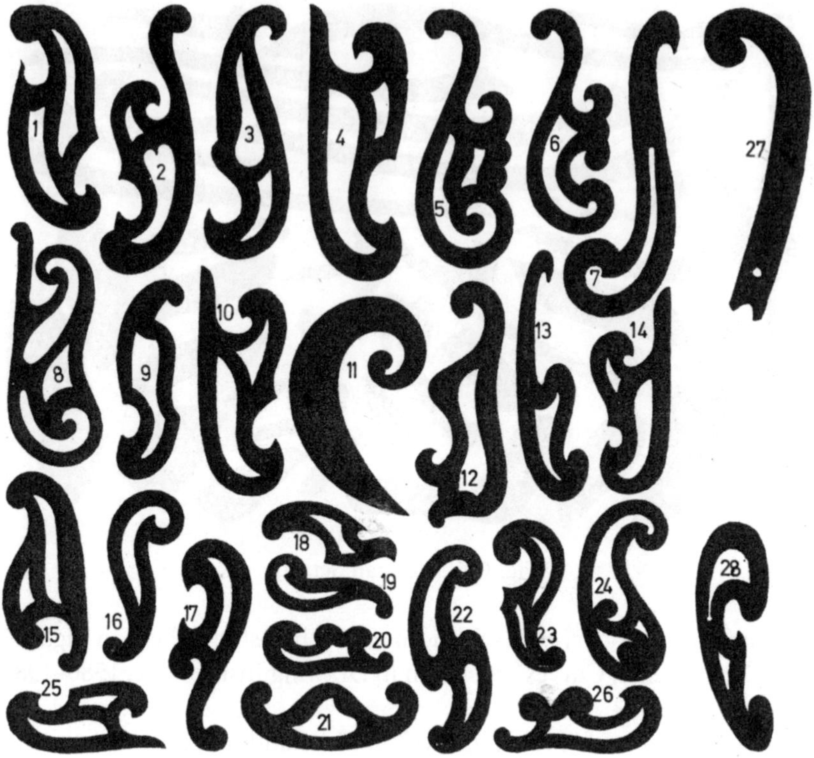

Fig. C.7: French Curves (not on uniform scale!) [9].

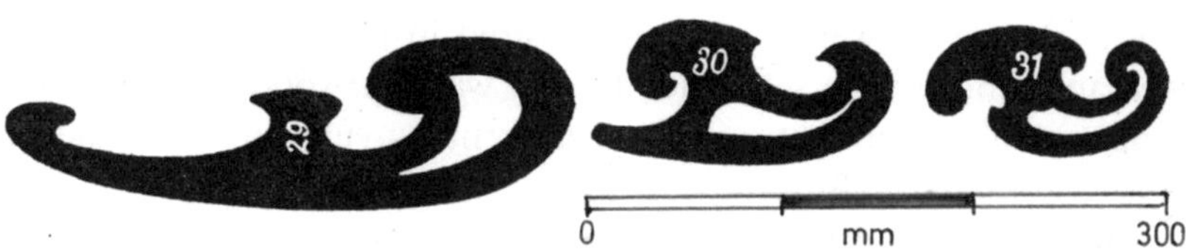

Fig. C.8: Burmester Curves [9].

to N° 26 in Fig. C.7 above) to show form and use of these curves. For further information he referred to the trade where perfectly cut French Curves of various contours and sizes ("...pistolets variés de contours et de grandeurs, parfaitement decoupés...") could be purchased. The way d'Étroyat dealt with this subject indicates that the use of French Curves in naval architecture had been a routine matter by 1850.

Burmester Curves

What seems to be a subset of three of the French Curves is in reality a special mathematical variation of these and was developed by Professor Ludwig Burmester (1840–1927), a mathematician from Munich [10]. Each of these three curves (see Fig. C.8) consists of a number of different quarter-ellipses blended into each other [9]. These curves serve a great variety of applications and are still available.

References

1. Nowacki, Horst: "Splines im Schiffbau", Proc. 21st Duisburg Colloquium on Ship and Ocean Technology: The Ship out of the Computer, pp. 27–53, Gerhard Mercator University, Institute of Ship Technology, Duisburg, May 2000.
2. Müller, Ch. G. D.: "Anfangsgründe der Schiffbaukunst" (translation of H. L. Duhamel du Monceau: Élémens de l'Architecture Navale, Paris 1758), bei Joachim Pauli, Berlin 1791, reprint Horst Hamecher, Kassel 1973.
3. Harris, Daniel G.: "F. H. Chapman, The First Naval Architect", Conway Maritime Press, London 1989.
4. Timmermann, Gerhard: "Skibskonstruktionstegningen gennem tiderne" i Årbog 1962, Handels- og Sjøfartsmuseet på Kronborg 1962.
5. Steel, David: "The Shipwright's Vade-mecum", P. Steel, London 1805, reprint Jean Boudriot Publications, Rotherfield 1995.
6. Steel, David: "Éléments et pratique de l'architecture navale 1805", Le Chasse-Marée/ArMen, Douarnenez Cedex 2001.
7. Hedderwick, Peter: "A Treatise on Marine Architecture", Edinburgh 1830.
8. MacGregor, David R.: "Schnellsegler 1775–1875, Konstruktion und Geschichte" (German translation of "Fast Sailing Ships"), Delius Klasing, Bielefeld 1974.
9. Gebr. Wichmann: "Hauptkatalog 1938", Berlin 1938, in possession of the library of Deutsches Schiffahrtsmuseum Bremerhaven sign. 83–1387. Figs. 2, 4, 5, 6, 7 and 8 reproduced by kind permission of Gebr. Wichmann G.m.b.H. Berlin.
10. Hambly, Maya: "Drawing Instruments 1580–1980", Sotheby's Publications, London 1988.
11. Jess Stein (ed.): "The Random House Dictionary of the English Language", New York 1967.
12. "Dictionnaire usuel illustré", Librairies Flammarson et Quillet, Paris 1982.
13. Larousse, Pierre: "Grand Dictionnaire universel du XIXe siècle", Vol XII 2e partie, Paris 1982.
14. d'Étroyat, Adrien: "Traité élémentaire d'Architecture Navale", 1re édition Gousset, Lorient 1846, 2e édition Mallet-Bachelier, Paris 1863, reproduction Ancre, Nice 1993.

APPENDIX D

BIBLIOGRAPHICAL NOTES ON HISTORICAL METROLOGY

Compiled by Jobst Lessenich

The following lists of references on historical metrology are intended to serve as an aid for comparisons and conversions of historical measures found in the two architectures in various periods and regions.

1. Sources from German Library Catalogues, collected by Jobst Lessenich

Agricola, Georgius: Schriften über Maße und Gewichte, Heidelberg, Berlin, 1959.

Alberti, Hans Joachim: Maß und Gewicht, Berlin, 1957.

Beschreibung der Gewichten und Massen der Stadt und Landschaft Zürich, Zürich 1765.

Bohn, Gottfried Christian: G. Ch. Bohns wohlerfahrener Kaufmann (europäische Münzen, Maße und Gewichte), Hamburg, 1762.

Böckh, August: Metrologische Untersuchungen über Gewichte, Münzfüße und Maße des Altertums in ihrem Zusammenhange, Berlin, 1838.

Brandis, J.: Münz-, Mass- und Gewichtswesen in Vorderasien bis auf Alexander den Großen, 2. Auflage Berlin 1866, Nachdruck Amsterdam, 1966.

Bünting, Heinrich: Vande Munte Ende Ghewichte der heyliger Schrifture (Altes und Neues Testament), Lateinischer Titel: de monetis et mensuris sacrae scripturae, ohne Ort, 1596.

Bußemer, Manfred: Bekannte Münzen, Maße und Gewichte, Ramstein-Miesenbach 1994.

Cagnazi, Luca de Samuele: Über den Werth der Maaße und Gewichte der alten Römer, Kopenhagen, 1828.

Dilke, Oswald Ashton Wentworth: Mathematik, Maße und Gewichte in der Antike, Stuttgart, 1991.

Dove, H. W.: Über Maaß und Messen, 2. Auflage, Berlin 1835.

Doursther, Horace: Dictionnaire universel des poids et mesures, Reprint Amsterdam, 1965.

Engel, Franz: Tabellen alter Münzen, Maße und Gewichte, Rinteln, 1965.

Eytelwein, Joh. Albert: Vergleichung der gegenwärtig und vormals in den Königlich preußischen Staaten eingeführten Maaße und Gewichte, Berlin, 1810.

Fischer, Ernst Gottfried: Tabellen zur Vergleichung der vornehmsten Münzen, Maasse und Gewichte in Europa, Berlin, 1787.

Gerechnet büchlein von Muntz, Gewicht, Elen und Maß aller Land gegen eynander verglichen. Gedruckt in Nürnberg bey Georg Wachler, 1530.

Gerhardt, Markus Rudolf Balthasar: Handbuch der Deutschen Münz-, Maaß- und Gewichtskunde, Berlin, 1788.

Hagen, Gotthilf: Zu Fragen über das deutsche Maass, Berlin, 1861.

Hameyer, Uwe: Messen und Maße, Kiel, 1986.

Haupt, Waldemar: Maße, Währungen, Werte, Stuttgart, 1939.

Heinrich, Placidus: Bestimmung der Maasse und Gewichte des Fürstenthums Regensburg, Regensburg, 1808.
Hellwig, Gerhard: Lexikon der Maße und Gewichte, Gütersloh, 1979.
Hertzer, Hugo: Reduktionstabellen zur praktischen Einführung der norddeutschen Maasse und Gewichte, Berlin, 1869–70.
Hippel, Wolfgang von: Maß und Gewicht im Gebiet des Königreiches Württemberg und der Fürstentümer Hohenzollern am Ende des 18. Jahrhunderts, Stuttgart, 2000.
Kahnt, Helmut / Knorr, Bernd: Alte Maße, Münzen und Gewichte. Ein Lexikon, Mannheim, Wien, Zürich, 1987.
Karsten, Gustav: Maass und Gewicht in alten und neuen Systemen, Berlin, 1871.
Kletke, G. M.: Preußische Maaß- und Gewichts-Ordnung, mit Inbegriff sämmtlicher seit dem 16. Mai 1816 bis zum Jahre 1842 incl. erschienenen Erläuterungen, Ergänzungen und Zusätze, Berlin 1844.
Kletke, G. M.: Maaß- und Gewichtsordnung vom 17. August 1868 nebst der Eichordnung vom 16. Juli 1869, Berlin, 1871.
Klimpert, Richard: Lexikon der Münzen, Maße, Gewichte, Zählgrößen und Zeitgrößen aller Länder der Erde, 2 Auflage, Berlin, 1869, Nachdruck Graz, 1972.
Kruse, Johann Elert: Allgemeiner und besonders Hamburgischer Contorist (Währungen, Münzen, Gewigten, Maaßen etc.), Hamburg, 1771 .
Löhmann, Friedrich: Tafeln zur Verwandlung des Längen- und Hohlmaasses sowie des Gewichts... aller Hauptstädte Europas...., Leipzig, 1822–1832.
Mäurer, Eduard: Deutsches Maass- und Gewichtsbuch, Weimar, 1872.
Manger, Heinrich Ludwig: Mass- und Gewichts-Vergleichstabellen, Potsdam, 1808.
Mausy, M.: Umrechnungs-Tabellen russischer Längenmaße, Gewichte und Hohlmaße in deutsche Maße und Gewichte, Halle (Saale), 1913.
Mendelsohn, M.: Handbuch der Münz-, Maß- und Gewichtskunde, Neustrelitz, 1994.
Neue Maas- und Gewichtsordnung für das Großherzogthum Baden, Carlsruhe, 1829.
Romé de L'Isle, Jean Baptiste Louis: Metrologische Tafeln über die alten Maaße, Gewichte und Münzen Roms und Griechenlands, Braunschweig, 1792.
Rottleuthner, Wilhelm: Alte lokale und nichtmetrische Gewichte und Maße und ihre Größen nach metrischem System. Bearb. von Wilhelm E. Rottleuthner, Innsbruck, 1985.
Schneider, Friedrich-Wilhelm: Taschenbuch der Maaß- und Gewichtskunde (deutsche und ausländische Maße), Berlin 1839.
Targé, P.: Das Maaß und Gewicht der Staaten des Norddeutschen Bundes, Magdeburg 1869.
Tavernier, Jean-Baptiste: Herrn J.-B. Taverniers Vierzigjährige Reise-Beschreibung durch Türkey, Persien, Indien.... allda gangbaren Münze, Maß, Gewicht und deren Wert ausführlich verzeichnend, Nürnberg ca 1675.
Trapp, Wolfgang: Kleines Handbuch der Maße, Zahlen, Gewichte und der Zeitrechnung, 3. Auflage, Stuttgart, 1998.
Verdenhalven, Fritz: Alte Meß- und Währungssysteme aus dem deutschen. Sprachgebiet, Neustadt an der Aisch, 1993.
Witthöft, Harald: Umrisse einer historischen Metrologie zum Nutzen der wirtschaftsgeschichtlichen Forschung, 2 Bde, Göttingen, 1979.
Witthöft, Harald: Handbuch der Historischen Metrologie, Band 1: Deutsche Bibliographie zur Historischen Metrologie, St. Katharinen, 1991.
Witthöft, Harald: Handbuch der Historischen Metrologie, Band 2: Deutsche Maße und Gewichte des 19. Jahrhunderts, Teil 1: Die Orts- und Landesmaße, St. Katharinen, 1993.
Witthöft, Harald: Handbuch der Historischen Metrologie, Band 3: Deutsche Maße und Gewichte des 19. Jahrhunderts, Teil 2: Die Maß- und Gewichtseinheiten, St. Katharinen, 1994.

Witthöft, Harald: Handbuch der Historischen Metrologie, Band 4: Deutsche Maße und Gewichte des 19. Jahrhunderts, Teil 3: Korpus der Maße und Gewichte nach den Rechtsquellen des 19. Jahrhunderts, St. Katharinen, 1994.

Witthöft, Harald: Handbuch der Historischen Metrologie, Band 5: "Der Nelkenbrecher", St. Katharinen, 2003.

Zupko, Ronald E.: British Weights and Measures, Madison, Wisc., 1977.

Zupko, Ronald E.: Italian Weights and Measures from the Middle Ages to the 19th Century, Philadelphia, 1981.

2. Sources from the INTERNET, contributed by Jens Høyrup

a) *Alphabetical index of units*

The user will find this index either via the website of the main index at

www.sizes.com/indexes.htm

and then clicking on INDEX TO UNITS & SYSTEMS OF UNITS or by immediately addressing

www.sizes.com/units/index.htm

This list contains a great number of units of weights, measures and many other properties, ancient and modern, domestic and foreign.

The terms are listed in alphabetical order. A menue ("A to Z" and "home") at the right side of the first page supports the search for the term in question. For each term the user will find its signification, its definition and its metric equivalent.

b) *List of metrological publications*

The user will find this list at the website

www.sizes.com/units/majsource.htm

The publications are arranged by the language they have been written in, not by the region whose metrology they discuss. Click on one of the underlined languages to see the publications in this language.

INDEX

History of Science and Medicine Library

ISSN 1872-0684

1. Fruton, J.S. *Fermentation. Vital or Chemical Process?* 2006. ISBN 978 90 04 15268 7
2. Pietikainen, P. *Neurosis and Modernity.* The Age of Nervousness in Sweden, 2007. ISBN 978 90 04 16075 0
3. Roos, A. *The Salt of the Earth.* Natural Philosophy, Medicine, and Chymistry in England, 1650-1750. 2007. ISBN 978 90 04 16176 4
4. Eastwood, B.S. *Ordering the Heavens.* Roman Astronomy and Cosmology in the Carolingian Renaissance. 2007. ISBN 978 90 04 16186 3 (Published as Vol. 8 in the subseries *Medieval and Early Modern Science*)
5. Leu, U.B., R. Keller & S. Weidmann. *Conrad Gessner's Private Library.* 2008. ISBN 978 90 04 16723 0
6. Hogenhuis, L.A.H. *Cognition and Recognition: On the Origin of Movement.* Rademaker (1887-1957): A Biography. 2009. ISBN 978 90 04 16836 7
7. Davids, C.A. *The Rise and Decline of Dutch Technological Leadership.* Technology, Economy and Culture in the Netherlands, 1350-1800 (2 vols.). 2008. ISBN 978 90 04 16865 7 (Published as Vol. 1 in the subseries *Knowledge Infrastructure and Knowledge Economy*)
8. Grellard, C. & A. Robert (Eds.). *Atomism in Late Medieval Philosophy and Theology.* 2009. ISBN 978 90 04 17217 3 (Published as Vol. 9 in the subseries *Medieval and Early Modern Science*)
9. Furdell, E.L. *Fatal Thirst.* Diabetes in Britain until Insulin. 2009. ISBN 978 90 04 17250 0
10. Strano, G., S. Johnston, M. Miniati & A. Morrison-Low (Eds.). *European Collections of Scientific Instruments, 1550-1750.* 2009. ISBN 978 90 04 17270 8 (Published as Vol. 1 in the subseries *Scientific Instruments and Collections*)
11. Nowacki, H. & W. Lefèvre (Eds.). *Creating Shapes in Civil and Naval Architecture.* A Cross-Disciplinary Comparison. 2009. ISBN 978 90 04 17345 3
12. Chabás, J. & B.R. Goldstein (Eds.). *The Astronomical Tables of Giovanni Bianchini.* 2009. ISBN 978 90 04 17615 7 (Published as Vol. 10 in the subseries *Medieval and Early Modern Science*)

PLATES I – LII

1.3a

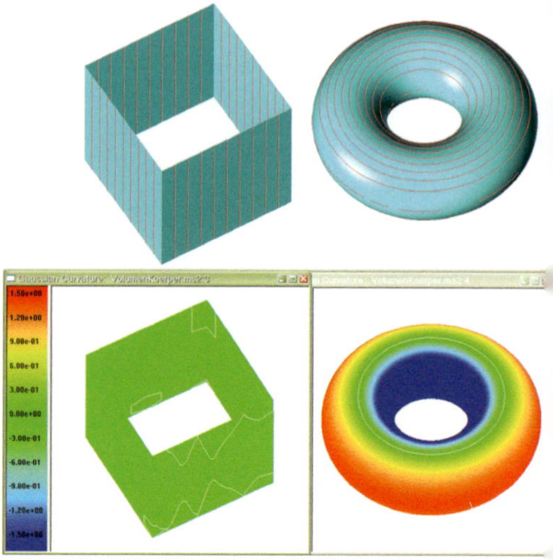

1.3b

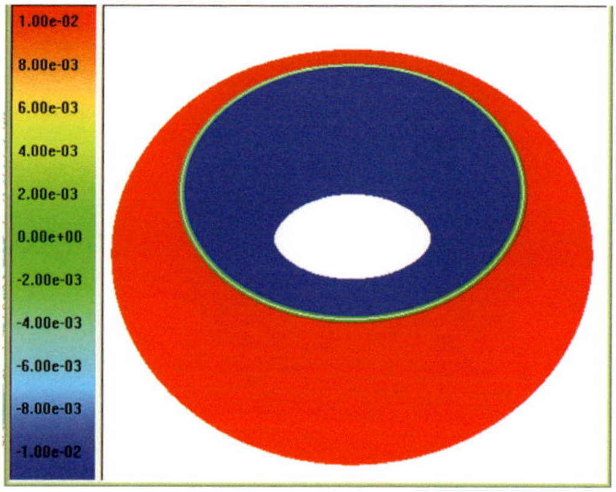

1.3c

Pl. I. Simple shapes: Perspective views and Gaussian curvature.
(Colors: red = positive, green = zero, blue = negative Gaussian curvature).
Fig. 1.3a: Sphere, cylinder, cone, Fig. 1.3b: Cube, torus, Fig. 1.3c: Torus with reset color table.

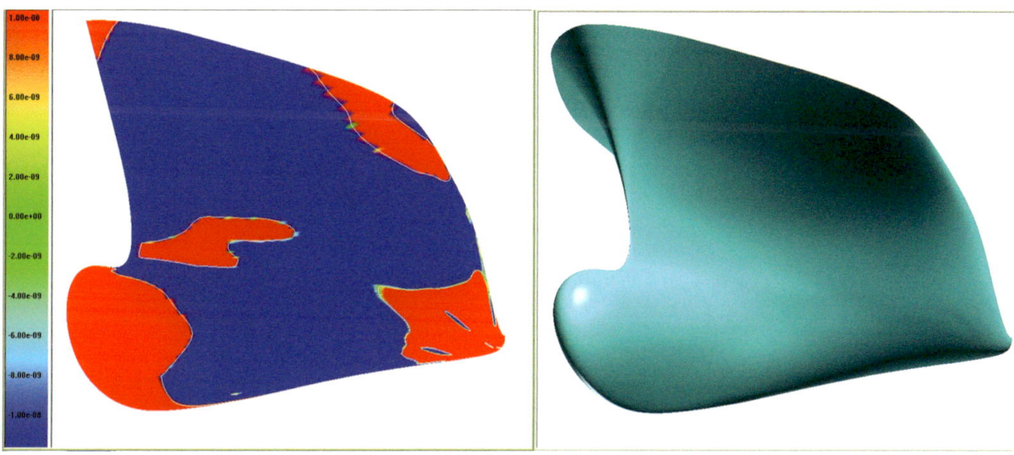

Pl. II. Fig. 1.4: Tanker forebody with bulbous bow: Gaussian curvature map and perspective view.

a

b

c

d

Pl. III. Ship hull shapes I: Figs 1.6a: Sailing log raft (Taiwan); 1.6b: Basket boat (Iraq, Vietnam); 1.6c: Bark canoe (North America); 1.6d: Fishing dugout (North America).

a

b

c

d

Pl. IV. Ship hull shapes II: Figs. 1.7a: Nile sailing boat; 1.7b: Indonesian outrigger; 1.7c: Hjortspring boat; 1.7d: Carvel planking construction.

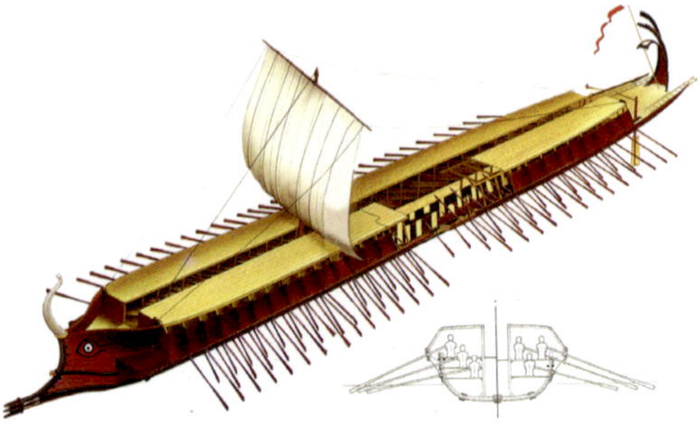

1.8a

1.8b

1.9a

1.10b

Pl. V. Ship hull shapes III: Figs. 1.8a: Greek trireme; 1.8b: Roman merchantman; 1.9a: Roskilde warship; 1.10b: A caravel.

Pl. VI. Fig. 1.12a: Frederik Henrik af Chapman. Portrait by L. Pasch the younger.

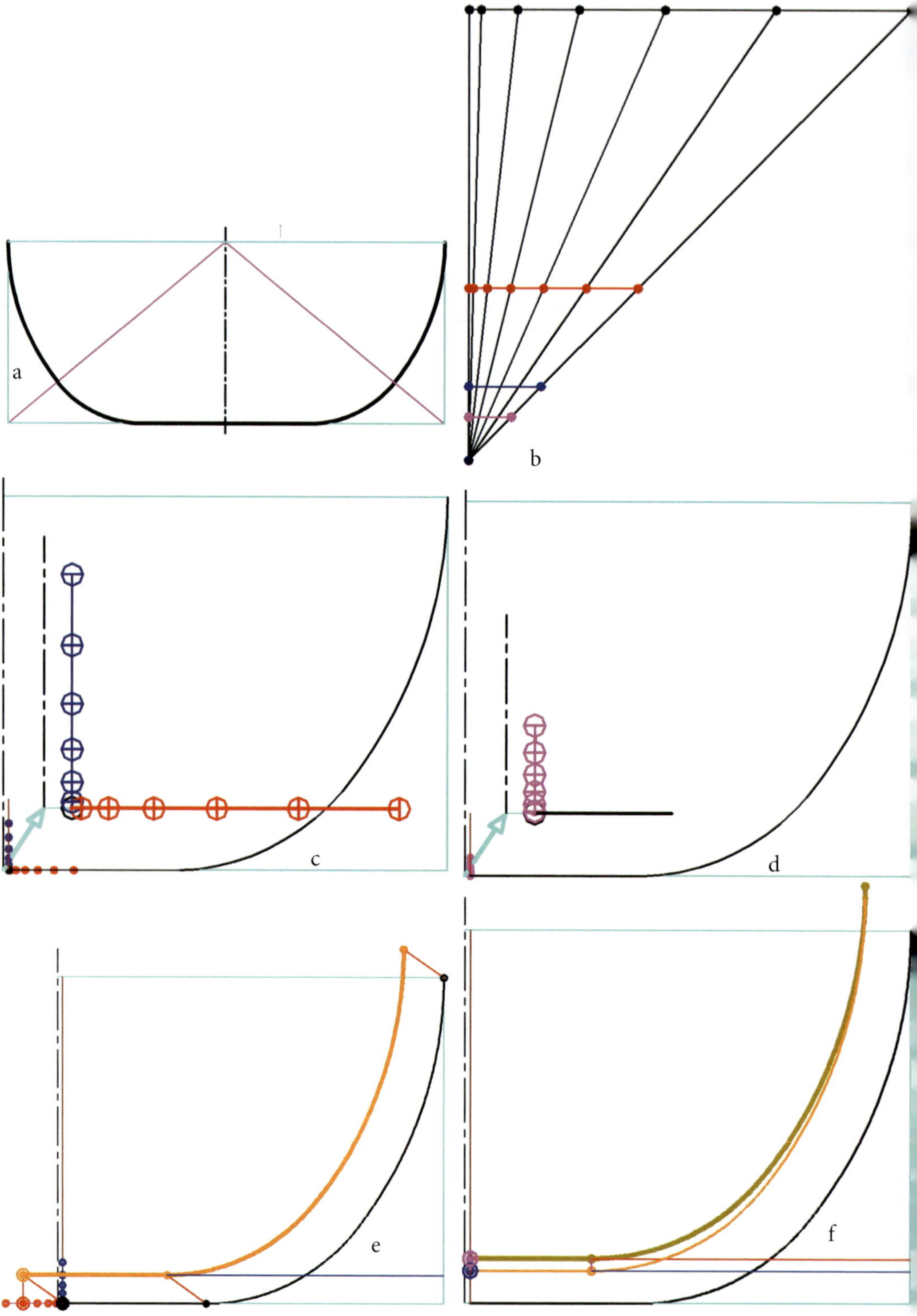

Pl. VII. Figs. 1.13a – 1.13f: Mediterranean lofting method: Procedural steps.

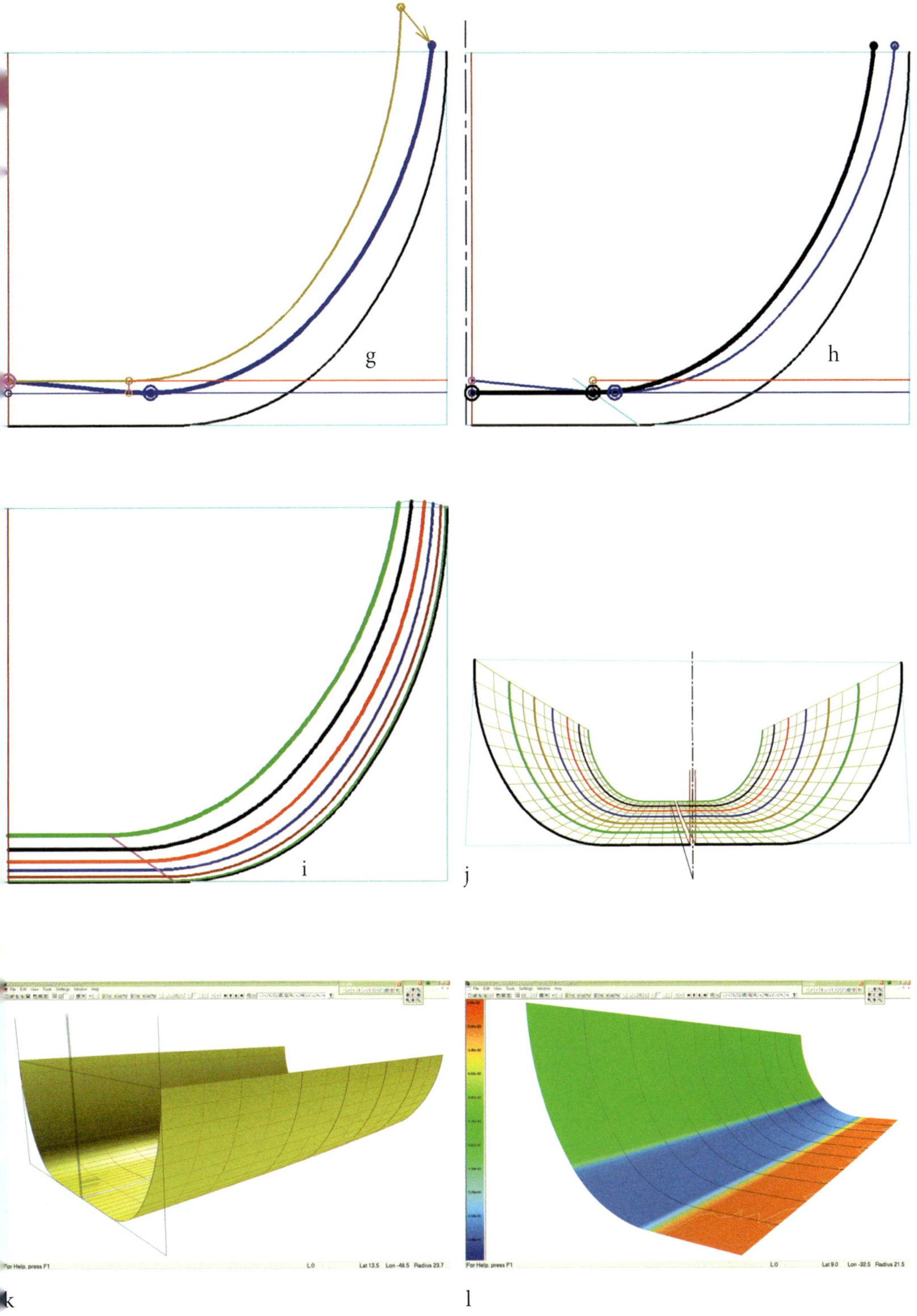

Pl. VIII. Figs. 1.13g – 1.13l: Mediterranean lofting method: Procedural steps.

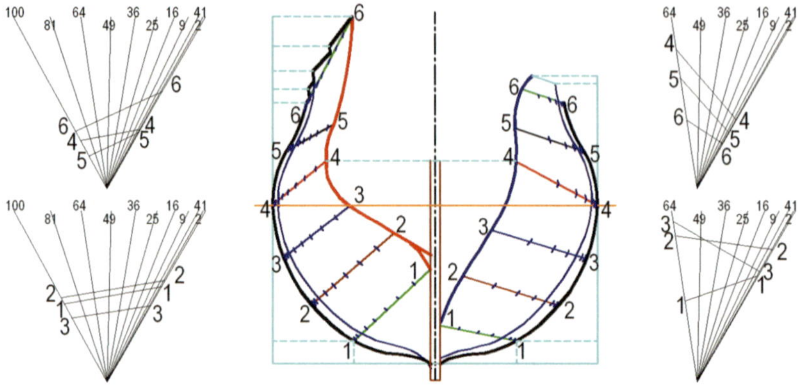

a

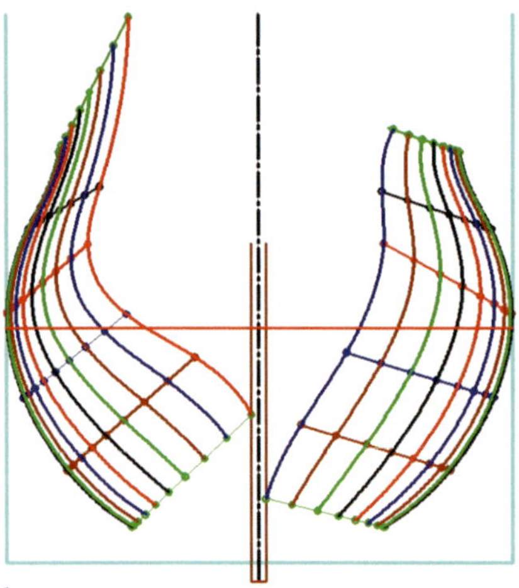

b

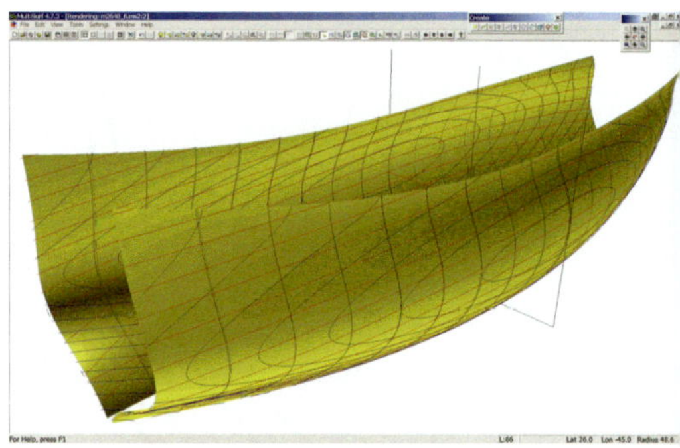

c

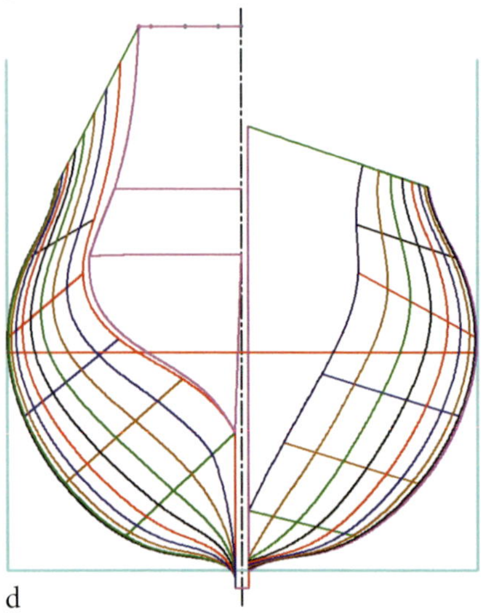

d

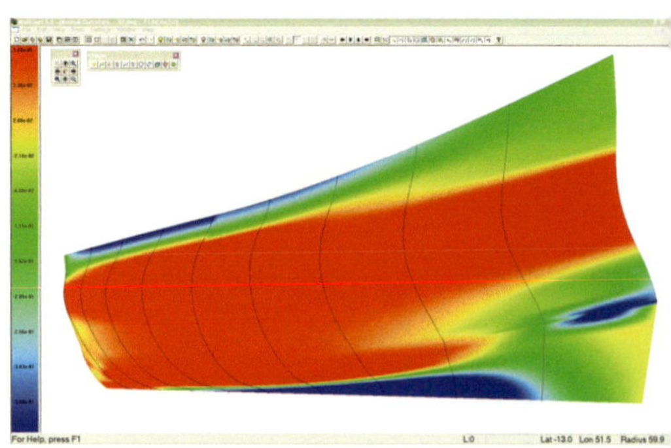

e

Pl. IX. Proportionate subdivision along diagonals, Example A (Boudriot): Figs. 1.14a: Lofting through two transverse sections per halfbody; 1.14b: lofted surface, forward and aft; 1.14c: rendered surface perspective view; 1.14d: complete body plan; 1.14e: normal curvature distribution.

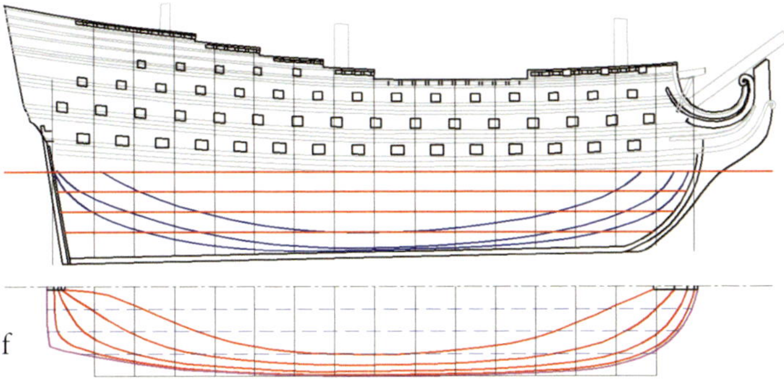

f

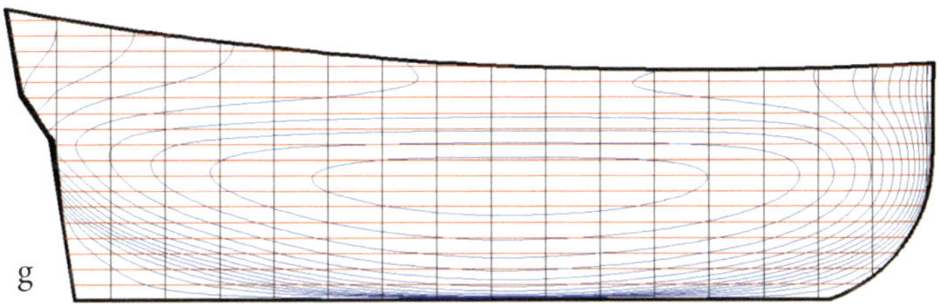

g

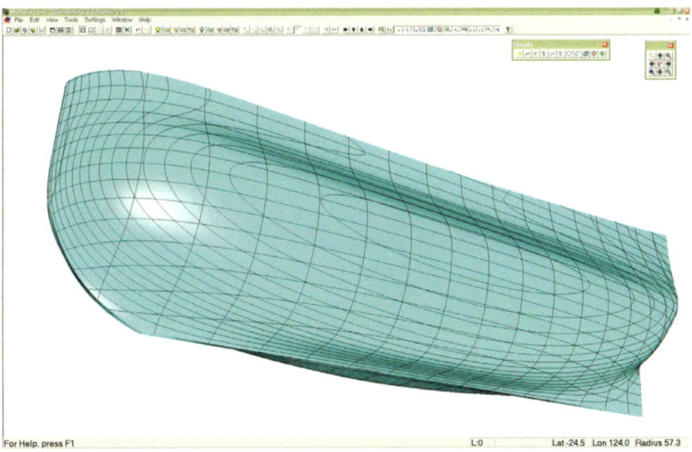

h

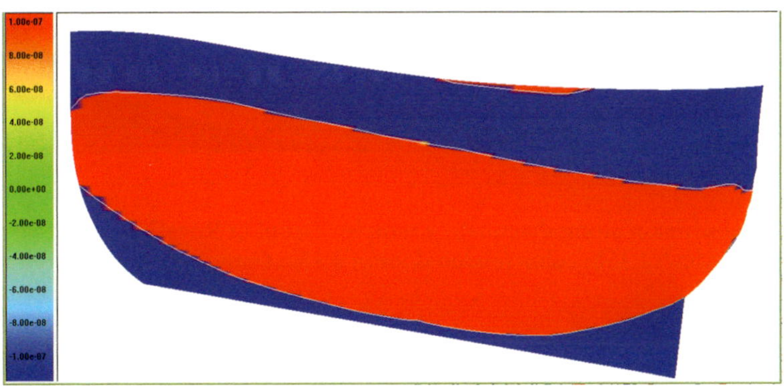

i

Pl. X. Proportionate subdivision along diagonals, Example B ("Royal Louis"): Figs. 1.14f: reconstruction by J. Sonntag; 1.14g: elevation, wireframe view; 1.14h: color rendering; 1.14i: Gaussian curvature distribution.

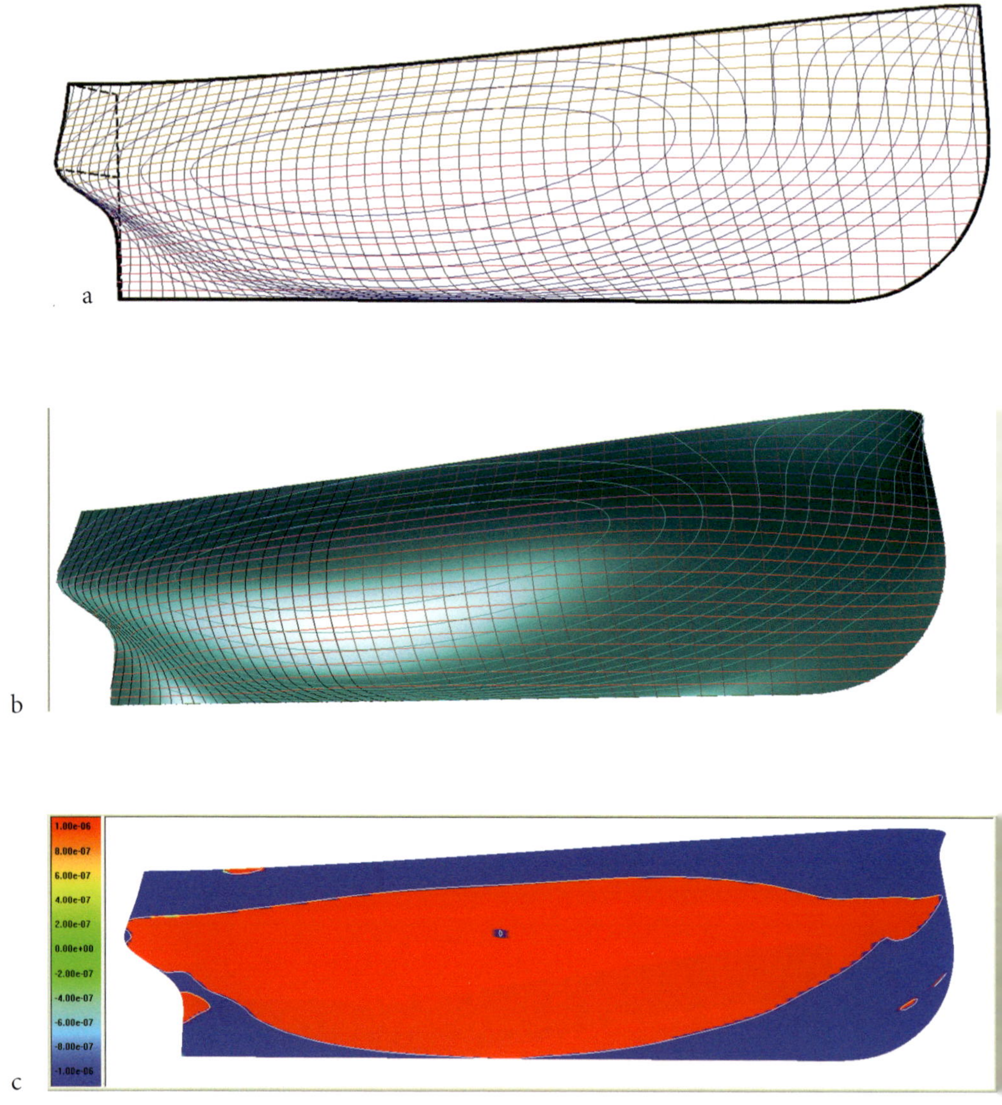

Pl. XI. Free subdivision along diagonals, Example “Dauphin Royal”: Figs. 1.15a: wireframe perspective view; 1.15b: color rendered view; 1.15c: Gaussian curvature rendering.

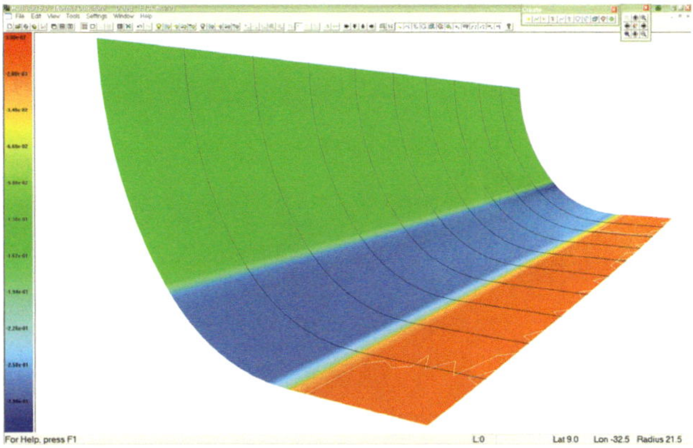

a

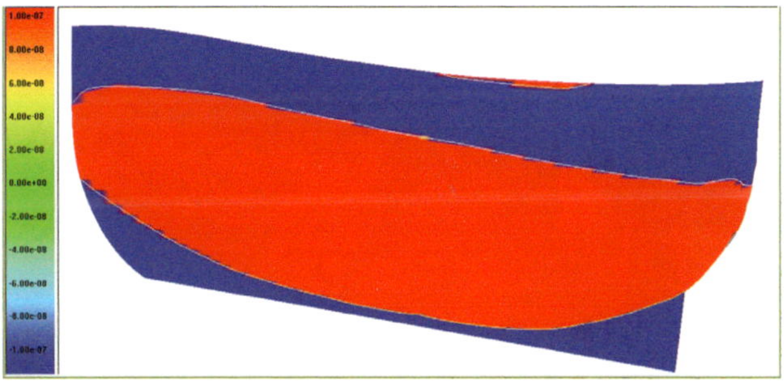

b

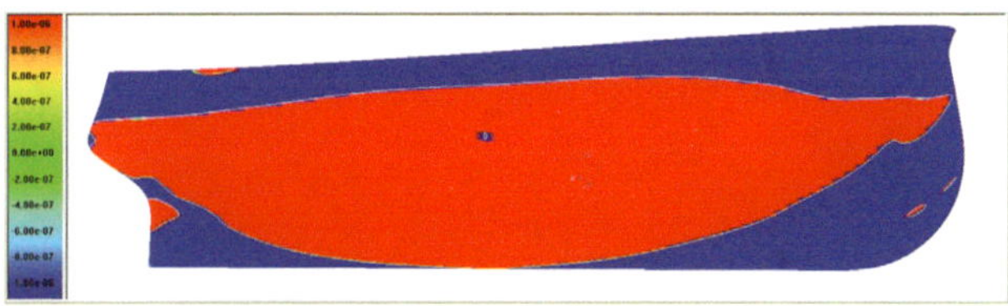

c

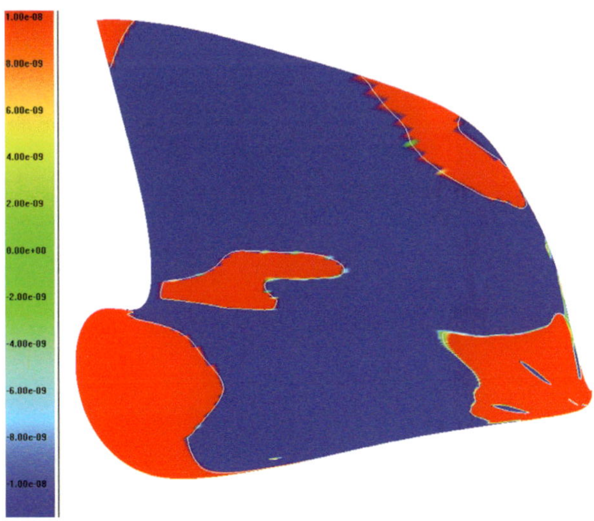

d

Pl. XII. Comparison of Gaussian curvature maps for various design methods: Figs. 1.16a: “La Madeleine”; 1.16b: “Royal Louis”; 1.16c: “Dauphin Royal”; 1.16d: “Modern tanker”.

Pl. XIII. Fig. 4.1: Mainz, wreck no. 5, in situ.

Pl. XIV. Fig. 4.10: Mainz, wreck no. 5, detail.

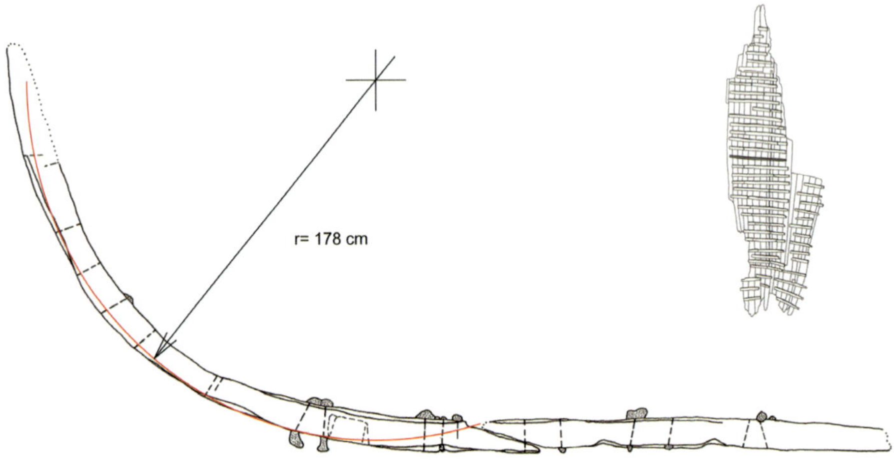

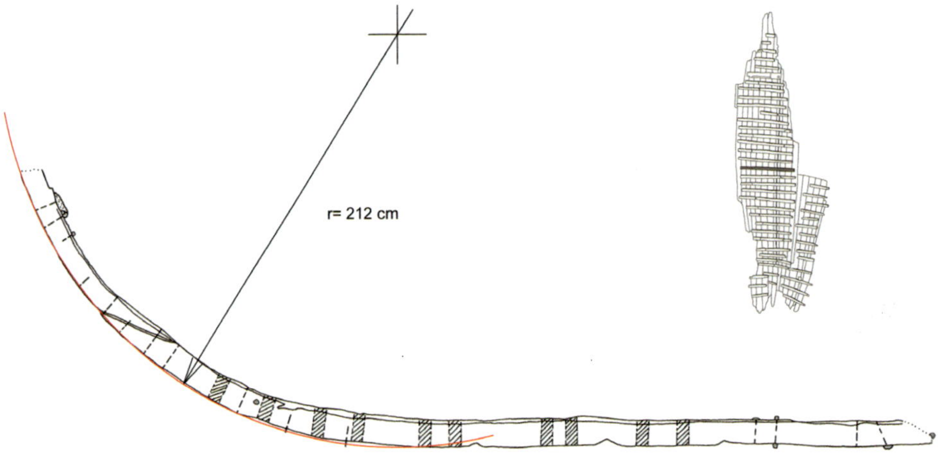

Pl. XV. Figs. 4.18a and 4.18b: Mainz, wreck no. 3. – Set of frame curvatures for frames 11 and 12 compared with projections of circular arcs.

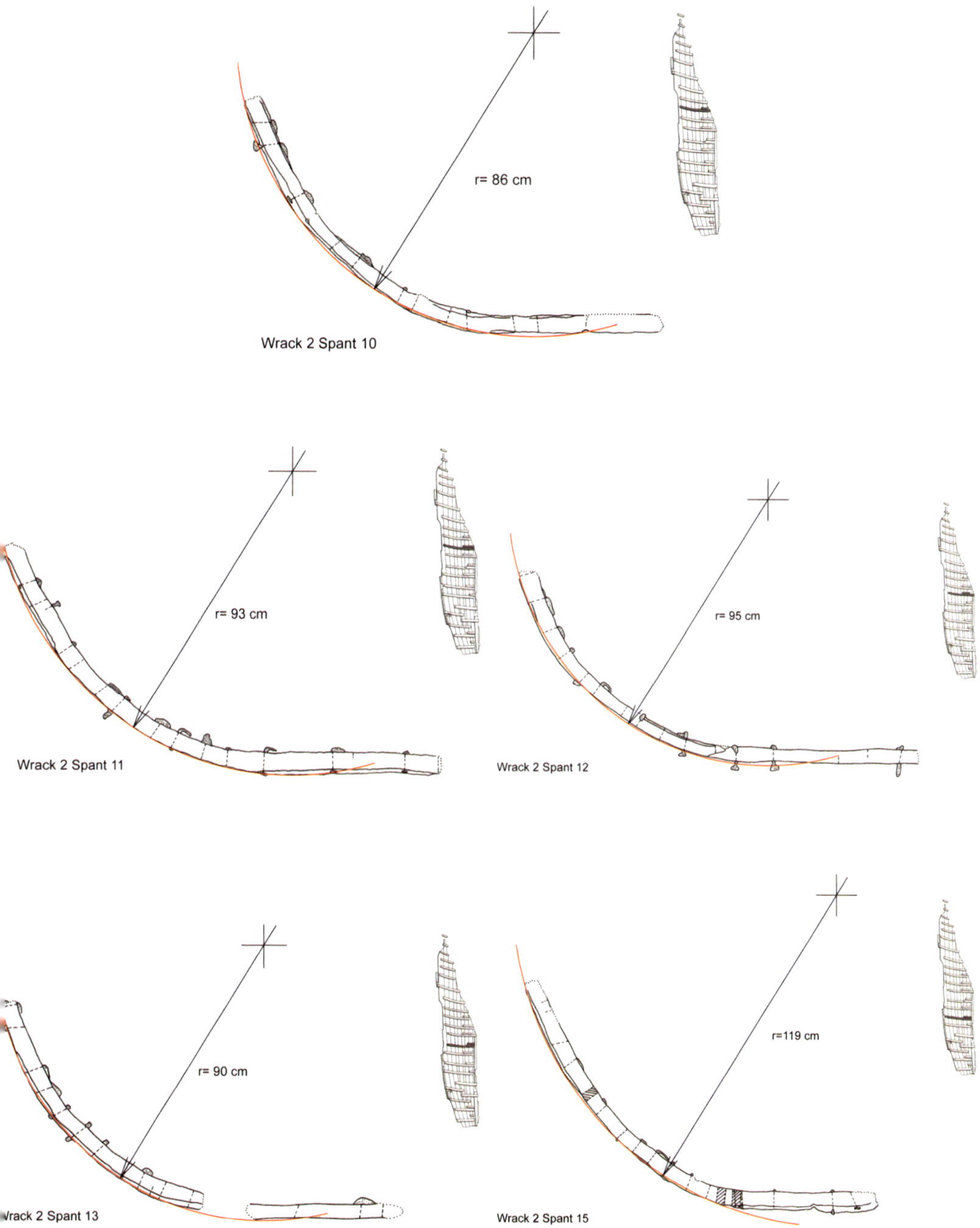

Pl. XVI. Fig. 4.19: Mainz, wreck no. 2. – Set of frame curvatures compared with projections of circular arcs for frames 10, 11, 12, 13 and 15.

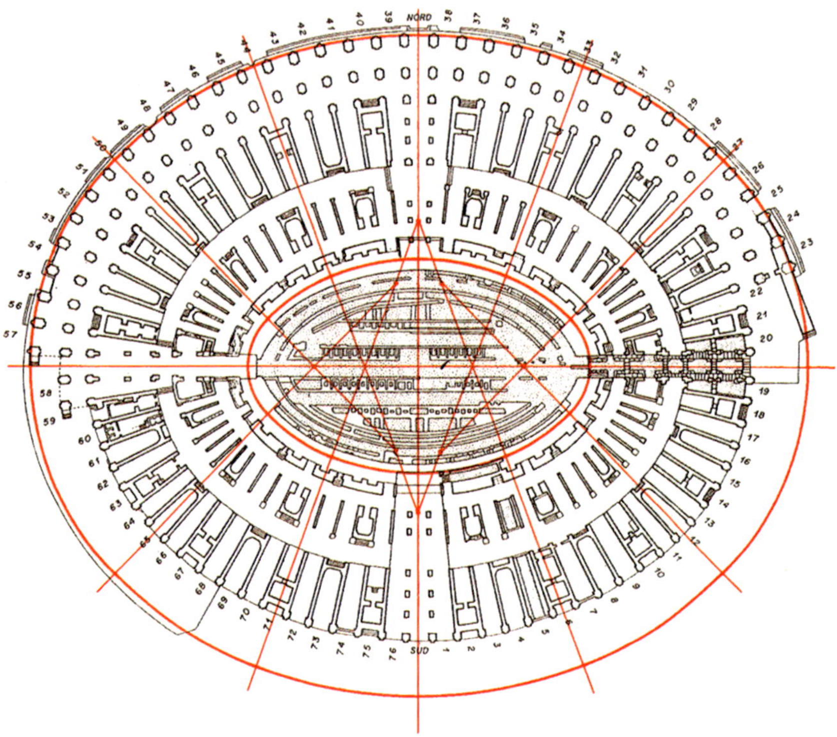

Pl. XIX. Fig. 5.12: Plan of the Colosseum, Rome.

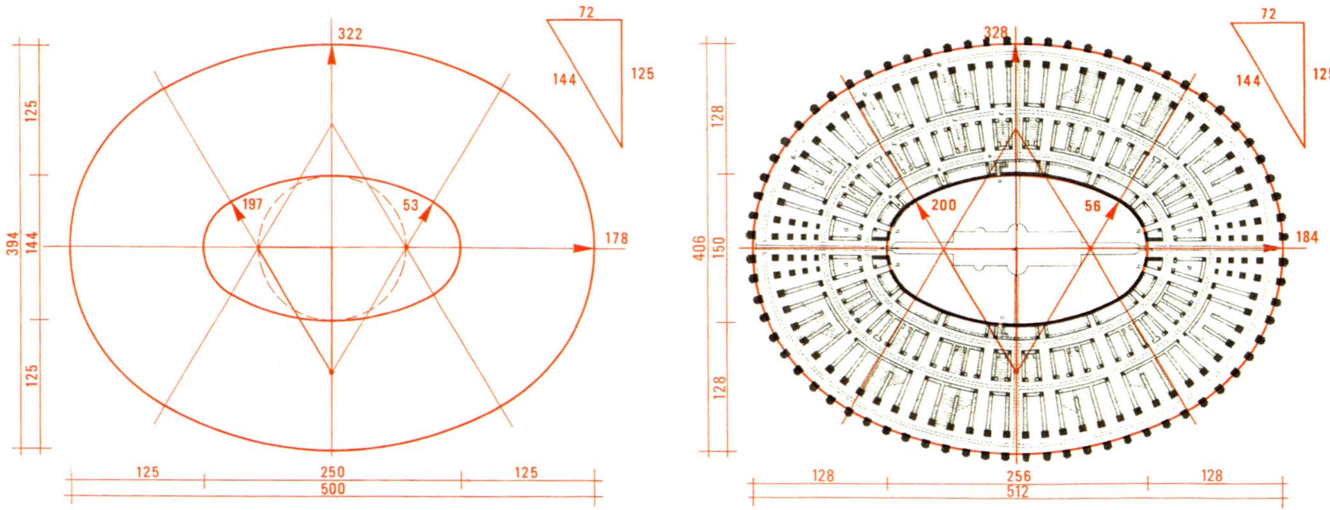

Pl. XX. Amphitheatre at Verona: Figs. 5.14a: hypothetical initial design; 5.14b: modified scheme.

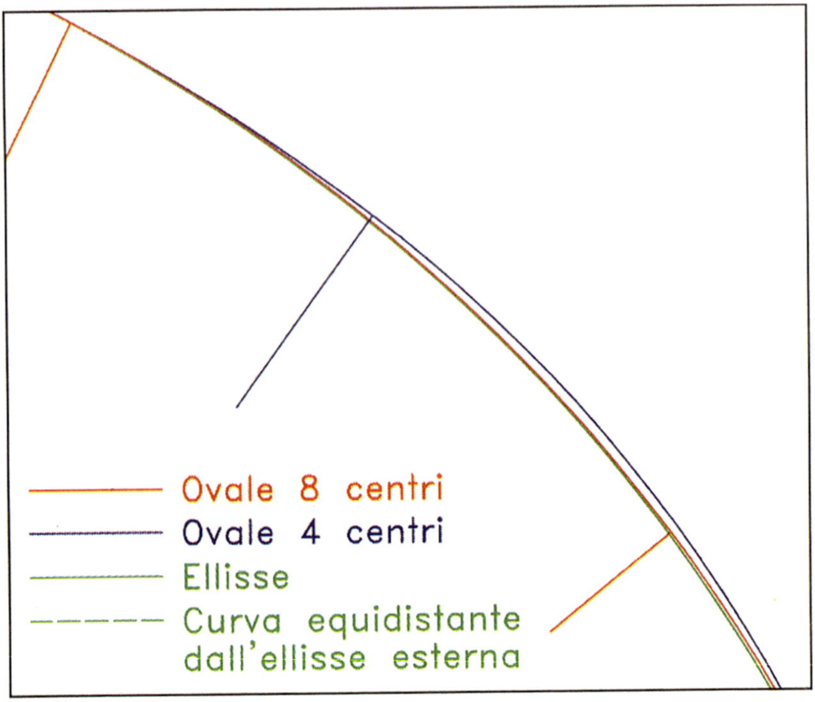

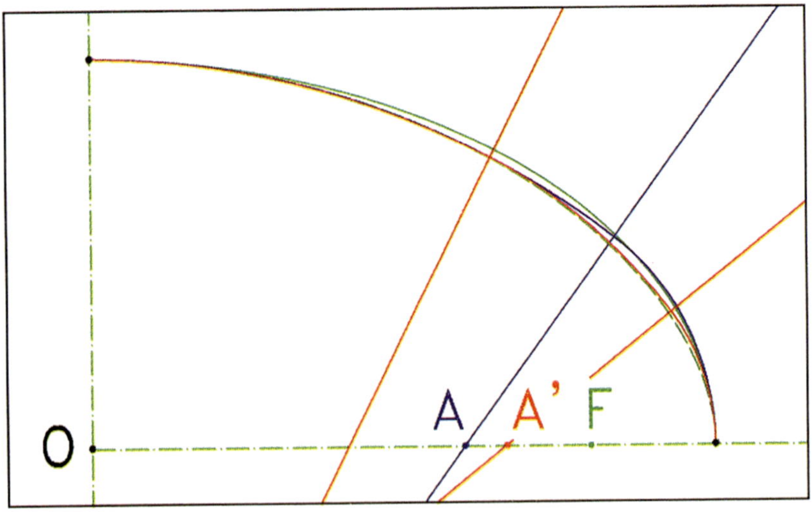

Pl. XXI. Figs 5.15a and 5.15b: Diagrammatic part-plans of the Colosseum, Rome.

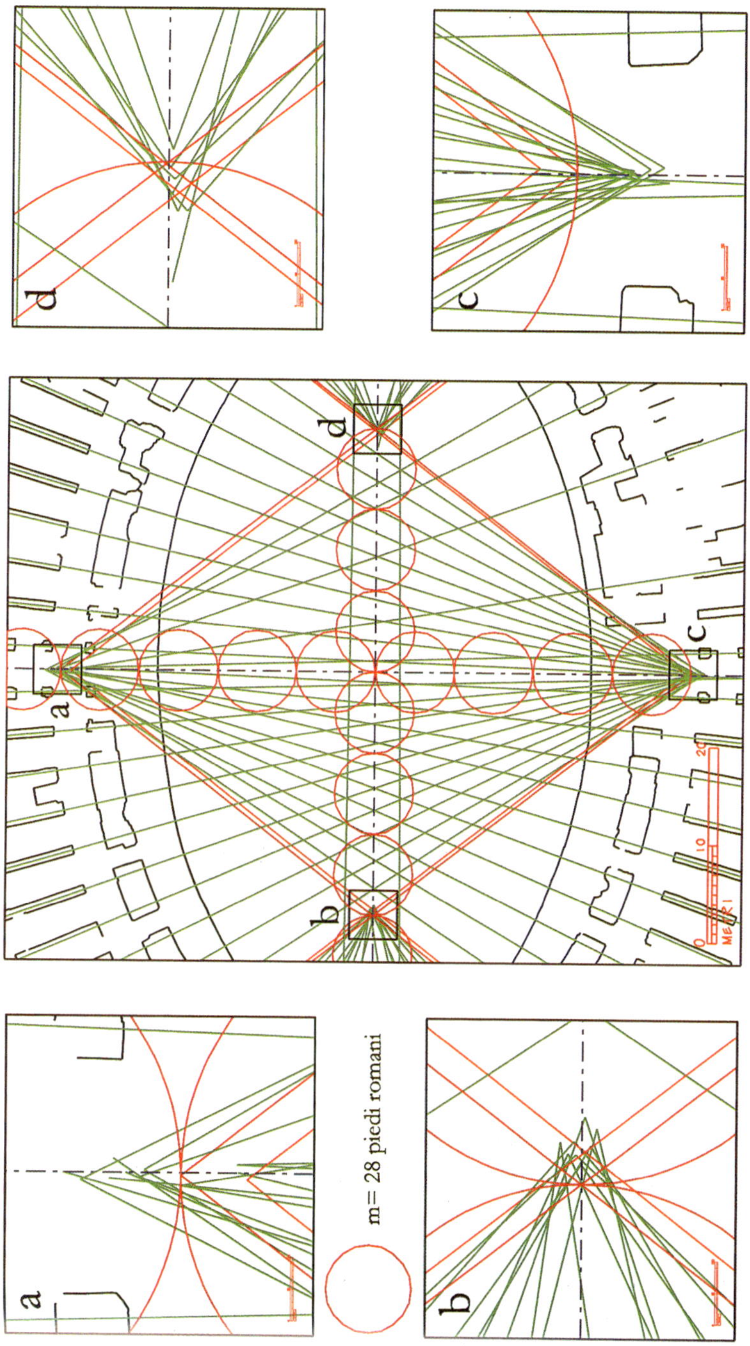

Pl. XXII. Fig. 5.16: Diagrammatic part-plans of the Colosseum, Rome.

Pl. XXIII. Fig. 7.2,3: Skuldelev 2 longship.

Pl. XXIV. Fig. 7.6: Launching of the *Roar Ege* reconstruction of Skuldelev 3 (1984).

Pl. XXV. Fig. 7.7: The shipbuilding scene on the Bayeux tapestry.

Pl. XXVI. Fig. 8.10: Building a rib vault in the Centro de los Oficios in León.

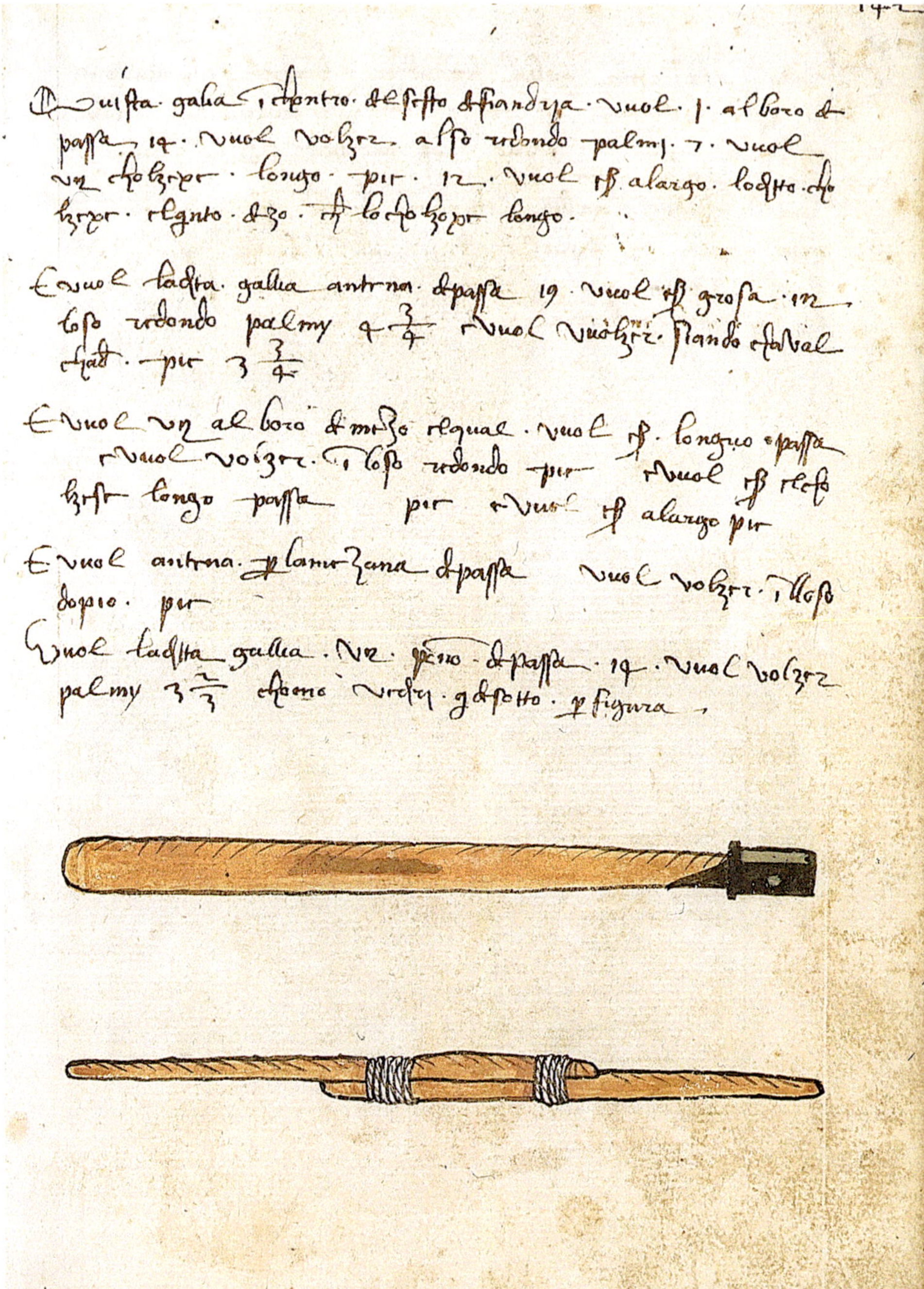

Pl. XXVII. Fig. 10.3: Illustrations of equipment contained in the Michael of Rhodes manuscript.

Pl. XXVIII. Fig. 10.4: Illustration of the galley of Flanders in the Michael of Rhodes manuscript.

Pl. XXIX. Fig. 10.5: Illustration of the galley of Romania in the Michael of Rhodes manuscript.

Pl. XXX. Fig. 10.6: Illustration of the *nave quadra* in the Michael of Rhodes manuscript.

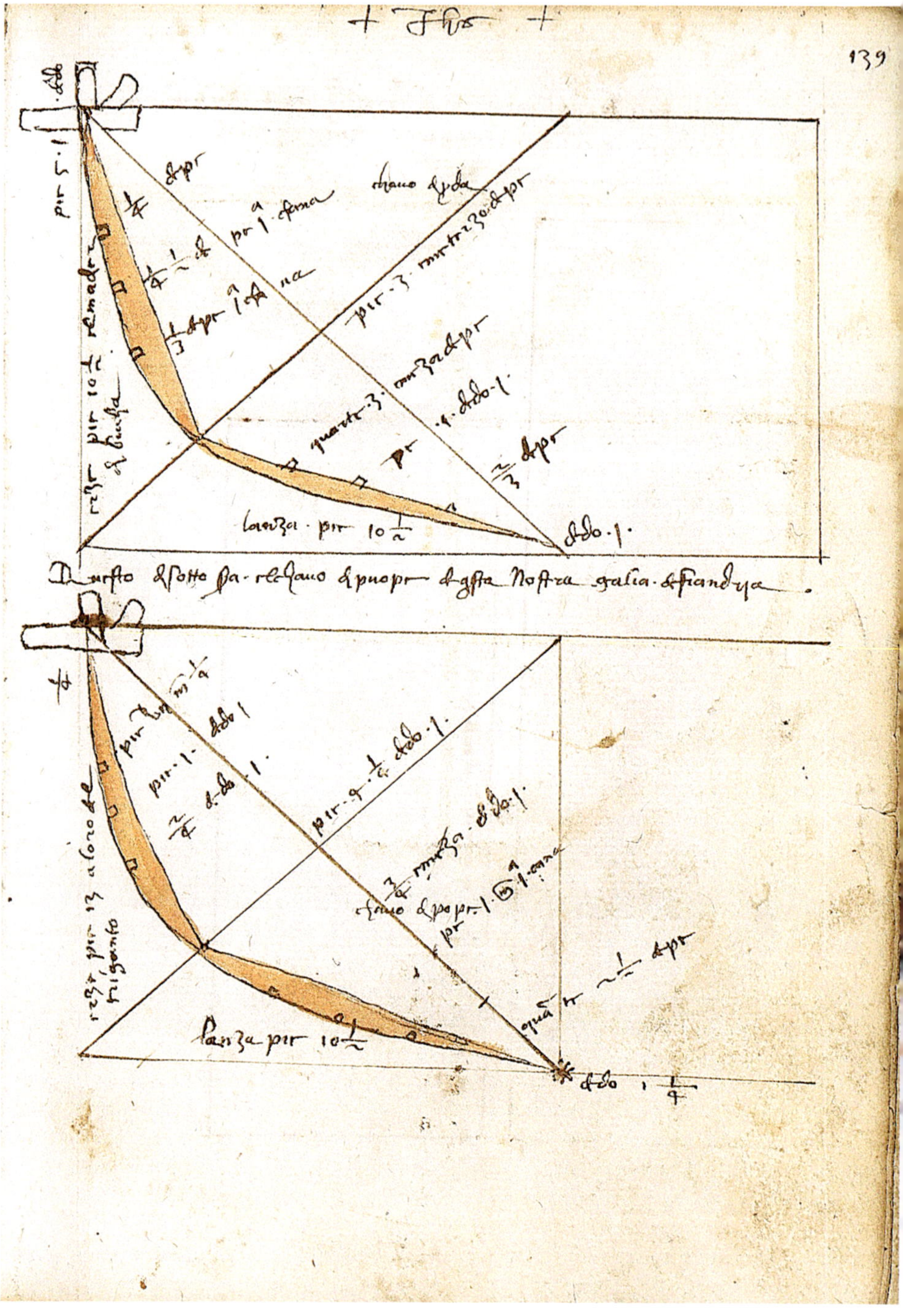

Pl. XXXI. Fig. 10.7: Bow and stern diagrams in the Michael of Rhodes manuscript.

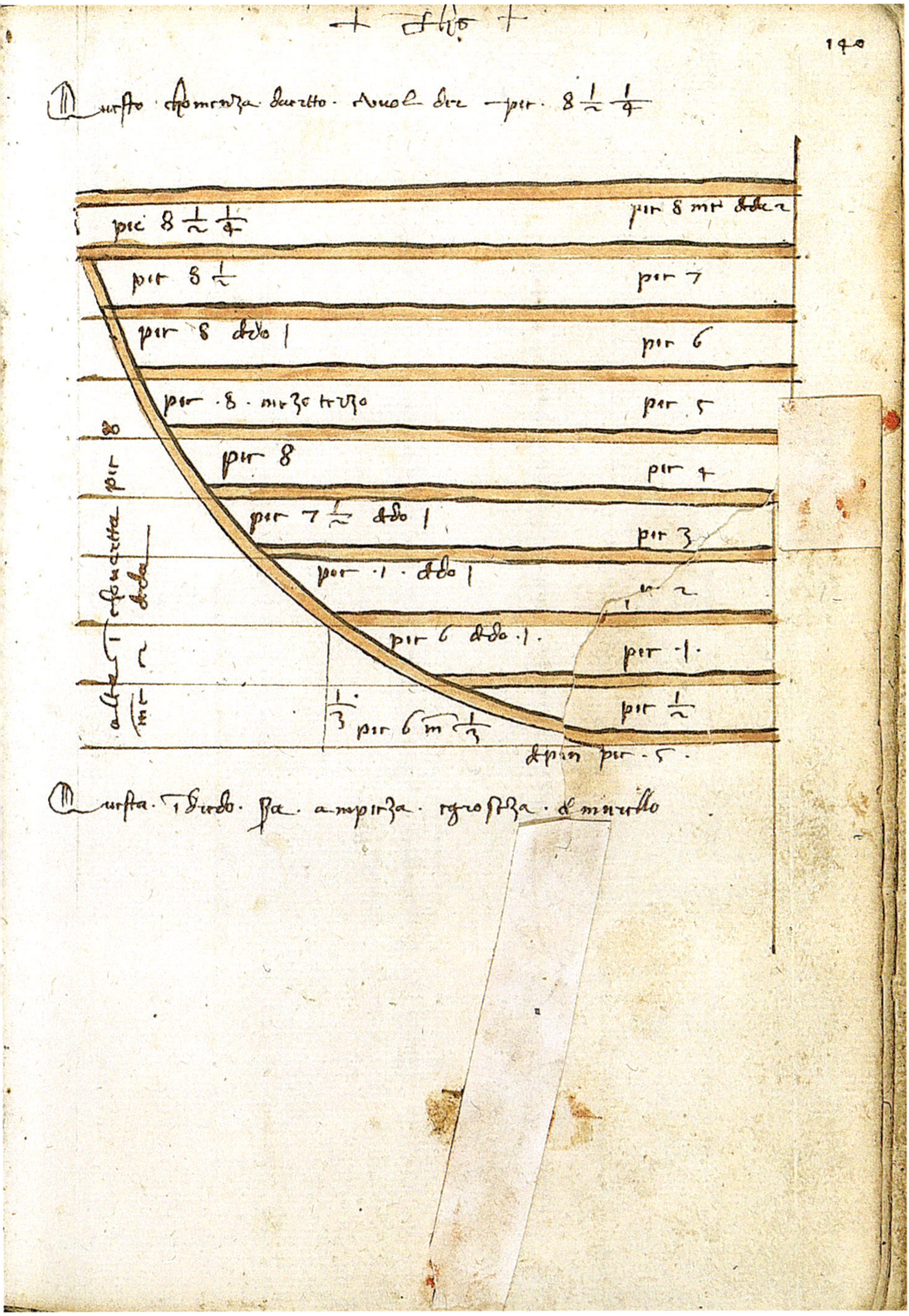

Pl. XXXII. Fig. 10.8: Midship bend of the galley of Flanders in the Michael of Rhodes manuscript.

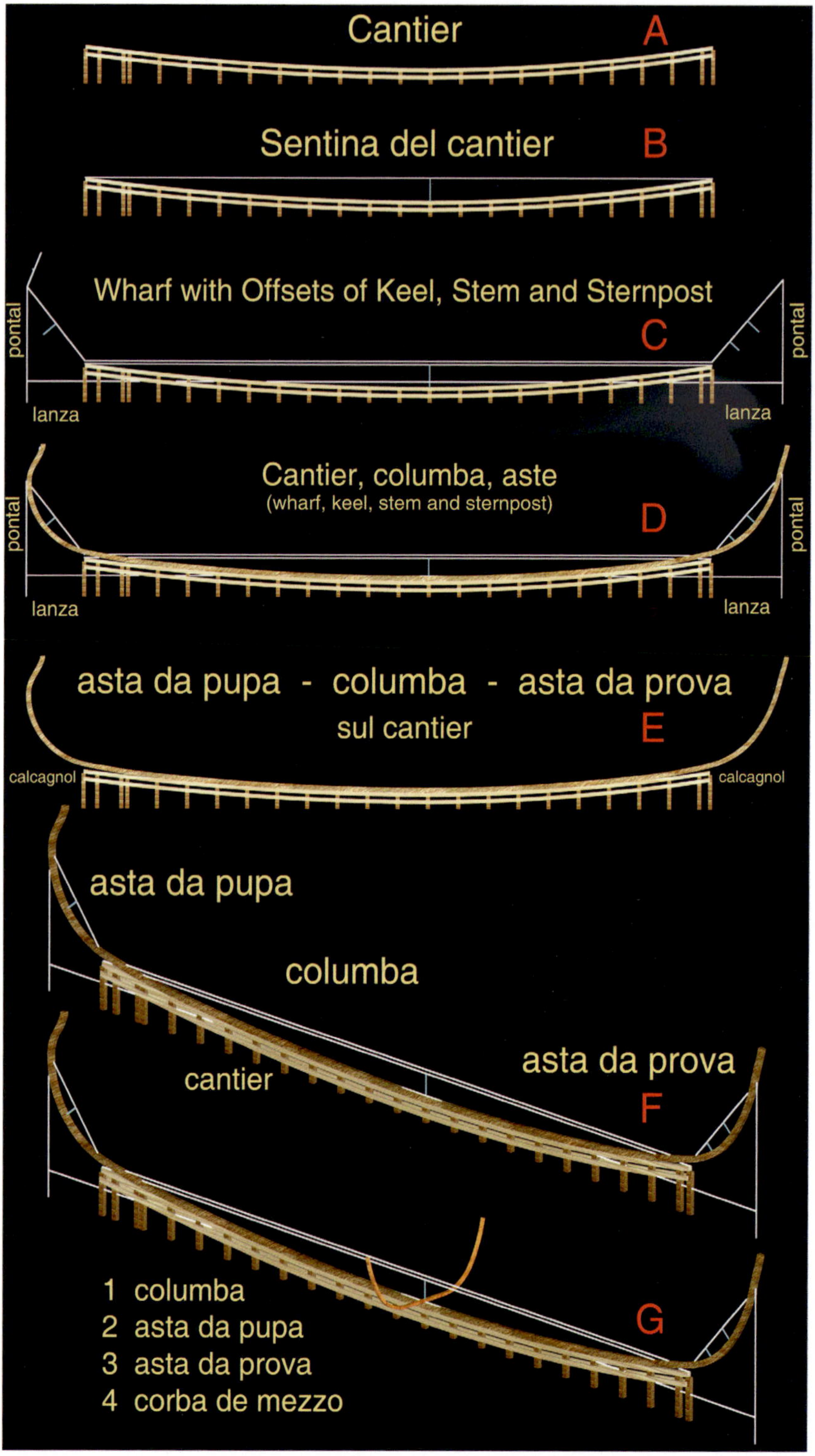

Pl. XXXIII. Figs. 11.9a and 11.9b: Shaping the lateral plan.

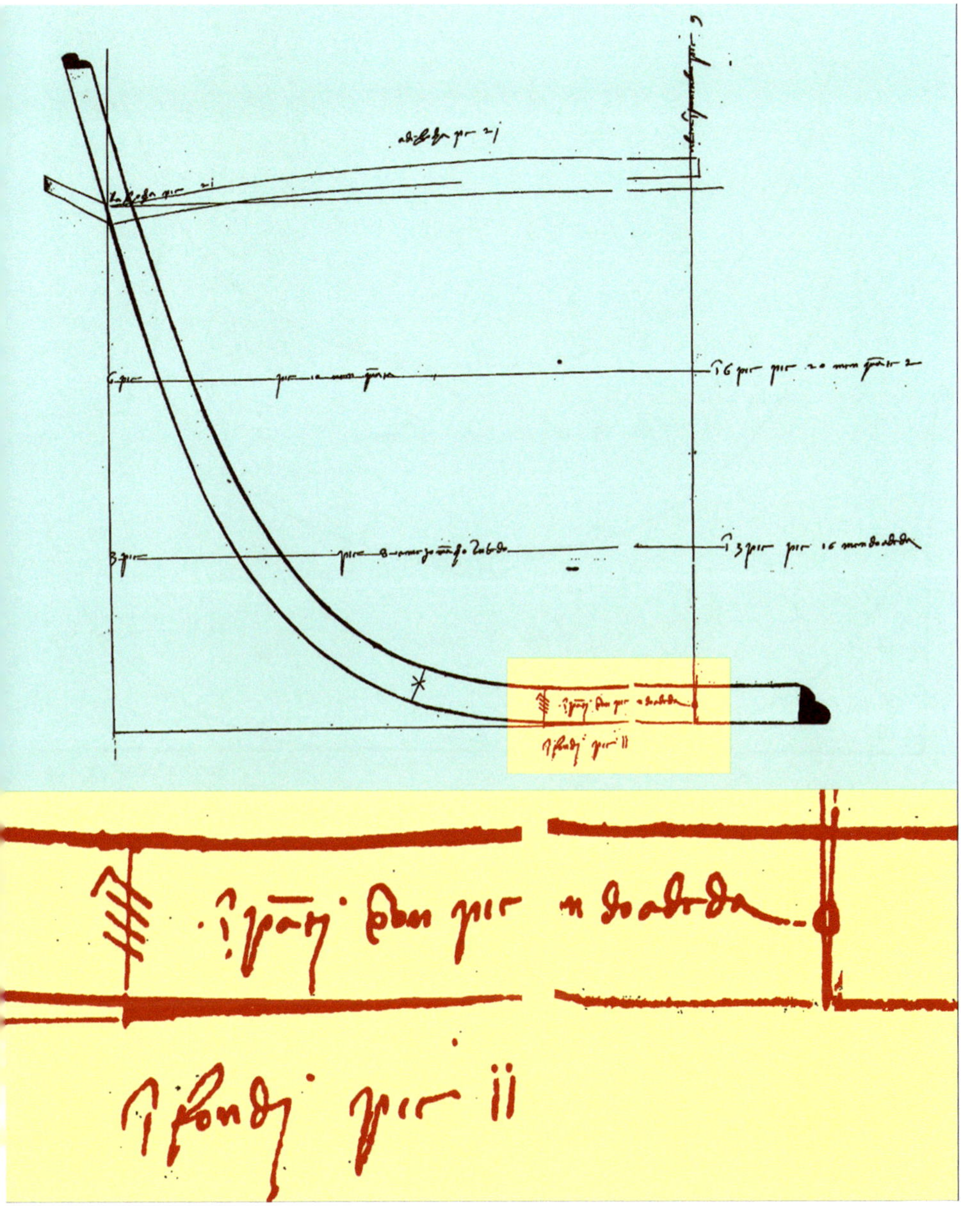

Pl. XXXIV. Fig. 11.10: Partison del fondo (Raggioni antique).

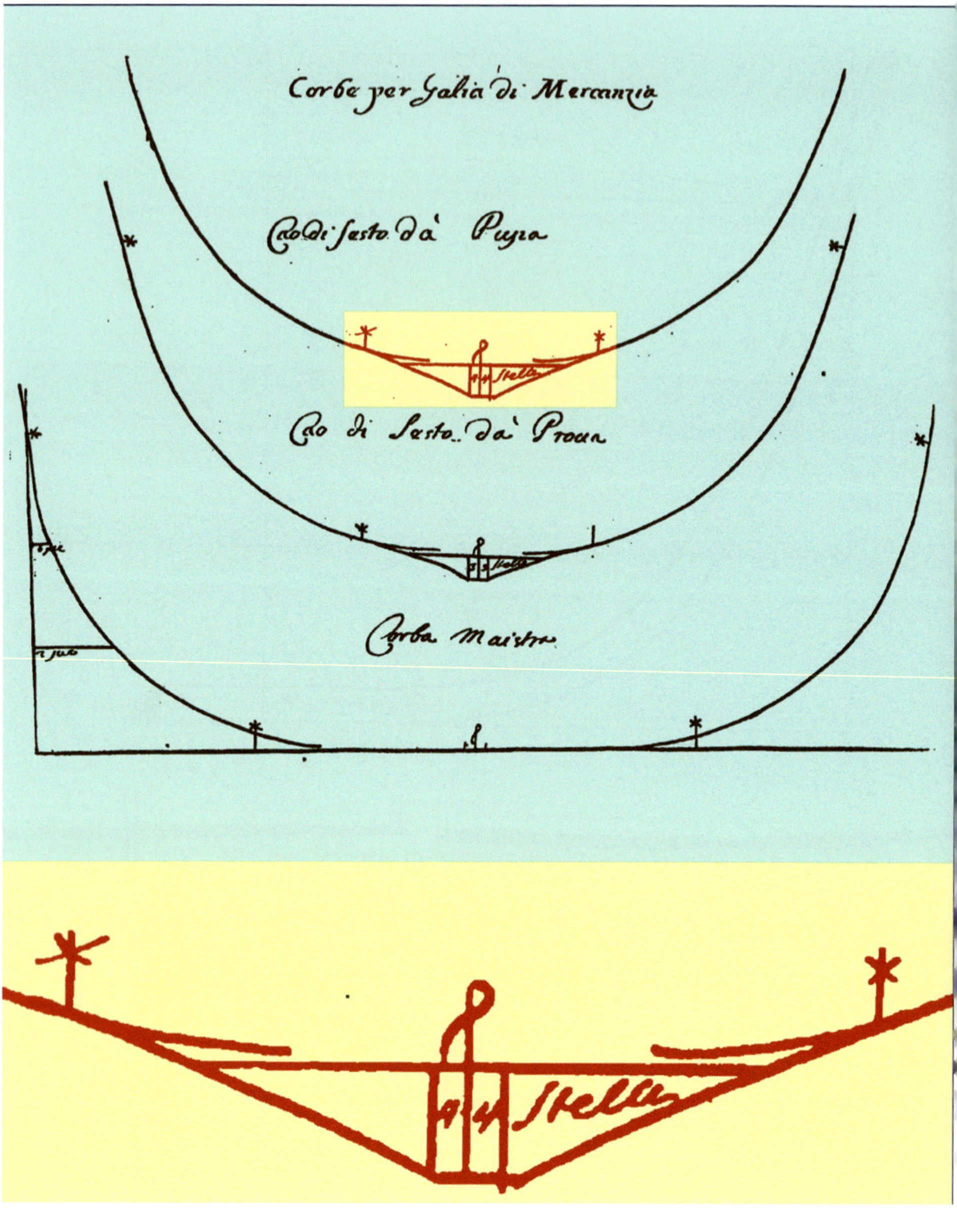

Pl. XXXV. Fig. 11.11: Stella (Zuanne).

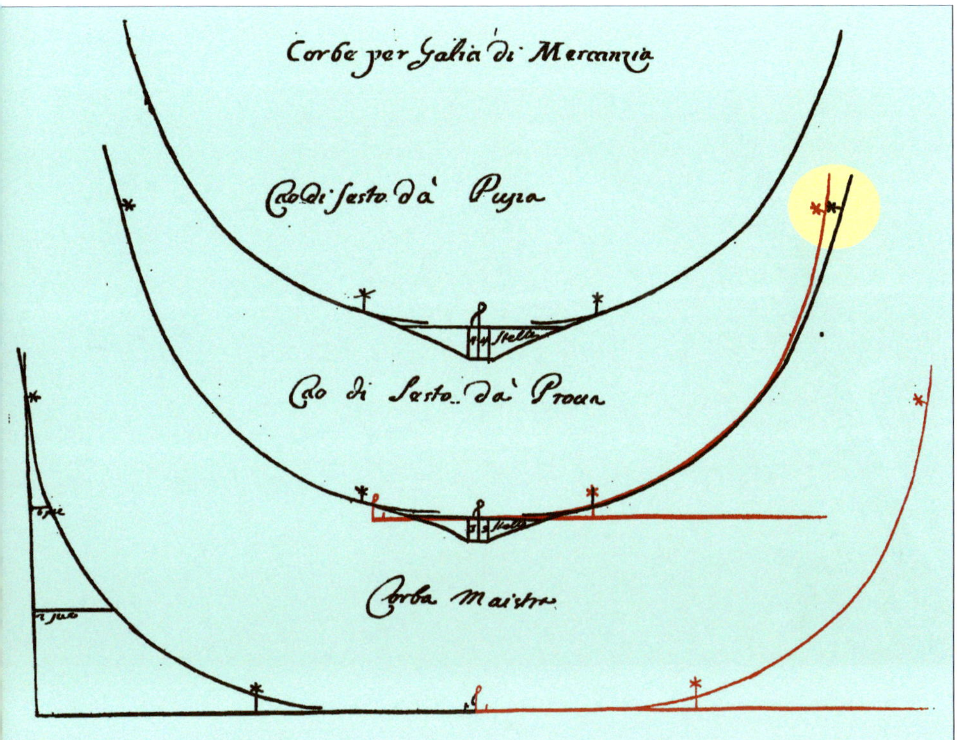

Pl. XXXVI. Fig. 11.12: Legno in ramo (Zuanne).

Pl. XXXVII. Fig. 11.14: Pavion.

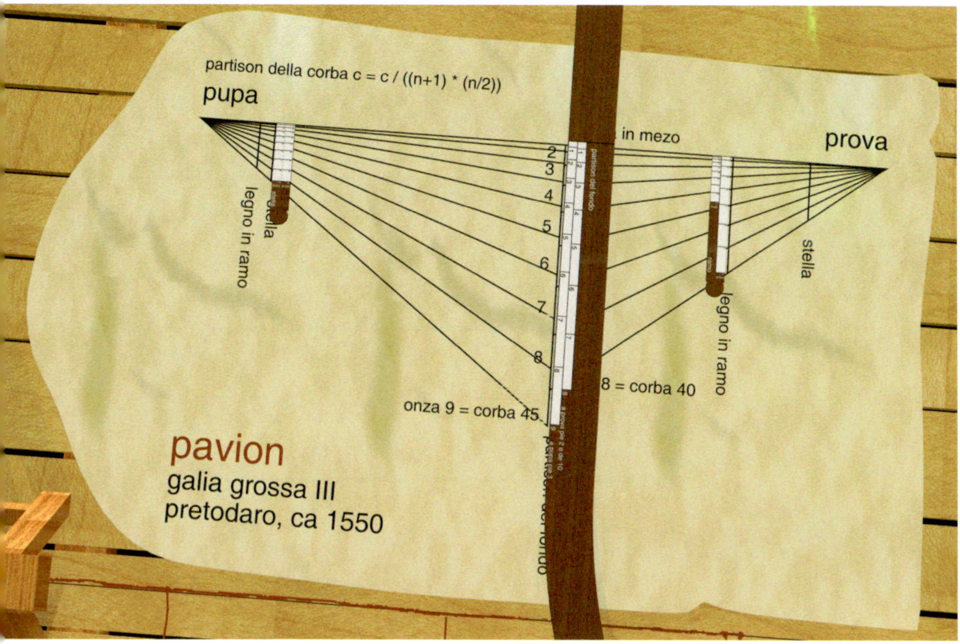

Pl. XXXVIII. Fig. 11.15: Morelli.

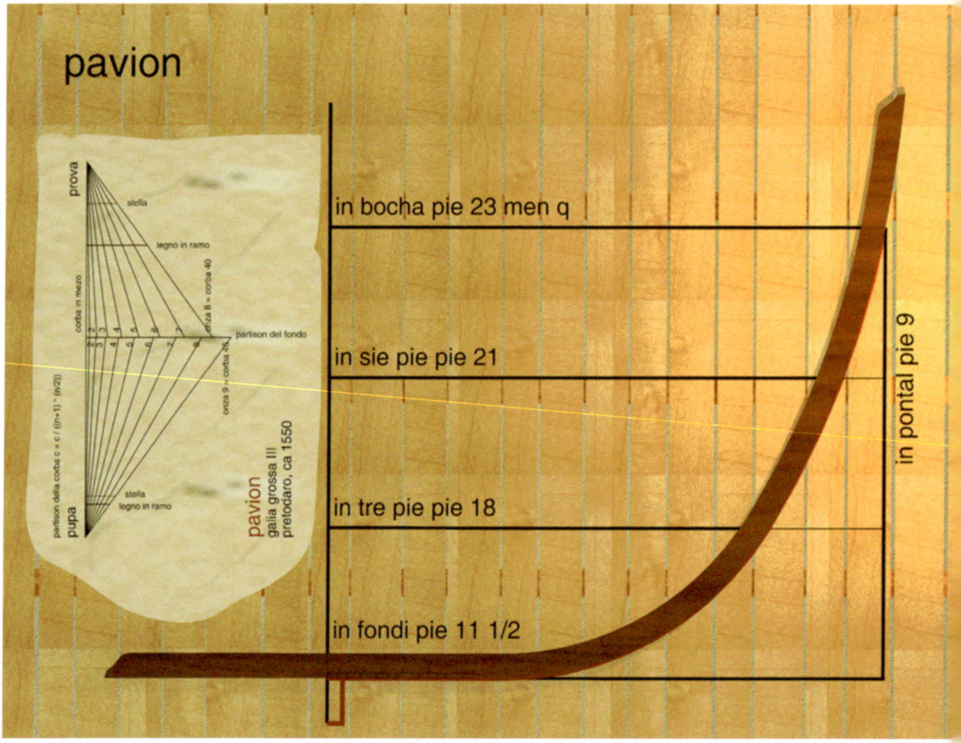

Pl. XXXIX. Fig. 11.16a: Pavion, sesto and its offsets.

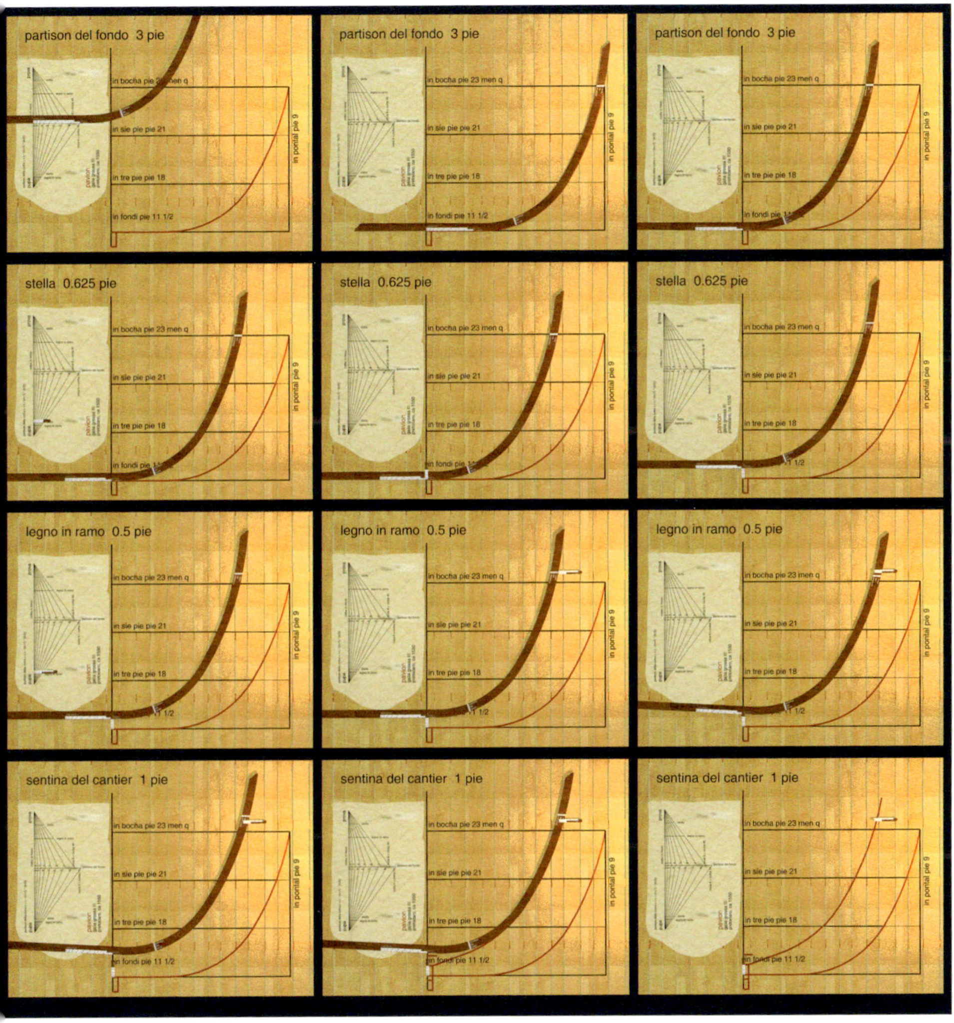

Pl. XL. Fig. 11.16b: The effects of: partison del fondo, stella, legno in ramo, sentina del cantier (from top).

Pl. XLI. Fig. 11.17: Sesti (moulds) in the Squero Tramontin.

Pl. XLII. Fig. 11.18: Tools in the mould loft.

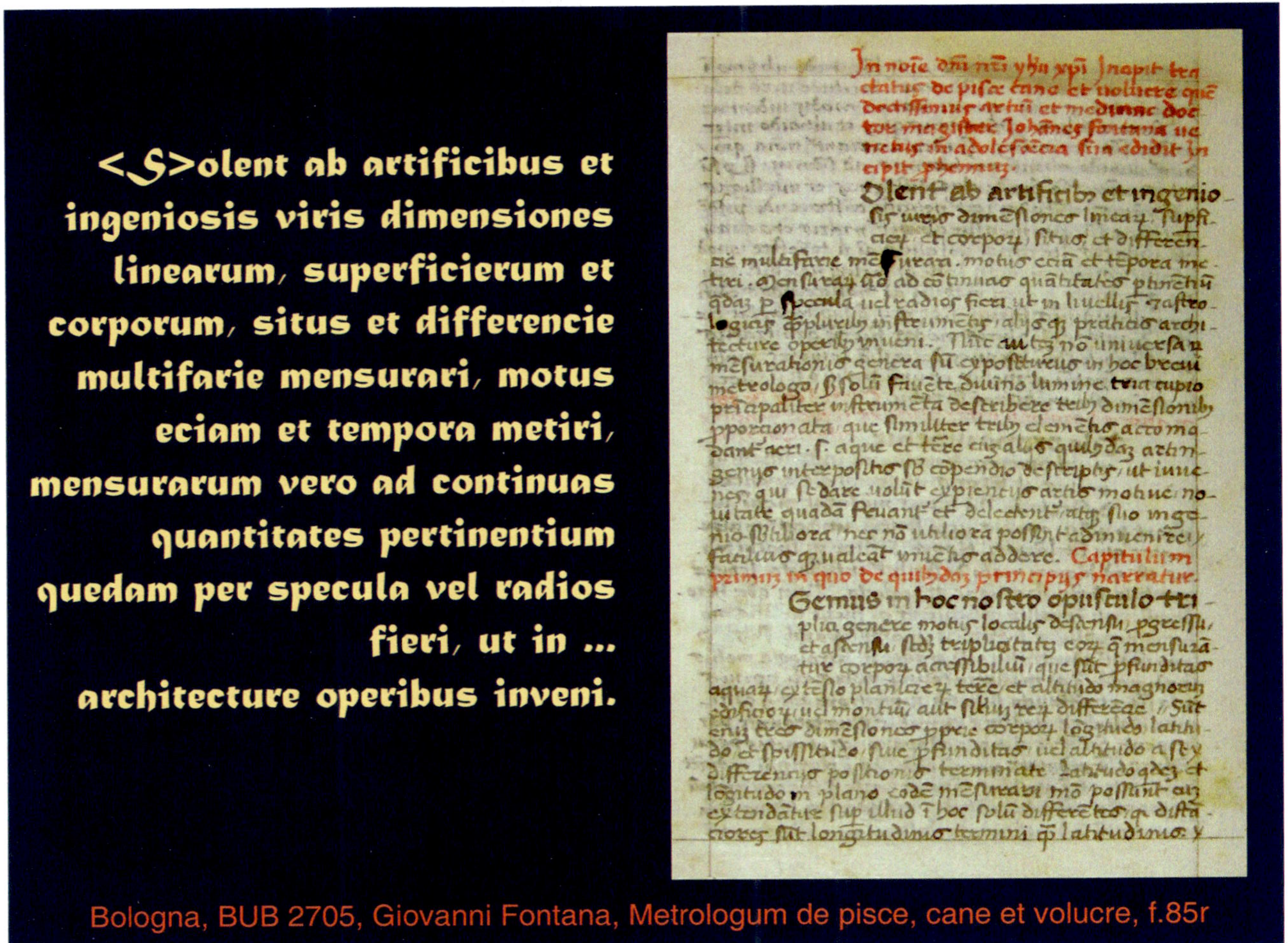

Pl. XLIII. Fig. 11.19: Folio 85r of Giovanni Fontana's Treatise on Metrology.

```
;*************************************************
;GALIA GROSSA C di Pre Theodoro
;in piedi x 10
;*************************************************
 (setq Titl "GALIA GROSSA C di Pre Theodoro")
;METHOD OF DESIGN
 (setq MezLun nil)                    ;'mezza luna' or triangle
 (setq KnikSp nil)                    ;defines the kind of frames
;HULL
 (setq LnRR    1380)                  ;length between perpendiculars
 (setq PM        10)                  ;number of parallel midship frames
 (setq KWL       80)                  ;waterline
;KEEL AND STRAKES
 (setq SenCan    10)                  ;'sentina del cantier'
 (setq Cantir '( 6.4    7   ))        ;thickness, height of 'cantier'
 (setq Columb '( 6.4    9   ))        ;thickness, height of 'columba' (keel)
 (setq Parmez '( 7.5    9.5 ))        ;thickness, height of 'paramezal'
;PLANK ABOVE THE CENTA (WALE)
 (setq MDB    '( 3.75 10))            ;thickness, height of 'magier de boca'
;PLANKS
 (setq fili    12)                    ;no of planks                (estimated)
;MAINFRAME
 (setq Fondo    115)                  ;'in fondo'
 (setq trepie   180  )                ;'in trepie'
 (setq siepie   210  )                ;'in siepie'
 (setq boca     227.5)                ;'in bocha'
 (setq Corb   '( 4.375 4.375))        ;dimension of frame
 (setq Lata   '( 3.75  4.375))        ;dimension of deckbeam
 (setq ponM      90  )                ;height of frame
 (setq bolson    12.5)                ;'bolson della coverta'
 (setq Colo   bolson)                 ;length of 'colo della lata'
;DECK
 (setq Corda  '( 4      5))           ;breadth, height of 'corda'    (estimated)
 (setq SCorda '( 5      7.5))         ;breadth, height of sotto corda
 (setq Spina  '( 3.75   7.5))         ;breadth, height of spina
 (setq SSpina '( 5      7.5))         ;breadth, height of sotto spina
 (setq Co-Co     40)                  ;distance 'corsia' - 'corda'
 (setq ScaLat    22)                  ;no of the 'lata' close to the 'schaza'
 (setq ScaLen     5)                  ;length of 'schaza' - in late  (estimated)
;VOGA
 (setq DZovo      4.75)               ;dimension of 'zovo'           (estimated)
 (setq HPave     28  )                ;height  of 'pavesata'
 (setq Post   '( 5      5.7  ))       ;breadth, height of 'postiza'
 (setq Banda  '( 3.33   9    ))       ;breadth, height of 'banda'
 (setq CadCol '( 4.375 6.875))        ;breadth, height of 'chadena da cholo'
 (setq Banchi    30)                  ;number of 'banchi'
 (setq fogoNo    27)                  ;no of banco at the 'fogon'
 (setq P-P       35)                  ;distance of 'postizzi'
 (setq S-S        7.143)              ;distance of 'schermi'
 (setq Z-PP (+ P-P S-S ))             ;distance: 'zovo' - 'primo postizzo'
 (setq BacBac     2)                  ;distance between 'baccalari' (in frames)
```

```
;-------------- different data fore and aft -------------
;LIST OF ONZE
 (setq Onze    '((-1 ONZE)
                 (O (list 0 5 10 15 20 25 30 35 40)
                    (list 0 5 10 15 20 25 30 35 40 45))
                 (Z (list 45 50 52.5 53.75)
                    (list 50 55 57.5 60 62.5)) ))
;CAO DI SESTO (ALIGNING FRAME)
 (setq CaoDS   '((-1 CAODISESTO)
                 (stel     6.25   6.25)  ;'stella'
                 (part    26.25  30   )  ;'particion dal fondi'
                 (ramo    12.5    5   )  ;'ramo'
                 (Fpa      2      2   )
                 (ferir  205    240   ))) ;'ferir'
;STEM AND STERPOST
 (setq Calcag '((-1 CALCAGNOL)           ;'calcagnol'
                 (H        5.625 5.625)
                 (L       90     110  )))
 (setq Asta    '((-1 ASTA)               ;stem and sternpost
                 (H      130    175   )  ;height
                 (Lanzo  130    105   )  ;'slanzo'
                 (HLot-    0    -60   )  ;intersection: perpendicular-stem below H
                 (LDeck    0    -25   )
                 (HDeck    0      0   )  ;deck above stem or sternpost
                 (L+      14    -24.8 )  ;angle of stem / sternpost
                 (H+      50      0   )  ;height of stem or sternpost above H
                 (Vala    28     30   )  ;'valangin'
                 (ValaO    0.3    3   )  ;upper 'valangin' 'alla Faustina'
                 (ValaOO   0      0   )  ;upper part of upper 'valangin'
                 (ValaUU  23      0   )));lower 'valangin'
 (setq Trigto '((-1 TRIGANTO)            ;randsom
                 (B        0      5)     ;thickness of the 'triganto'
                 (H        0      6)     ;height
                 (L        0    140)     ;length or breadth
                 (BOLM    -1      0.3)   ;reduces 'bolzon' for 'triganto', 'zovo'
                 (voga          nil) ))  ;width of poop platform
;CENTA
 (setq Centa   '((-1 CENTA)
                 (B        6.25)         ;thickness 'per il morso'
                 (H        5   )         ;height
                 (MDB+     0     18  )   ;enlarges distance 'mdb'-'centa'
                 (Ende     S      S) ))  ;end of 'centa' (sternpost or deck)
;VOGA
 (setq Voga    '((-1 VOGA)
                 (B      296  )          ;breadth between 'postizze'
                 (H       32.5)          ;height above deck
                 (B-       0      0)     ;diminishes breadth
                 (H-       1.25   5)     ;diminishes height
                 (Colo+   10     10)     ;enlarges length of 'colo'
                 (BaPo     4.2    4.2)   ;distance: 'banda'-'postizza'
                 (palm   100    145)))   ;'palmeta'                    (estimated)
 (setq Corsia '((-1 CORSIA)              ;gangway
                 (S        5.625)        ;thickness of 'corsia' side planking
                 (B       27.5)          ;breadth of the gangway
                 (H       25)            ;height above deck
                 (B-       0      0)     ;diminishes breadth of the 'corsia'
                 (H-       3.75   0) ))  ;diminishes height
```

Pl. XLV. Fig. 11.23: Galere grosse and their boats.

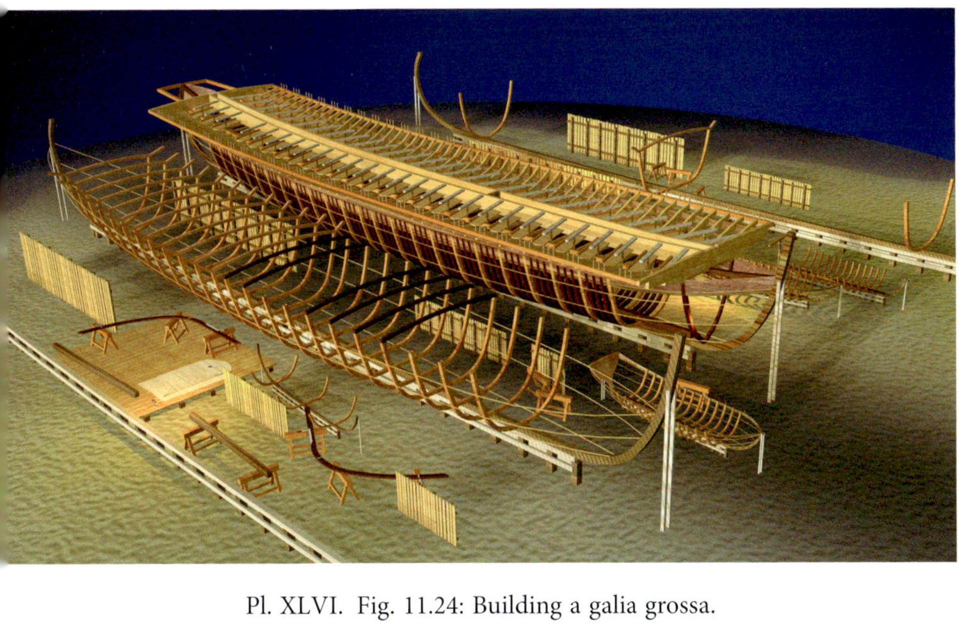

Pl. XLVI. Fig. 11.24: Building a galia grossa.

Pl. XLVII. Fig. 11.25: Merchant galleys of 1434, 1550, and 1686 (from right to left).

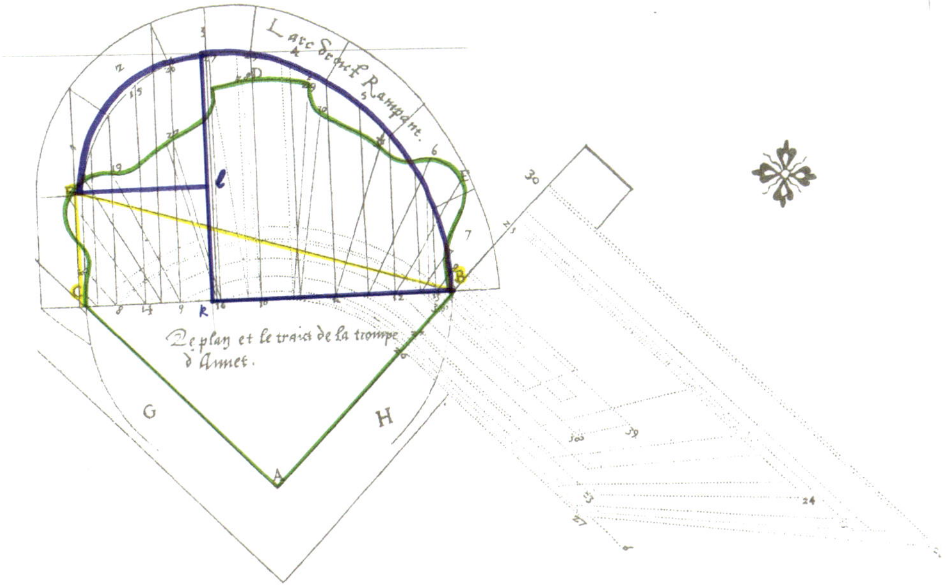

Pl. XLVIII. Fig. 14.9: De l'Orme's second drawing of the Trompe d'Anet (Green: Integrated ground plan. Yellow: Height of the inclination. Blue: Construction of the flattened arch).

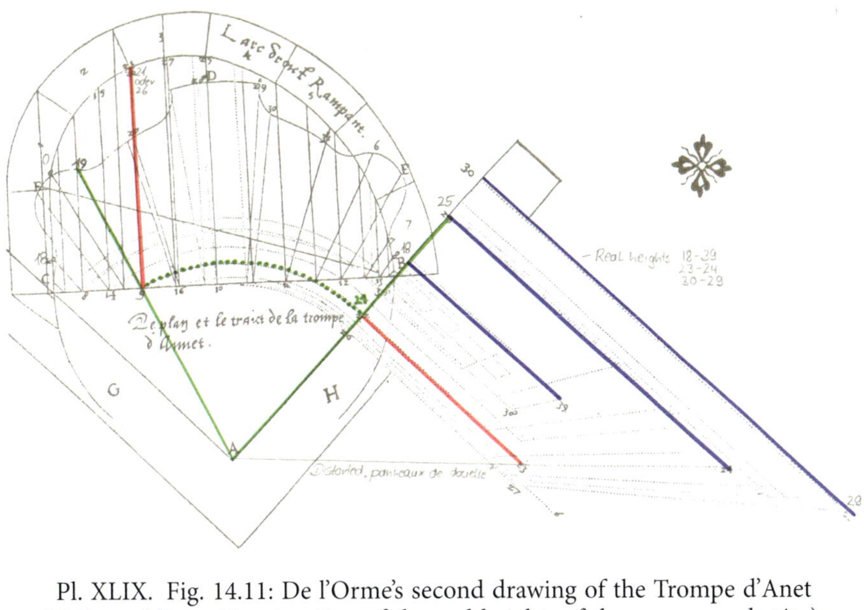

Pl. XLIX. Fig. 14.11: De l'Orme's second drawing of the Trompe d'Anet (Coloured lines: Construction of the real heights of the panneaux de tête).

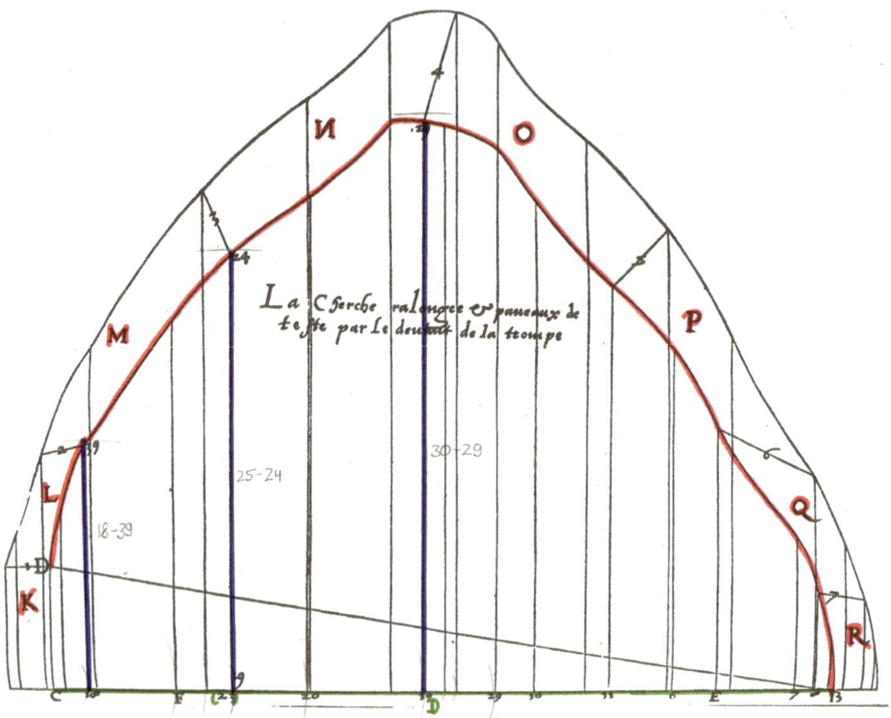

Pl. L. Fig. 14.15: De l'Orme's third drawing of the Trompe d'Anet (Green line CB: Unfolding of the ground plan; blue: real heights; red: Unfolded real contour).

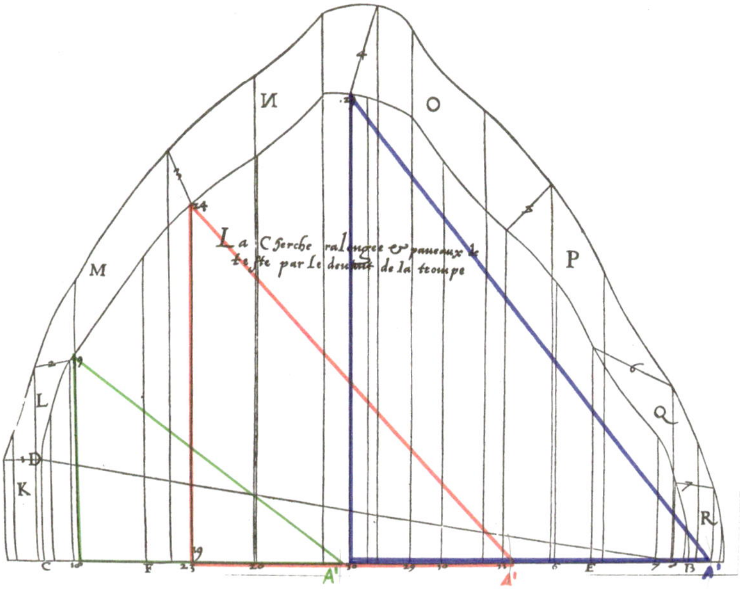

Pl. LI. Fig. 14.18: De l'Orme's fourth drawing of the Trompe d'Anet.

Pl. LII. Fig. 14.19: Three-dimensional model of the squinch of Anet.

Birkhäuser

Systems & Control: Foundations & Applications